Conservation and Mobile Indigenous Peoples

STUDIES IN FORCED MIGRATION
General Editors: Stephen Castles, Dawn Chatty, and Chaloka Beyani

Volume 1
A Tamil Asylum Diaspora: Sri Lankan Migration, Settlement and Politics in Switzerland
Christopher McDowell

Volume 2
Understanding Impoverishment: The Consequences of Development-Induced Displacement
Edited by Christopher McDowell

Volume 3
Losing Place: Refugee Populations and Rural Transformations in East Africa
Johnathan B. Bascom

Volume 4
The End of the Refugee Cycle? Refugee Repatriation and Reconstruction
Edited by Richard Black and Khalid Koser

Volume 5
Engendering Forced Migration: Theory and Practice
Edited by Doreen Indra

Volume 6
Refugee Policy in Sudan, 1967–1984
Ahmed Karadawi

Volume 7
Psychosocial Wellness of Refugees: Issues in Qualitative and Quantitative Research
Edited by Frederick L. Ahearn, Jr.

Volume 8
Fear in Bongoland: Burundi Refugees in Urban Tanzania
Marc Sommers

Volume 9
Whatever Happened to Asylum in Britain? A Tale of Two Walls
Louise Pirouet

Volume 10
Conservation and Mobile Indigenous Peoples: Displacement, Forced Settlement and Sustainable Development
Edited by Dawn Chatty and Marcus Colchester

Volume 11
Tibetans in Nepal: The Dynamics of International Assistance among a Community in Exile
Ann Frechette

Volume 12
Crossing the Aegean: An Appraisal of the 1923 Compulsory Population Exchange between Greece and Turkey
Edited by Renée Hirschon

Conservation and Mobile Indigenous Peoples

DISPLACEMENT, FORCED SETTLEMENT AND SUSTAINABLE DEVELOPMENT

Edited by

Dawn Chatty and *Marcus Colchester*

Berghahn Books
New York • Oxford

First published in 2002 by **Berghahn Books**

www.BerghahnBooks.com

Library of Congress Cataloging-in-Publication Data
Conservation and Mobile Indigenous Peoples : Displacement, Forced Settlement and Sustainable Development / edited by Dawn Chatty and Marcus Colchester
    p. cm. -- (Studies in forced migration : v. 10)
    Includes bibliographical references and index.
    ISBN 1-57181-841-3 (cloth: alk. paper). -- ISBN 1-57181-842-1 (pbk. : alk. paper)
    1. Nature--Effect of human beings on.   2. Forced migration--Case studies.
3. Land settlement--Case studies.   4. Indigenous peoples--Land tenure.
5. Conservation of natural resources   I. Chatty, Dawn.   II. Colchester, Marcus.   III. Series.

GF75 .D57 2002
333.7'2--dc21   2002018270

British Library Cataloguing in Publication Data

A catalogue record for this book is available
from the British Library.

Printed in the United Kingdom on acid-free paper

ISBN 1-57181-841-3 (hardback)
ISBN 1-57181-842-1 (paperback)

To Jill, Kito and Merry
Nick, Adrian and Miranda

Contents

List of Tables and Figures

Tables

Figures

Preface

It is estimated that 10 million people are displaced from their homes and communities each year through a combination of civil unrest, armed conflict, development projects (especially dam construction) and other interventions. Over the past few years, the Refugee Studies Centre at the University of Oxford has undertaken to study many of these aspects of forced migration. The disruption to human lives and livelihoods which has resulted from recent wildlife conservation and other environmental protection projects has not, however, been studied systematically. Each year hundreds of thousands of mobile people, located in difficult-to-reach, marginal areas are displaced and often forced into permanent settlements in order to set aside land for the conservation of wildlife

This book has emerged from these two separate concerns: recent local, national and international efforts to protect the biodiversity of our planet; and the sustainable livelihoods of marginal communities around the world. For most of the past decade, I had observed first hand in Oman and later in Syria, how these two important dimensions of universal interests often collide and create distress and upheaval in the lives of indigenous and local peoples as well as in the work of social and natural scientists. Recognizing the global dimensions of this collision, I set about organizing a conference which would bring together social and natural scientists – anthropologists, ecologists, and wildlife conservation specialists – to examine the impact conservation and other environmental protection projects have on the lives and livelihoods of the peoples who inhabit the same territory and ecological niches. A call for papers went out in 1998 under the auspices of the Refugee Studies Centre. Over 80 abstracts were received of which 36 were invited to proceed to full papers. In September 1999, the conference *Displacement, Forced Settlement and Conservation* was held at St. Anne's College, University of Oxford. All but two of the original papers delivered at that conference appear in the on-line version of this book with Berghahn Publishers (www.berghahnbooks.com).

Marcus Colchester, Director of the Forest Peoples Programme, kindly agreed to co-edit this volume with me. His extensive knowledge and experience of forest peoples in Latin America and Southeast Asia gives a balance to my own expertise in nomadic pastoral systems in the Middle East and North Africa. Together we have selected the twenty papers to make up this volume in which the interface between conservation and indigenous peoples is analysed in Latin America, East Africa, Southern Africa, the Middle East and the Mediterranean, South Asia, Southeast Asia, and the Pacific. In some cases, papers critically examine the contemporary efforts to bring indigenous peoples into the management and running of conservation efforts. In others, the plight of indigenous communities who are ignored or treated dismissively is revealed. All these papers highlight the need for a wide multi-disciplinary platform upon which indigenous peoples can voice their concerns alongside those of natural and social scientists in the effort to protect the biodiversity of the planet we all share.

Dawn Chatty

Acknowledgements

It is sometimes very difficult to separate out those who deserve special mention in a set of projects which involved so many people. There are all the marginal indigenous people, researchers, and conservation professionals to thank for the time, energy and courtesy they extended to aspects of this project. There are at the closing end of the project those who helped put together the September 1999 conference. My sincere thanks to Sean Loughna, my able research assistant, who helped pull together the final logistics for the 36 paper presentations over a period of three days and who put together the conference report. Also to Adam Hundt, my thanks for his having volunteered to be an extra set of hands at the conference and helping out with problems as they cropped up, however small or large.

My heartfelt thanks and appreciation go to Dominique Attala, who not only managed all the correspondence, accommodation and travel details for 42 participants at the 1999 conference, but has also taken over the logistical challenge of keeping tabs on 35 authors and their papers in preparation for this volume in both its hard copy and on-line edition. I don't know how this project would have been managed without her efficiency, good cheer and stamina.

I would also like to acknowledge the financial support which the Summit Foundation (USA), the Wenner-Gren Foundation for Anthropological Research (USA), and the Foreign and Commonwealth Office (UK) provided to the September 1999 conference upon which this collection is based.

Notes on Contributors

George Arab has a BSc in Agricultural Sciences from the University of Aleppo and worked as an agricultural engineer in the Syrian Ministry of Agriculture and Agricultural Reform from 1970 to 1993. For several years he also acted as a consultant to the Pasture, Forage and Livestock Programme at the International Center for Agricultural Research in Dry Areas (ICARDA) and is currently a research assistant to the Natural Resource Management Programme at ICARDA.

Aref Abu-Rabia has a BA in Education from Ben-Gurion University, a Masters in Public Health from the Hebrew University of Jerusalem and a PhD in Anthropology from Tel-Aviv University. At present he is a lecturer at the Department of Middle East Studies at Ben-Gurion University. He is the author of *The Negev Bedouin and Livestock Rearing*, (1994) and *The Traditional Bedouin Medicine*, (1999); and is co-author with S. Bar-Zvi, and G. Kressel, of *The Charm of Graves: Mourning Rituals and Tomb Worshipping Among the Negev Bedouin* (1998).

Sue Armstrong is a freelance writer and broadcaster specializing in science, health and development. She works regularly on assignment for the World Health Organization (WHO), and the United Nations Joint Programme on HIV/AIDS (UNAIDS). She researched and wrote 'AIDS: Images of the Epidemic' and 'Action for Children Affected by AIDS', both published by WHO in 1994. Sue has worked as a foreign correspondent in Brussels (1983–86) and South Africa (1988–96), and her work has appeared in New Scientist, The Economist Development Report, Scotland on Sunday, BBC Wildlife magazine, and on BBC radio. Sue is consultant to the Panos Oral Testimony Project among the San in Botswana.

Olivia Bennett has worked on development issues since 1976. As a freelance writer, she published many books, primarily in the field of development education. In 1990 Olivia joined the Panos Institute, an independent information organization which works with NGOs and media specialists in Africa, Asia and South America to raise the level of debate and understanding of key environment and development issues. She edited and co-wrote a series of Panos books and also worked with the Sahel Oral History Project. Using this experience, she conceived and now directs Panos' Oral Testimony Programme, which has been exploring the uses and value of testimony collection within the development process, as well as gathering testimony on specific themes. The programme combines detailed work and training with NGOs and community-based groups with wide dissemination of their views and experiences.

Randall Boone is a scientist working with the Natural Resource Ecology Laboratory, Colorado State University, USA on African ecosystem modelling and management. Dr Boone is a wildlife ecologist adept at computational ecology, and has written about integrated assessment using ecosystem models, species–habitat associations and habitat modelling, error assessments in habitat modelling, habitat fragmentation, animal dispersal, telemetry and spectrophotometry. His current work deals with using climate forecasts with ecosystem models, integrated assessments in East Africa, and dispersal in amphibians and large herbivores.

Dawn Chatty is Dulverton Senior Research Fellow and Deputy Director of the Refugee Studies Centre, Queen Elizabeth House, University of Oxford. She is a social anthropologist whose ethnographic interests lie in the Middle East, particularly with nomadic pastoral tribes. Among her recent publications are *Mobile Pastoralists: Development Planning and Social Change in Oman* (1996) and *Organizing Women: Informal and Formal Women's Groups in the Middle East* (with Annika Rabo (eds), 1997). She is currently examining the impact which conservation schemes have on the mobility and livelihoods of pastoral populations, focusing on the recent animal reintroduction schemes in Oman, Jordan and Syria.

Marcus Colchester received his doctorate in anthropology at the University of Oxford. As Projects Director of Survival International his work focused on the human rights impacts of imposed development schemes, especially in Amazonia and South and South East Asia. He sat on the International Labour Organization's expert committee on the revision of Convention 107. He is a founder member of the World Rainforest Movement and set up the Forest Peoples Programme which has developed into a well-known NGO active in the field of indigenous rights and the environment. He is currently Director of the programme. He has strongly advocated reforms in conservation policies to respect indigenous peoples' rights. In 1994 he was awarded a Pew Foundation Conservation Fellowship in recognition of his work in this field. He has acted as a consultant for the International Commission on International Humanitarian Issues, the United Nations Research Institute on Social Development, the World Bank, the World Commission on Dams and the Biodiversity Support Programme. He has published extensively in academic and NGO journals and is the author and editor of numerous books including *The Struggle for*

Land and the Fate of the Forests (1993) with Larry Lohmann and *Guyana: Fragile Frontier – Loggers, Miners and Forest Peoples* (1997).

Chris de Wet is Professor and Head of Department of Anthropology at Rhodes University, South Africa. He has twenty years research experience on population resettlement arising from development projects and from political programmes. Recently he has acted as a consultant for the World Commission on Dams, and served on the Environmental Review Panel for the Maguga Dam in Swaziland. Currently he is coordinator of a project on development-induced displacement and resettlement, funded by the Department for International Development (UK) and carried out at the Refugee Studies Centre, University of Oxford. He is the author of a monograph on resettlement in South Africa, and the co-editor of three books.

Christopher R. Duncan is the Royal Anthropological Institute Fellow in Urgent Anthropology at Goldsmiths College, University of London. His research focuses on conversion to Christianity, resettlement and changing resource use patterns among the Forest Tobelo, a group of forest-dwelling foragers on the island of Halmahera in eastern Indonesia. He is currently editing a book on the various policies that Southeast Asian governments have for resettling and developing peripheral minorities.

Cristina Eghenter studied philosophy at the University of Florence, Italy (BA) and anthropology and human ecology at Rutgers University, USA (MA and PhD). Her doctoral research focused on the causes and circumstances of long-distance migrations among the indigenous people of the interior of East Kalimantan, Indonesia. In 1995, she was appointed Field Director of Culture and Conservation, an interdisciplinary research programme on the cultural dimensions of forest management sponsored by the Ford Foundation and linked to the Kayan Mentarang project, WWF Indonesia, in East Kalimantan, Indonesia. After that initial assignment, she continued to be involved in the management of the project, a large conservation and development project in the Kayan Mentarang National Park. As Director of the Community Development Programme for the last three years, she designed and co-ordinated activities in support of community-based management for the National Park, including: community mapping; participatory planning for the management of the park; training of community representatives and project staff; advocacy of indigenous rights in the conservation area; analysis of economic potential. Her affiliation as a research fellow with the Centre for South-East Asian Studies, University of Hull, has allowed her to carry on writing and publishing on environmental and development issues in Borneo. She is also a member of the Advisory Board of the Center for Social Forestry at Mulawarman University, East Kalimantan, Indonesia.

Jim Ellis is Senior Research Scientist at the Natural Resource Ecology Laboratory, Colorado State University. He is an ecosystem ecologist who has worked in arid and semi-arid pastoral ecosystems for the past twenty years. He has worked extensively in Africa and is currently working in Central Asia, Mongolia and in China. Current research includes examining the effects of a strong global warming trend on the steppes of Mongolia and Inner Mongolia. He is also exploring the extent of rangeland degradation and paths to economic and ecological recovery in Central Asia. In Africa he is leading a team to provide information and understanding to African policy makers and land managers on balancing the needs of wildlife, ecosystem integrity, and pastoral food security. He has written extensively on African rangelands ecosystems including the seminal publication on pastoral non-equilibrium systems in the *Journal of Range Management* (1988).

Christo Fabricius is Associate Professor and Head of the Environmental Science Programme at Rhodes University, Grahamstown, South Africa, where he directs a growing group of young scientists specializing in people–environment interactions. With a PhD from the University of Cape Town he has extensive local and international experience in linkages between social and ecological systems, the role of biodiversity in rural livelihoods, and environmental policy making and planning. Before entering academia he spent 12 years as a conservation scientist in South Africa, and two years as a Research Associate with the International Institute for Environment and Development (IIED) in London, where he became involved in natural resource management projects in Latin America, Asia, South-East Asia and throughout the SADC region. He regularly advises South African forestry and conservation agencies on policy matters.

Eleanor Fisher is a Social Anthropologist based in the Centre for Development Studies, University of Wales, Swansea. Her research includes studies on livelihood issues, forced resettlement, natural resource management, African apiculture and ethical trade (Europe, Africa and the Caribbean).

Kathleen A. Galvin is Associate Professor in the Department of Anthropology and Senior Research Scientist at the Natural Resource Ecology Laboratory, Colorado State University. She has published on issues of African pastoral adaptation, health, nutrition and strategies of coping with climate variability and conservation policy. She is exploring the effects of climate variability on land use and the use of climate forecasts among indigenous and commercial ranchers in South Africa. She is examining the same issues among farmers and ranchers in the US Great Plains. She has also been investigating pastoral land use, well-being and conservation issues in northern Tanzania. Strategies for balancing pastoral food security, biological conservation and ecosystem integrity in East Africa with use of an integrated modelling and assessment system is also a research focus.

Chris Griffin studied sociology as an undergraduate before turning to postgraduate studies in social anthropology. Following on fieldwork in France which led to a PhD from Sussex University, Chris moved to the University of the South Pacific, in Fiji, where from 1975 to 1982 he taught in the sociology department. In 1982 he returned to the UK and spent a year at London University's Institute of Commonwealth Studies. Later he became a Gypsy Site Warden in London, working mainly with Irish Travellers but also some Romany families. In 1987 he moved to Western Australia to take up a lecturing position in what is now Edith Cowan University. His teaching and research interests include 'travel', political anthropology, indigenous issues, ethnicity, and applied anthropology. Trying to balance his interests in Fiji with those in Travellers and Gypsies (including Gypsies in Australia) has proved a perpetual challenge. His relationship with Daniel Meshack grew from years of earlier correspondence and in 1999 he was his guest at the Rural Education and Development Association in Chennai.

Graham Griffin is a Senior Research Scientist with the Commonwealth Scientific and Industrial Research Organization's Centre for Arid Zone Research at Alice Springs in Central Australia. Trained in geology, he worked as an exploration geologist across Australia in the early 1970s. Following postgraduate studies in ecology, he worked for over 20 years as a systems and landscape ecologist in arid Australia. Graham has undertaken wide-ranging studies in the ecology and management of fire in desert ecosystems, re-establishing traditional Aboriginal burning management regimes in vast areas of desert grasslands. He has undertaken extensive work in the fields of natural research inventory, landscape and process heterogeneity, remote sensing image analysis, species distribution modelling, analysis and application to regional planning, focused on sand desert and arid mountain range ecosystems. Graham has been undertaking ecological studies, particularly fire research and human impact studies, at Uluru Kata Tjuta National Park and in other arid regions since the early 1980s. He was appointed by the Federal Minister for the Environment as a member of the Aboriginal owned and managed National Park Board of Management in 1987 and has been active in strategic development and policy setting in joint management with Aboriginal owners since that time.

Jim Igoe conducted ethnographic fieldwork in several Maasai communities on the borders of Tarangire National Park in Tanzania, between 1993 and 1997. His research focused on community-based land rights movements and Maasai NGOs, with special interest in issues of community conservation. His dissertation examines the historical and contemporary links between these movements and global institutions of money, power, and ideas. In a separate project, he examined the impacts of conservation and development on local resource management systems in Tarangire. Jim Igoe is now an assistant professor at the University of Colorado at Denver.

Stacy Lynn is a PhD student in the Graduate Degree Program in Ecology, Colorado State University. Her work focuses on the triangular interface of pastoral populations and their livestock, local ecology, and conservation policy.

Her Master's thesis 'Conservation Policy and Local Ecology: Effects on Maasai land use patterns and human welfare in northern Tanzania' (CSU 2000) investigated the relative effects of policy and ecology on human welfare in the Ngorongoro Conservation Area, Tanzania and used a geographic information system to study the effects of ecology on cattle herd migration distance and settlement patterns. Her PhD research is part of a 'Global-Livestock Collaborative Research Support Program' grant to investigate the pastoralism–ecology–policy interface in and around protected areas of East Africa.

Ann Magennis is Associate Professor in the Department of Anthropology, Colorado State University. A biological anthropologist, Ann is interested in children's growth, both for extant and skeletal populations. Most of her work has been in the New World but she has recently worked with pastoral populations in Tanzania.

J. Terrence McCabe received his MA and PhD (1985) in Anthropology from the State University of New York at Binghamton. He was an Assistant Professor at the University of Georgia from 1985 to 1989. He has been at the University of Colorado at Boulder since 1989. He is a Faculty Research Associate in the Environment and Behavior Program of the Institute for Behavioral Science. He is also Associate Professor of Anthropology in the Department of Anthropology. For the past twenty years he has been studying the processes by which people use the land and manage natural resources in the arid and semi-arid savannas of East Africa. In particular, Dr McCabe worked among the Turkana of Northwest Kenya during the 1980s and early 1990s. He was a contributing researcher on the South Turkana Ecosystem Project, a large multi-disciplinary project involving anthropologists and ecologists. The STEP has been referred to as the most detailed study of a human population conducted within an ecosystem framework. An ethnography is currently being written by Dr McCabe based upon his sixteen years of work among the Turkana. In the last ten years, Dr McCabe has also been working in northern Tanzania. He has been examining the impact of conservation policy on the economy and land use practices of the Maasai living in the Ngorongoro Conservation Area (NCA). This research began in 1989 with funding from the International Union for the Conservation of Nature. It has now become another multi-disciplinary project involving many of the natural and social scientists who worked on STEP. Over the next five years, Dr McCabe's research will continue to focus on population processes, land use change, and the conservation of natural resources with a special emphasis on the relationships between wildlife and indigenous peoples.

Pamela McElwee is a PhD candidate in the Departments of Forestry and Environmental Studies and Anthropology at Yale University. She also holds an MSc in Forestry from Oxford. Since 1996, she has been conducting research in Viet Nam on the effects of biodiversity conservation on local livelihoods.

Daniel Meshack holds degrees in theology and development sociology and is the founder, Director, Secretary and principal field officer of the Rural Education and Development Association (READA), Madras, a community development agency. For twenty years READA has focused its attention on the plight of Dalits (or 'untouchables'), Fisher peoples, Narikuravas Gypsies and Tribals in Tamilnadu; its modus operandi, participatory action research. READA projects include Narikuravas leadership training, the organization and promotion of women's associations, urban land acquisition, fresh water supply, infant education, adult literacy, AIDS awareness, public education programmes for the greater tolerance of Narikuravas, networking other ethnic communities and public figures in support of Narikuravas' rights, on-going advocacy and legal representation. Daniel is himself a Dalit, and a member of the Church of South India; his development work is informed by Freire and other liberation theological-sociologists. He has helped organize numerous workshops and conferences on development issues, and writes poetry when time allows. He is married with two children.

Miguel Montoya, PhD is a researcher at the Department of Social Anthropology, Stockholm University. He is currently involved in a study of settlers in the Ticoporo Forest Reserve in Venezuela, and is also working on a project in economic anthropology about stock market investors in the emerging markets. He has previously written about peasants, state agencies and involuntary migration in Western Venezuela in connection with the construction of a hydroelectric scheme. He spent the 1997–98 academic year doing postdoctoral work at the University of Texas at Austin.

Tom Nordblom holds a PhD in Agricultural and Resource Economics from Oregon State University. He served at ICARDA (Aleppo, Syria) from 1981 to 1998, and at Charles Sturt University (Wagga Wagga, Australia) where he has been a Senior Research Fellow in the School of Agriculture from 1998. His professional experience includes research, training and outreach in domestic and international agricultural development contexts: benefit-cost analyses *a priori* and *ex post*, short and long-run project impacts on investment, livelihoods, environment. His interests include decision modelling, resource allocation, development of institutions for sustainable resource management and, in particular, the challenges of quantifying the trade-offs among the conflicting goals of economic efficiency, social equity and environmental sustainability.

Jonathan Rae who has recently completed his DPhil at Oxford, is Human Geography lecturer at the University of Brighton. His research focuses on institutions for natural resource management in common property situations and in particular dryland pastoral systems. He continues to contract with the Land Tenure Service of the Food and Agriculture Organization of the United Nations on projects in the Syrian Steppe, and as a communications consultant with non-government organizations.

Jin Sato is associate professor of environmental politics at the Institute of Environmental Studies, Graduate School of Frontier Sciences at the University of Tokyo. He has a BA in Anthropology and MA and PhD in International Relations from the University of Tokyo. He also has a Master's Degree in Public Policy from the Kennedy School at Harvard. He was a visiting scholar at Kasetsart University in Thailand (1995–97) and Post-Doctoral Fellow at the Agrarian Studies Program, Yale University (1998–99). His recent publication in English is: 'People in Between: Conversion and Conservation of Forest Lands in Thailand', *Development and Change*, 31(1).

Nicole Smith did her Masters degree at Colorado State University in the Anthropology Department. She did her thesis fieldwork in northern Tanzania. Her thesis, entitled 'Maasai Household Economy: A comparison between the Loliondo Game Controlled Area and the Ngorongoro Conservation Area, (NCA) Northern Tanzania' looked at the effects of conservation policy on the economic status of the NCA Maasai relative to the economic welfare of Maasai living outside the conservation area. She currently is living in southern Colorado.

Sian Sullivan (PhD Anthropology, London 1998) is a British Academy Post-Doctoral Research Fellow in the Department of Anthropology at the School of Oriental and African Studies (London University). Her academic interests include cultural landscapes, dryland ecology and resource use, 'community'-based conservation, environment and development discourses, gender, dance and 'the body'. Among her publications are *Political Ecology: Science, Myth and Power* (co-edited with P. Stott, 2000), articles in *Anthropos, Journal of Biogeography, Cimbebasia*, and *Economic Botany*, and contributions to several edited volumes (see references cited). She dances and performs with a small London-based group called Gravitas Dance Company (www.gravitas-dance.com).

Dimitrios Theodossopoulos is a lecturer in the Anthropology Department at the University of Wales, Lampeter. In the early nineties he carried out fieldwork on environmental politics and the indigenous perceptions of the environment in rural Greece. Despite his early involvement in ecological conservation, his work is a sustained critique of environmentalism, inspired by the axiom that a thorough study of indigenous cultures is a fundamental step to understanding conflicts over the environment. He is currently teaching and writing anthropology on a variety of themes, ranging from the human–environmental relationship to the ethnography of conflict and nationalism in the Balkans. His most recent field of interest focuses on the anthropology of Lower Central America and, in particular, the Garifuna (Black Carib) society.

Philip Thornton is programme co-ordinator of the Systems Analysis and Impact Assessment Programme at ILRI, the International Livestock Research Centre in Nairobi. His training is in farm management and agricultural economics. In addition to work in priority setting and impact assessment, he is involved in modelling land-use change and crop–livestock interactions in smallholder systems in the tropics and subtropics. Current research activities include poverty mapping in East Africa, modelling household economics of pastoral systems, and assessing the possible impacts of climate change on smallholders' agricultural systems over the next 50 years.

David Turton (BSc (Soc.), PhD) is a social anthropologist who retired as Director of the Refugee Studies Centre at the University of Oxford in December 2000. He previously taught in the Department of Social Anthropology at the University of Manchester where he helped to establish the Granada Centre for Visual Anthropology. His field research has been mainly in southwestern Ethiopia amongst the Mursi, a group of cattle herders and cultivators, and his theoretical interests have been in responses to drought and long-term ecological change, warfare, ethnicity, and ethnographic film. Apart from many articles on the Mursi, his publications include, as editor, *War and Ethnicity: Global Connections and Local Violence* (1997) and, as co-editor, *Warfare amongst East African Herders* (1978), *Film as Ethnography* (1990), *Cultural Identities and Ethnic Minorities in Europe* (1999), and *Ethnic Diversity in Europe: Challenges to the Nation State* (2000). He has made six programmes for television on the Mursi.

Reed L. Wadley is Assistant Professor of Anthropology at the University of Missouri-Columbia (USA). He has been Research Fellow at the International Institute for Asian Studies (Netherlands), Associate Scientist with the Center for International Forestry Research (Indonesia), Instructor in Anthropology at Arizona State University (USA), and Consultant with the Asian Wetlands Bureau (now Wetlands International) and with the Center for International Forestry Research. His publications include 'Disrespecting the Dead and the Living: Iban Ancestor Worship and the Violation of Mourning Taboos', *Journal of the Royal Anthropological Institute* (1999), 5(4): 595–610; 'Understanding Local People's Use of Time: A Pre-condition for Good Co-management', *Environmental Conservation* (1999), 26: 41-52 (with Carol Colfer and P. Venkateswarlu); and 'Hunting Primates and Managing Forests: The Case of Iban Forest Farmers in West Kalimantan, Indonesia', *Human Ecology* (1997), 25: 243–271 (with Carol Colfer and Ian Hood).

1

Introduction

CONSERVATION AND MOBILE INDIGENOUS PEOPLES

Dawn Chatty and Marcus Colchester

The close of the twentieth century has witnessed an upsurge in international concern about people's impact on the natural environment. As pressure on natural resources has intensified, the conventional means of protecting habitat and preventing species extinctions, through the establishment of 'protected areas', has increasingly come into question. Conventional conservation approaches have been accused of ignoring the wider forces causing environmental damage and, even, of being part of the same mindset, which imposes land use categories from the 'top-down', classifying lands as protected areas or zones. This, say the critics, has only legitimized and encouraged unsustainable land use outside protected areas, placing further pressure on natural resources and the beleaguered protected areas themselves. Some have, thus, demanded broader changes in national and global economies and focused attention on the underlying causes of environmental destruction – social injustice, the lack of secure land tenure, the enclosure of the commons, consumerism, the rise of corporations, global trade, and government collusion or indifference (WRM 1990; IUCN 1991; Colchester and Lohmann 1993; Ecologist 1993; Verolme and Moussa 1999; Barraclough and Ghimire 2000; Wood et al. 2000).

The classic conservation approach has also been challenged from a different but related quarter. As our appreciation of the value of traditional knowledge and community-based natural resource management has grown (Posey 1999; Roe et al. 2000), there has been a corresponding growth of concern about the social impacts of the imposition of protected areas on indigenous peoples (West and Brechin 1991; Wells and Brandon 1992; Kemf 1993; Colchester 1994; Ghimire and Pimbert 1997).

Indigenous Peoples and Forced Migration

The linking themes, which bring these chapters together, are those of displacement and forced migration, which have emerged as a central area of research at the Refugee Studies Centre at the University of Oxford. The literature on forced resettlement in general and of indigenous peoples in particular is now very extensive and will not be reviewed here (see for example Cernea and Guggenheim 1993; Cernea 1999). Yet the sheer scale of the process is still little appreciated. For example, it is estimated that between 40 and 80 million people worldwide have been displaced by large dams alone (WCD 2000: 104) and our ignorance of the exact numbers of people displaced itself provides shocking evidence of the degree to which local community interests can be ignored by planners. Indigenous peoples have suffered disproportionately from this process (Colchester 1999; WCD 2000). Summarizing the impacts of forced resettlement on rural communities, the World Bank (1994: iii–iv) notes:

> When people are forcibly moved, production systems may be dismantled, long-established residential settlements are disorganized, and kinship groups are scattered. Many jobs and assets are lost. Informal social networks that are part of daily sustenance systems – providing mutual help in childcare, food security, revenue transfers, labour exchange and other basic sources of socio-economic support – collapse because of territorial dispersion. Health care tends to deteriorate. Links between producers and their consumers are often severed, and local labour markets are disrupted. Local organizations and formal and informal associations disappear because of the sudden departure of their members, often in different directions. Traditional authority and management systems can lose leaders. Symbolic markers, such as ancestral shrines and graves, are abandoned, breaking links with the past and with peoples' cultural identity. Not always visible or quantifiable, these processes are nonetheless real. The cumulative effect is that the social fabric and economy are torn apart.

We lack accurate statistics about just how many people, indigenous or otherwise, have been displaced to make way for protected areas. One estimate suggests that as many as 600,000 'tribal' people have been displaced by protected areas in India alone (PRIA 1993). In recent years, with involuntary resettlement becoming increasingly questioned, the trend in India has been to place so many restrictions on indigenous peoples to limit their movements and livelihoods as to make their continued residence almost impossible, obliging them to relocate 'by choice' (JBHSS 2000).

The chapters in this volume show just how widespread this process of forced resettlement continues to be. The case studies also illustrate a very wide range of social situations resulting from the imposition of protected areas on indigenous peoples and other local communities ranging from forced removals, serious impoverishment, through strategies of resistance and conflict management, to 'best practice' examples of co-management, as in Australia.

The papers reveal how hard it is to impose clear categories on the manifold local experiences. Some marginalized peoples like the 'Gypsies' of Tamilnadu described by Meshack and Griffin (this volume) are obliged to be so mobile that isolating a single cause of displacement makes little sense. Yet many other mobile or 'nomadic' peoples in fact have quite well-defined territories (Chatty 1996). Not all protected areas that have led to the displacement of prior residents have directly caused 'involuntary resettlement'. Rather in many cases such as that of the Bedouin of the Negev described by Abu Rabia (this volume), the impositions of protected areas have limited the extent of subsistence or grazing lands, leading inexorably to out-migration from the contested areas. Displacement, intended or not, has been the result. The clarification is important at a time when the World Bank is engaged in revising its policy on 'Involuntary Resettlement' and is seeking to discriminate between those whom its projects oblige to move to make way for development and those whose livelihoods are curtailed due to imposed conservation areas. Controversially, benefits accorded the former – such as being informed about their options and rights, being consulted about alternatives, provided with prompt compensation, ensured the timely sharing of information, infrastructural support, provisions of alternative livelihoods, and (where possible) replacement land for land lost – are denied the latter group (World Bank 2001).

Recent Historical Precursors to Current Conservation Paradigms

Government organized parks and protected areas first made their appearance in America and Europe during the nineteenth century. Significant areas of land were set aside as 'wilderness', to be preserved 'untouched by humans', for the good of humankind. In 1872 a tract of hot springs and geysers in northwestern Wyoming was set aside to establish Yellowstone National Park. The inhabitants of the area, mainly Bannock, Crow, Sheepeater and Shoshone native American Indians, were driven out by the army, which took over management of the area (Morrison 1993).

In the United Kingdom, conservationists were mainly foresters whose philosophy stressed that the public good was best served through the protection of forests and water resources, even if this meant the displacement of local communities (McCracken 1987: 190). This expertise and philosophy was transferred abroad to all of Great Britain's colonial holdings. However, whereas in Britain National Parks, in the main, recognize existing rights and established farming systems, in the colonies the customary rights of native peoples were often denied (Harmon 1991; Colchester 1994). Now, more than a century later, most national parks in Latin America, Asia, Africa and the rest of the developing world have been, and to an extent still continue to be, created on the model pioneered at Yellowstone and built upon by the early British colonial conservationists. The fundamental principle of operation remains to protect the park or reserve from the damage which the indigenous or other local communities are supposed to inflict.

In much of the developing world, conservation efforts during the past century have been largely based on the assumption that human actions negatively affect the physical environment. Problems like soil erosion, degradation of rangelands, desertification, and the destruction of wildlife have been viewed as principally due to local, indigenous misuse of resources. Recent studies have clearly shown that models of intervention developed in the West, in its particular historical context, have been transferred to the developing world with no regard for the specific contexts of the actual receiving environments or peoples (e.g. Sanford 1983; Anderson and Grove 1987; Manning 1989; Behnke et al. 1991). For example, the common Western, urban notion of wilderness as untouched or untamed land has pervaded conservation thinking. Many policies are based on the assumption that such areas can only be maintained without people. They do not recognize the importance of local management and land-use practices in sustaining and protecting biodiversity. Nearly every part of the world has been inhabited and modified by people in the past, and apparent wildernesses have often supported high densities of people (Colchester 1994; Pimbert and Pretty 1995). In Kenya, for example, the rich Serengeti grassland ecosystem was, in part, maintained by the presence of the Maasai and their cattle (Adams and McShane 1992). There is good evidence from many parts of the world that local people do value, utilize and efficiently manage their environments (Nabhan et al. 1991; Oldfield and Alcorn 1991; Scoones et al. 1992; Novellino 1998; Abin 1998) as they have done for millennia. These findings suggest, in complete reversal of recent conservation philosophy, that it is when local or indigenous people are excluded that degradation is more likely to occur. 'It suggests that the mythical pristine environment exists only in our imagination' (Pimbert and Pretty 1995: 3).

Rangeland management has had a similar history of Western philosophies and technologies being transposed onto the developing world. The concept of sustainable yield and the goal of improved productivity originated in Germany and North America, respectively, and were rapidly adopted in Australia. All these territories were organized on a system of privately-owned land. For the last fifty years, policy makers have defined the major concern of pastoral regions of the developing world to be overstocking that leads to certain ecological disaster. In this view the problem (too many livestock) has a technical solution (destocking). However, the central assumption being made is that pastoral ecosystems are potentially stable and balanced, and become destabilized by overstocking and overgrazing. This bias has led to the establishment of a multitude of development projects that promoted group ranching, grazing blocks and livestock associations. But these schemes have failed, leading to a fundamental questioning of the basic assumptions underlying this tradition of range management. Behnke et al. (1993) have admirably shown that pastoral systems are not equilibrium systems. Instead they are continuously adapting to changeable conditions, and their very survival depends upon this capacity to adapt. It is, in fact, the 'conventional development practices themselves that are the destabilizing influences on pastoral systems, as they have prevented traditional adaptive systems from being used' (Pimbert and Pretty 1995: 5).

Early Colonial Policy towards Indigenous People in Protected Areas

As has been briefly summarized above, in the late nineteenth century and throughout the first half of the twentieth century, conservation meant the preservation of flora and fauna and the exclusion of people. As was the case in the formation of Yellowstone National Park, armies or colonial police forces in Latin America, Africa, Asia and much of the developing world have been employed to expropriate and exclude local communities from areas designated as 'protected', often at great social and ecological costs. Forced removal and compulsory resettlement, often to environments totally inadequate for sustainable livelihood, were common practices.

Accompanying this forced removal was the view that indigenous people who rely on wild resources are 'backward' and so need help to be developed. Occasionally the 'primitive' or 'backward' habits of the indigenous people were regarded as attractive for tourism and, in carefully regulated circumstances, a limited number of groups, such as the San in areas of the Kalahari, were allowed to remain in or near traditional lands. The situation of the Maasai in Kenya and Tanzania is another example (Jacobs 1975; Lindsay 1987). In 1904, in an effort to pacify the Maasai and to clear preferred land for European settlers, the British government created the Northern and Southern Maasai Reserves. Subsequently, over the next ten years, the colonial government abolished the Northern Reserve and forced its resident population to move, effectively denying them access to much productive rangeland. It prohibited all hunting of wild animals on the reserve, although many authorities apparently felt that the Maasai could continue to coexist with the wildlife population. These reserves served the purpose of preserving primitive Africa where 'native and game alike have wandered happily and freely since the Flood' (Cranworth 1912: 310 quoted in Lindsay 1987: 152).

Post-colonial Policy

By the 1940s and 1950s, late colonial policies and early independent government policies began to change. The image of the harmless, pristine native was replaced by that of a dangerous and uncivilized local. Meanwhile indigenous peoples, already highly constrained if not prohibited from pursuing their livelihoods as they had in previous centuries, became more often regarded as backward primitives, and as impediments not only to the state's conservation policy, but also to its general desire to modernize and develop. Subsistence systems were denigrated, and policies were adopted aimed at forcibly settling not only nomadic pastoralists, but also swidden farmers and hunters and gatherers. As Fisher (this volume) demonstrates, resettlement of native peoples in colonial Tanganyika was often justified in the colonial mind as a means of promoting development, easing administration, and providing essential services such as health care. Little thought was given to indigenous priorities, perspectives or even systems of resource use and the administrators' discourse evolved to legitimize these manipulations. In the Middle East, the newly-independent countries set about settling Bedouin in an effort to 'modernize' them and, as in Israel, a sometimes repressive state took little account of local knowledge or pastoral regimes (Abu Rabia this volume). In Latin America, India and South East Asia similar policies took hold, either settling indigenous communities to better control them, or pushing them off land which the state deemed important for its own economic development (Ewers 1998; Dangwal 1998). These kinds of impositions led to a denial of land rights (Turton this volume; Rae, Arab and Nordblom this volume) and were often shaped by quite explicit prejudices against mobile peoples (Sato this volume; Meshack and Griffin this volume). The same arguments about the need to provide government services to such backward peoples are still deployed in Indonesia (Li 1999b) but the underlying purposes are geopolitical and economic, a means of freeing up access to resources for other interests (Duncan this volume). As in Vietnam, the establishment of protected areas, and concomitant expulsion of local residents, is also motivated by government needs for foreign exchange (McElwee this volume).

In East Africa, post-colonial policy was directed at sedentarizing the Maasai and shifting their livestock economy from a subsistence to a market basis. As Lindsay summarizes (1987: 152–5), the government constructed dams and boreholes, and tapped watering holes. Livestock numbers appeared to increase rapidly over subsequent decades, and by the 1960s conservationists began to perceive that wildlife in the reserves was being threatened by the Maasai herders and their livestock. The reserve boundaries were redrawn and talk began to centre on the possible exclusion of Maasai livestock from the reserves. With growing tourist revenues, the government declared a livestock-free area in the middle of the reserves to protect wildlife. Local Maasai elders began to demand formal ownership of all the land in the region. A confrontation between resident pastoralists and government/conservationists was inevitable. Maasai began killing wildlife such as rhinoceroses and elephants in protest against the threatened loss of more grazing land. More recently there has been some evidence of collaboration with poachers in response to the trade in ivory and horn (Douglas-Hamilton 1979).

Experiences with Protected Areas

Indigenous spokespersons reviewing their experience with protected areas in Latin America have remarked that conservation and development are just two sides of the same coin (Gray et al. 1998). They experience both as top-down impositions which deny their prior rights to land and devalue their indigenous knowledge and systems of land use (Colchester and Erni 1999).

This experience has been a worldwide phenomenon. In East Africa, for example, people have long been forced off their lands in order to create parks and sanctuaries for wildlife (Turton 1987; Howell 1987; McCabe et al.

1992; Kwokwo Barume 2000). Authors in this volume detail these impacts on indigenous peoples in the Middle East (Chatty), South Africa (Fabricius and de Wet) and East Africa (McCabe). Galvin and her colleagues (this volume) provide detailed evidence of the real impoverishment caused by the progressively tightening restrictions imposed on the Maasai in the Ngorongoro conservation area.

The assumption was that local communities overstocked, overgrazed or otherwise overused the natural environment and were thus obstacles to effective natural resource management. 'Scientific' management of these areas was assumed to require the removal of the indigenous peoples for the long-term benefit of these wildlife preserves or the imposition of strict limits on their livelihoods. However, the capacity of State institutions to regulate better these resources is not always clear. Montoya (this volume) describes how a rapidly degrading area of open access in Venezuela was nominally converted into protected State property as the Ticoporo Reserve without taking into account either the current occupants or further pressure from colonists. Using the 'weapons of the weak', the colonists have been able to take over most of the reserve and some have even gained secure title to their holdings. Rent seeking by forest guards has undermined efforts of protection, allowing logging to strip out the best timber. The author recommends reducing the role of the State except to further secure peasant tenure rights and encourage agroforestry as a way of improving forest cover. Likewise Turton (this volume) argues that the main threat to the 'sustainability' of the local ecology in the Omo valley in Southern Ethiopia comes not from the Mursi but from top-down conservation initiatives which will cause forced resettlement or severely curtail traditional rights of use and access. Nearer home, Theodossopoulos (this volume) details local resistance on the island of Zacynthos in Greece where imposed conservation laws will extinguish land titles, without compensation.

These cases also reveal clearly how conservation objectives are undermined when local people's needs and rights are ignored. As Chatty (this volume) demonstrates in her examination of conservation initiatives in Jordan and Syria, although initially the curtailment of local rights may, for a time, allow projects to be portrayed as 'successes', in the longer term conservation goals have been defeated by local resentment and resistance.

Recent Alternatives to the Traditional Conservation Paradigm

The near universal model of protected areas and natural parks, which was derived from a Western, positivist approach to science, has lately shown signs of accepting alternative paradigms. Until quite recently scientific investigation was dominated by the Cartesian positivist or rationalist paradigm. This assumes the existence of only one reality, and that the aim of science is to discover, predict, and control that reality. This approach reduces the complex aspects of a problem into discrete parts that can be analysed, so that predictions can be made on these discrete parts. It is then assumed that knowledge can be summarized into universal laws or generalizations. Conservation science is firmly set within this paradigm, and so too are the inherently ethnocentric basic values and assumptions of its professionals. This has produced a body of work and industry based on a top-down transfer of technology model of conservation that has consistently ignored the complexity of ecological and social relationships at the local level (Pimbert and Pretty 1995: 13; Jensen 1998).

For several decades now, however, a minority opinion has grown that argues for a more pluralistic way of thinking about the world and how to change it (e.g. Kuhn 1962; Checkland 1981; Vickers 1981; Pretty et al. 1994). It is becoming increasingly clear that ecological systems of plants and animals exist as a function of their unique pasts. Understanding the particular history of a community or ecosystem is critical for its current management. The old, conventional view of ecosystems as a function of their current operating mechanisms, and the assumption that human interference caused depletion of biological diversity, formerly justified the removal of people from national parks and reserves. But as ecosystems are now more clearly regarded as dynamic and continuously changing, the importance of people in their development and functioning is being acknowledged. Recent studies, for example, indicate that Amerindians played a far greater role in manipulating scrub savannas than had previously been suspected (Anderson and Posey 1989). In southeast Asia, it is now seen that the intermittent clearances of patches of land in the forests of the Pwo Karen never caused forest degradation, but rather encouraged the larger wild herbivores to enter by creating sporadic open spaces in the forest (Ewers 1998). In Africa, a few conservationists now realize that some biodiversity loss in protected areas actually stems from the restrictions placed on the activities of local communities. For example, the Serengeti grassland ecosystem is now understood to have been maintained in the past by the presence of the Maasai and their cattle. With the expulsion of the Maasai, the Serengeti is increasingly being taken over by scrub and woodland, leaving less grazing for antelope (Adams and McShane 1992). A similar lesson was learned in Tsavo National Park (Botkin 1990) where resource management to protect and control the elephants caused severe deterioration of the land within the park, while the inhabited area outside the park remained forested.

In the closing decade of the twentieth century, there was thus a change of heart, and international conservationist circles now reverberate with conceptual discussion of 'conservation with a human face' (Bell 1987), and the need for community participation (Cernea 1991; IIED 1994; Beltran 2000). Even conservation biologists who have given some the impression they are holding out for the old protectionist approach (Brandon et al. 1998), now agree that 'traditional and indigenous people can claim incontrovertible rights to their land' and as 'morally responsible humans we must support their struggle' (Redford and Sanderson 2000: 1362).

The preferred means for achieving a reconciliation of social justice and conservation goals is through 'community-based natural resource management' (CBNRM). These efforts seek to promote 'the collective use and management of natural resources in rural areas by a group of people with a self-defined, distinct identity' (Fabricius 2002: 2). A few promising examples of Latin American, Asian and African conservation efforts are now emerging where indigenous peoples are beginning to be effectively integrated into conservation and development projects. In Bolivia, for example, participatory research by the local indigenous community in the Balfor project has permitted members to harvest caiman and peccaries at sustainable levels, based on their own monitoring (Fabricius 2002). In Africa, the CAMPFIRE scheme in Zimbabwe has been widely promoted as allowing for the sharing of benefits – however small – by the community and at the same time giving indigenous peoples a voice in rural politics (Fabricius 2002). Fabricius and de Wet (this volume) argue that the community-based approach, combined with the provision of land security, is allowing conservation and development goals to be met simultaneously. In India, proposals are now being discussed to allow an indigenous people to become key actors in designing and implementing the management plan for the Rajaji National Park in Uttar Pradesh (Dangwal 1998: Colchester and Erni 1999). Encouragingly, some of the major international conservation organizations have responded to this challenge and are seeking out new forms of collaboration to secure indigenous peoples' rights and long-term conservation goals simultaneously (BSP 2000; Margoluis et al. 2000; Weber et al. 2000).

However, some authors in this volume warn us against facile 'solutions'. Understanding the complex histories of land use systems is important if current conflicts of interest over resources are to be resolved, as Wadley (this volume) demonstrates by examining the Danau Sentarum Wildlife Reserve in West Kalimantan (Indonesian Borneo). There are also dilemmas for conservationists and others promoting participation and the indigenous voice in deciding who speaks for the group, as Armstrong and Bennett (this volume) describe in their examination of the situation of the San of the Kalahari Game Reserve in Botswana. Sullivan (this volume) critically examines the record of community-based natural resource management in Namibia. She finds that the 'new' conservation policy has favoured large (mainly white settler) farmers and even in communal areas has failed to secure ownership rights or substantial benefits. Heavy costs are also borne by the villagers in terms of loss of crops, and even lives, to wildlife. She doubts the conservancies are sustainable without continuing flows of grant aid or a resort to heavy-handed police methods little different from the 'fortress conservation' of the past.

Indeed, a central concern of conservation science today is to find practical ways of putting people back into conservation, or, as Bell argues, to give conservation a human face (Bell 1987). This does not mean substituting the assumption that all indigenous peoples are environmental destroyers with an equally simplistic notion that all indigenous peoples are 'noble ecologists'. Indeed, there are doubts about the extent to which it is appropriate to represent indigenous knowledge through the discourse of 'biodiversity conservation' (Ellen et al. 2000; Lawrence et al. 2000). Long-term field studies suggest that some, perhaps many, indigenous peoples regulate their impact on the environment not so much through consciously limiting direct pressure on resources but through their own political processes which space out communities over wide territories (Colchester 1981; Harms 1981, 1999; Hames 1991). As pressure on land intensifies and wildlife habitats shrink, and as indigenous peoples are increasingly drawn into the market, overhunting of game is increasingly a problem (Robinson and Bennett 2000). Moreover, some ecologists have warned that the elimination of top-predators may lead to much wider impacts on ecosystems than might be predicted (Terborgh 1999 cited in Schwartzman et al. 2000). Many indigenous peoples recognize that changes in their economies, social organization and values may now pose a threat to the natural environments that they depend on, and they seek the assistance of conservationists in addressing these imbalances (International Alliance 1996). Eghenter (this volume) in her examination of Dayak land use in the Kayan Mentarang National Park in East Kalimantan demonstrates clearly the need for detailed studies of the environmental and social impacts of community resource use. However, the study by Galvin and her colleagues (this volume) warns us not to expect to find 'win–win' scenarios everywhere. Their long-term study of declining Maasai livelihoods and nutrition in Ngorongoro revealed no viable options for restoring Maasai standards of living, based on their current land use systems, which do not imply costs to conservation.

Beyond 'Participation'

'Participation' has now become part of the normal language of development theory. It has become so fashionable that almost everyone claims participation to be part of their work. In the world of conservation, the term has been used to justify the extension of control by the State, or to justify external decisions. In the 1970s 'participation' was often a scheme for achieving the voluntary submission of people to protected area schemes (*passive participation, and participation for material incentives*). Often it was no more than a public relations exercise in which local people were passive actors (*participation in information giving*). In the 1980s it was defined as local interest in natural resource protection (*participation by consultation*). In the 1990s, some agencies saw it as a means of involving people in protected area management (*functional participation and interactive participation*). All too often 'participation' in protected area management is quite nominal as Turton (this volume) documents for the Mursi in Ethiopia. Imposed management systems, while nominally 'participatory' may take little account of indigenous institutions and processes of decision-making, as the Maasai have found in Ngorongoro (McCabe this volume).

At last, and at least, we now recognize that without local involvement there is little real chance of protecting wildlife. Encouraging though such initiatives are they remain weak judged against the demands of indigenous peoples themselves and their rights as recognized in international law. Genuine recognition of indigenous rights to their lands and to self-determination, requires that conservationists engage as advisers to indigenous land owners, implying the need for a further transfer of power and resources in favour of marginalized groups and new mechanisms of accountability between 'outsiders' and indigenous people (Colchester 1996).

Conservation and Indigenous Rights in International Law

The rights of indigenous peoples in conservation concerns have long antecedents and have even found their way, albeit ambiguously, into international law. Notably, the global agreements negotiated at the Earth Summit

Table 1.1 A Typology of Participation

Typology	*Components of each type*
Passive participation	People participate by being told what is going to happen or what has already happened. It is a unilateral announcement by project management; people's responses are not taken into account.
Participation in information giving	People participate by answering questions posed by extractive researchers. People do not have the opportunity to influence proceedings.
Participation by consultation	People participate by being consulted, and external agents listen to views. Professionals are under no obligation to take on board people's views.
Participation for material incentives	People participate by providing resources, for example labour, in return for food, cash or other material incentives. It is very common to see this called participation, yet people have no stake in prolonging activities when incentives end.
Functional participation	People participate by forming groups to meet predetermined objectives. Such involvement tends to be after major decisions have been made. These institutions tend to be dependent on external initiators and facilitators.
Interactive participation	People participate in joint analysis, which leads to action plans. It tends to involve interdisciplinary methodologies that seek multiple perspectives. These groups take control over local decisions, and so people have a stake in maintaining structures or practices.
Self-mobilization	People participate by taking initiatives independent of external institutions to change systems.

Adapted from Pretty *et al.* 1994

in 1992 gave a prominent place to 'indigenous peoples'. For example the 'Rio Declaration' in Article 22 explicitly noted that:

> Indigenous peoples and their communities and other local communities have a vital role in environmental management and development because of their knowledge and traditional practices. States should recognize and duly support their identity, culture and interests and enable their effective participation in the achievement of sustainable development (cited in International Alliance 1997).

The Convention on Biological Diversity (CBD), which was also finalized at the Earth Summit and which has now been ratified by some 174 countries, also makes provisions relevant to indigenous peoples. Article 8(j) obliges States that are party to the convention 'as far as possible and as appropriate': 'Subject to its national legislation, to respect, preserve and maintain knowledge, innovations and practices of indigenous and local communities embodying traditional lifestyles relevant for the conservation and sustainable use of biological resources ...' Similarly, with reference to *in situ* conservation practices, article 10(c) obliges States, 'as far as possible and as appropriate' to: '*protect and encourage customary use of biological resources in accordance with traditional cultural practices that are compatible with conservation or sustainable use requirements*'.

In 1994, the World Conservation Union adopted a revised set of categories of protected areas which accept that indigenous peoples may own and manage protected areas (IUCN 1994). In 1996, following several years of intensive engagement with indigenous peoples' organizations, the World-Wide Fund for Nature-International adopted a *Statement of Principles on Indigenous Peoples and Conservation*, which endorsed the UN Draft Declaration on the Rights of Indigenous Peoples, accepts that constructive engagement with indigenous peoples must start with a recognition of their rights, upholds the rights of indigenous peoples to own, manage and control their lands and territories and to benefit from the application of their knowledge (WWF 1996).

The same year the World Conservation Congress, the paramount body of the World Conservation Union, adopted seven different resolutions on Indigenous peoples (IUCN 1996). These resolutions *inter alia*:

- Recognize the rights of indigenous peoples to their lands and territories, particularly in forests, in marine and coastal ecosystems, and in protected areas
- Recognize their rights to manage their natural resources in protected areas either on their own or jointly with others
- Endorse the principles enshrined in the International Labour Organization's Convention 169, Agenda 21, the CBD and the Draft Declaration on the Rights of Indigenous Peoples
- Urge member countries to adopt ILO Convention 169
- Recognize the right of indigenous peoples to participate in decision-making related to the implementation of the CBD

- Recognize the need for joint agreements with indigenous peoples for the management of protected areas and their right to effective participation and to be consulted in decisions related to natural resource management.

In 1999, the World Commission on Protected Areas adopted guidelines for putting into practice the principles contained in one of these six resolutions. These guidelines place emphasis on co-management of protected areas, on agreements between indigenous peoples and conservation bodies, on indigenous participation and on a recognition of indigenous peoples' rights to 'sustainable, traditional use' of their lands and territories (WCPA 1999).

The prominence given to them in this debate is the result of a spectacular resurgence of indigenous peoples, who have strategically and quite consciously mobilized to occupy political space at national and global levels to claim recognition of their human rights (Veber et al. 1993; Barnes et al. 1995; Wearne 1996; MRG 1999). A culmination of this sustained advocacy has been a re-interpretation of international human rights instruments in the light of the particular circumstances of indigenous peoples. A body of jurisprudence has resulted which effectively recognizes indigenous peoples' rights to the ownership, use and management of their lands and territories, to represent themselves through their own institutions, to exercise their customary law in conformity with other human rights standards, to their intellectual property, to a measure of self-governance, and to self-determination (Simpson 1997; Pritchard 1998; Roulet 1999; Kambel and MacKay 1999).

These rights have been consolidated in a number of human rights instruments including: the International Labour Organization's revised Convention on Tribal and Indigenous Peoples, No. 169, adopted in 1989; the draft United Nations Declaration on the Rights of Indigenous Peoples, which was adopted by the Sub-Commission on the Prevention of Discrimination and Protection of Minorities in 1993 and is now being reviewed by a working group in the United Nations Human Rights Commission; and the Proposed American Declaration on the Rights of Indigenous Peoples being developed by the Organization of American States, which amounts to a contextualized restatement of rights already recognized in the American Convention on Human Rights. The recent establishment of a United Nations Forum on Indigenous Issues at the level of the Economic and Social Council has further consolidated these gains (Garcia Alix 1999).

The importance of this growing recognition of indigenous peoples' rights has yet to be widely appreciated. It constitutes a significant shift in the evolution of international law, in which growing recognition is given to the *collective* rights of human groups to maintain their distinctive identities, customs and relations with their natural environments (Crawford 1988). Historians may look back on this era of growing recognition of indigenous peoples' rights as a sea-change as significant as the anti-slavery movements of one hundred and fifty years earlier.

Disputes remain, of course. A complex question that has yet to be addressed, in a way that satisfies all, is the *definition* of 'indigenous peoples' –

something which is conceptually impossible until the very notion of 'peoples', a basic element of international law, is itself defined (Kingsbury 1998). 'Objective' criteria that help identify who indigenous peoples are include the following, offered by the Chairperson of the UN's Working Group on Indigenous Populations and the World Bank:

- Priority in time with respect to the occupation and use of a specific territory
- Voluntary perpetuation of cultural distinctiveness
- Self-identification, as well as recognition by other groups, or by State authorities as a distinct collectivity
- An experience of subjugation, exclusion or discrimination (Daes 1996)
- Vulnerability to being disadvantaged in the development process
- Close attachment to ancestral territories and to natural resources in these areas
- Presence of customary social and political institutions
- Primarily subsistence oriented production (World Bank 1991)

What these 'check lists' of 'indigenousness' duly recognize is the principle of self-identification, a principle that is also incorporated into the ILO's Convention No. 169 (Article 1c) and the UN's Draft Declaration on the Rights of Indigenous Peoples. The editors of this volume accept both approaches, interpreting the term 'indigenous peoples' as both a polythetic and a self-ascribed class of human societies. Human groups collectively choose to identify themselves as 'indigenous peoples' in order to secure control of their lands and natural resources, to renegotiate their political relations with nation-states and to overcome discrimination and marginalization. As anthropologist Tania Murray Li (1999a: 151) notes: 'My argument is that a group's self-identification as tribal or indigenous people is not natural or inevitable, but neither is it simply invented, adopted or imposed. It is, rather, a *positioning* which draws upon historically sedimented practices, landscapes and repertoires of meaning and emerges through particular patterns of engagement and struggle.'

Sustainable Conservation and Development

Sustainable conservation requires, above all else, the acceptance of outsiders *by* the indigenous peoples. McCabe and others (1992: 353–66) suggest that linking conservation with human development offers the most promising course of action for long-term sustainability of nature and human life. McCabe argues that nature reserves and other protected areas must be placed into their own regional context. If the economy of the local communities is not vigorous, or is in a serious state of decline, the establishment of a wildlife reserve in its midst does not promote long-term sustainability. The population is unlikely to see any benefit from such a scheme and cooperation is unlikely. If, on the other hand, the problems of the human population are addressed and the community anticipates benefit from a combined conservation/development scheme, then cooperation and long-term sustainability are possible.

Measures that address the needs of wild and domesticated animals as well as the human group, such as veterinary care and prophylactic health campaigns for both animals and humans, water wells and water distribution, seed distribution and the extension of fodder crop growing, are a few of the broad array of programmes that can draw conservation closer to development.

Policy and principles, however, need to be translated into practice. As long as national governments are driven by global economic concerns, structural adjustment policies and debt repayment, small-scale, marginal and often illiterate indigenous communities have little chance of becoming equal partners in any participative exercise. Furthermore, multi-national corporate activities in tourism, mining, logging, and other extractive activities are powerful actors in the conservation discourse. It is of great importance that leading international conservation organizations take a more progressive stand and move beyond words into action. The study by Griffin (this volume) of the Anangu people in Uluru-Kata Tjuta National Park in Australia provides an example of how this is possible. The Anangu are recognized as landowners of the park, have accepted joint management and have begun to benefit from the income stream generated by tourism.

Along with the emerging forces and vibrant influences that many indigenous people themselves are now exerting at the local, national and international level, policy-makers, planners, managers and researchers need to remain vigilant that indigenous peoples are equal, if not more-than-equal, partners in the search for biodiversity conservation. Protecting the cultural diversity of our planet – in its continuous adaptation to its environment – is, after all, also part of the broader biodiversity that we all seek to preserve.

References

Abin, R. 1998. 'Plantations: Village Development Threatens the Survival of Indigenous Dayak Communities in Sarawak'. *Indigenous Peoples*, 4: 15–23.

Adams, J. and McShane, T. 1992. *The Myth of Wild Africa: Conservation without Illusion.* New York: W.W. Norton and Co.

Anderson, D. and Grove, R. (eds). 1987. *Conservation in Africa: People, Policies and Practice.* Cambridge: Cambridge University Press.

Anderson, D. and Posey, D. 1989. 'Management of a tropical scrub savanna by the Gorotire Kayapò of Brazil'. In *Advances in Economic Botany*, 7: 159–73.

Barnes, R.H., Gray, A. and Kingsbury, B. (eds). 1995. *Indigenous Peoples of Asia.* Ann Arbor: Association for Asian Studies.

Barraclough, S.L. and Ghimire, K.B. 2000. *Agricultural Expansion and Tropical Deforestation: Poverty, International Trade and Land Use.* London: Earthscan.

Behnke, R., Scoones, I., and Kerven, C. (eds). 1991. *Redefining Range Ecology: Drylands Programmes.* London: IIED

Behnke, R., Scoones, I. and Kerven, C. (eds). 1993. *Range Ecology at Disequilibrium: New Models of Natural Variability and Pastoral Adaptation in African Savannas.* London: Overseas Development Institute.

Bell, H. 1987. 'Conservation with a Human Face: Conflict and reconciliation in African land use planning'. In D. Anderson and R. Grove (eds). *Conservation in*

Africa: People, Policies and Practice, pp. 79–101. Cambridge: Cambridge University Press.

Beltran, J. (ed.). 2000. *Indigenous and Traditional Peoples and Protected Areas: Principles, Guidelines and Case Studies.* Gland: WCPA and IUCN.

Botkin, D. 1990. *Discordant Harmonies: A New Ecology for the Twenty-First Century.* New York: Oxford University Press.

Brandon, K., Redford, K.H. and Sanderson, S.E. 1998. *Parks in Peril: People, Politics and Protected Areas.* Washington DC: The Nature Conservancy.

BSP. 2000. *Lessons from the Field: Linking Theory and Practice in Biodiversity Conservation.* Washington DC: Biodiversity Support Program.

Cernea, M. 1991. *Putting People First: Sociological Variables in Rural Development.* New York: Oxford University Press.

Cernea, M. (ed.). 1999. *The Economics of Involuntary Resettlement: Questions and Challenges.* Washington DC: World Bank.

Cernea, M. and Guggenheim, S. (eds). 1993. *Anthropological Approaches to Resettlement: Policy, Practice and Theory.* Boulder: Westview Press.

Chatty, D. 1996. *Mobile Pastoralists: Development, Planning and Social Change in Oman.* New York: Columbia University Press.

Checkland, P. 1981. *Systems Thinking, Systems Practice.* Chichester: John Wiley.

Colchester, M. 1981. 'Ecological Modelling and Indigenous Systems of Resource Use: Some examples from the Amazon of South Venezuela'. *Antropologica,* 55: 51–72.

—— 1994. *Salvaging Nature: Indigenous Peoples, Protected Areas and Biodiversity Conservation.* Geneva: United Nations Research Institute for Social Development.

—— 1996. 'Beyond "Participation": Indigenous peoples, biological diversity conservation and protected area management'. *Unasylva,* 186(47): 33–9.

—— 1999. 'Sharing Power: Dams, indigenous peoples and ethnic minorities'. *Indigenous Affairs* (Special Double Issue), 3–4: 4–55.

Colchester, M. and Erni, C. (eds). 1999. *Indigenous Peoples and Protected Areas in South and Southeast Asia: From Principles to Practice.* Forest Peoples Programme and the International Work Group for Indigenous Affairs, Document 97, Copenhagen.

Colchester, M. and Lohmann, L. (eds). 1993. *The Struggle for Land and the Fate of the Forest.* London and Penang: Zed Books and World Rainforest Movement.

Crawford, J. (ed.). 1988. *The Rights of Peoples.* Oxford: Clarendon Press.

Daes, E. 1996. Supplementary report of the Special Rapporteur on the Protection of the Heritage of Indigenous Peoples. United Nations Sub-Commission on Prevention of Discrimination and Protection of Minorities, forty-eighth session. E/CN.4.Sub.2/1996/22.

Dangwal, P. 1998. 'Van Gujjars at Apex of National Park Management'. In *Indigenous Affairs,* 4: 24–31.

Douglas-Hamilton, I. 1979. *The African Elephant Action Plan.* Nairobi: IUCN/WWF/NYZS Elephant Survey and Conservation Programme.

Ecologist, The. 1993. *Whose Common Future? Reclaiming the Commons.* London: Earthscan.

Ellen, R., Parkes, P. and Bicker, A. 2000. *Indigenous Environmental Knowledge and Transformations: Critical Anthropological Perspectives.* Amsterdam: Harwood Academic Publishers.

Ewers, K. 1998. 'The Politics of Conservation: Pwo Karen Forest People of Thailand'. In *Indigenous Affairs,* 4: 32–5.

Fabricius, C. (in press) 'Community-Based Natural Resource Management'. In *Encyclopaedia of Life Support Systems.* EULSS Publishers and UNESCO. Paris.

Garcia Alix, L. 1999. *The Permanent Forum for Indigenous Peoples: The Struggle for a new Partnership.* Copenhagen: International Work Group for Indigenous Affairs.

Ghimire, K.B. and Pimbert, M.P. (eds). 1997. *Social Change and Conservation. Environmental Politics and Impacts of National Parks and Protected Areas.* London: Earthscan Publications.

Gray, A., Newing, H. and Padellada, A. 1998. *Indigenous Peoples and Biodiversity Conservation in Latin America: From Principles to Practice.* Copenhagen: International Work Group for Indigenous Affairs and Forest Peoples Programme.

Hames, R. 1991. 'Wildlife Conservation in Tribal Societies'. In M. Oldfield and J. Alcorn (eds). *Biodiversity, Culture, Conservation and Ecodevelopment,* pp. 172–99. Boulder: Westview Press.

Harmon, D. 1991. 'National Park Residency in Developed Countries: The example of Great Britain' In P.C. West and S.R. Brechin (eds). *Resident Peoples and National Parks: Social Dilemmas and Strategies in International Conservation,* pp. 33–9. Tucson: University of Arizona Press.

Harms, R. 1981. *River of Wealth, River of Sorrow: The Central Zaire Basin in the Era of the Slave and Ivory Trade, 1500–1891.* New Haven: Yale University Press.

—— 1999. *Games Against Nature: An Eco-Cultural History of the Nunu of Equatorial Africa.* Cambridge: Cambridge University Press.

Howell, P. 1987. 'Introduction'. In D. Anderson and R. Grove (eds). *Conservation in Africa: People, Policies and Practice,* pp. 105–09. Cambridge: Cambridge University Press.

ILO. 1989. *Indigenous and Tribal Peoples Convention No. 169.* Geneva: International Labour Organization.

International Alliance. 1996. *Indigenous Peoples, Forests and Biodiversity: Indigenous Peoples and the Global Environmental Agenda.* Copenhagen: International Alliance of Indigenous-Tribal Peoples of the Tropical Forests and International Work Group for Indigenous Affairs.

—— 1997. *Indigenous Peoples' Participation in Global Environmental Negotiations.* London: International Alliance of Indigenous-Tribal Peoples of the Tropical Forests.

International Institute for Environment and Development (IIED). 1994. *Whose Eden? An Overview of Community Approaches to Wildlife Management.* London: IIED.

IUCN. 1991. *Caring for the Earth: A Strategy for Sustainable Living.* London: Earthscan.

—— 1993. *The World Conservation Union Bulletin,* no. 3/93: 10–12.

—— 1994. *Guidelines for Protected Area Management Categories.* Commission on National Parks and Protected Areas. Gland: IUCN.

—— 1996. *World Conservation Congress: Resolutions and Recommendations.* Gland: IUCN.

JBHSS. 2000. 'Nagarahole: Adivasi Peoples' Rights and Ecodevelopment'. Paper presented by the Janara Budakattu Hakku Stapana Samithi to the workshop on *Indigenous Peoples, Forests and the World Bank: Policies and Practice* organized by the Forest Peoples Programme, 9–10 May 2000, Embassy Suites Hotel, Washington D.C.

Jacobs, A. 1975. 'Maasai Pastoralism in Historical Perspectives'. In T. Monod (ed.). *Pastoralism in Tropical Africa,* pp. 406–25. London: Oxford University Press.

Jensen, M. 'Editorial'. In *Indigenous Affairs,* 4: 2–3.

Kambel, E.-R. and MacKay, F. 1999. *The Rights of Indigenous Peoples and Maroons in Suriname.* Forest Peoples Programme and International Work Group for Indigenous Affairs, Document 96, Copenhagen.

Kemf, E. (ed.). 1993. *Indigenous Peoples and Protected Areas: The Law of Mother Earth.* London: Earthscan Publications.

Kingsbury, B. 1998. '"Indigenous Peoples"' in International Law: a constructivist approach to the Asian Controversy'. *The American Journal of International Law*, 92 (3): 414–57.

Kothari, A., Singh, N. and Suri, S. (eds). 1996. *People and Protected Areas: Towards Participatory Conservation in India.* New Delhi: Sage Publications.

Kuhn, T. 1962. *The Structure of Scientific Revolution.* Chicago: Chicago University Press.

Kwokwo Barume, A. 2000. *Heading Towards Extinction: Indigenous Rights in Africa – the case of the Twa of the Kahuzi-Biega National Park, Democratic Republic of Congo.* Copenhagen: Forest Peoples Programme and International Work Group for Indigenous Affairs.

Lawrence, A., Ambrose-Oji, B., Lysinge, R. and Tako, C. 2000. 'Exploring Local Values for Forest Biodiversity on Mount Cameroon'. *Mountain Research and Development,* 20(2): 112–15.

Li, T.M. 1999a. 'Articulating Indigenous Identity in Indonesia: Resource Politics of the Tribal Slot'. *Comparative Studies in Society and History,* 42(1): 149–79.

—— 1999b. 'Compromising Power: Development, Culture, and Rule in Indonesia'. *Cultural Anthropology,* 14(3): 295–322.

Lindsay, W. 'Integrating Parks and Pastoralists: Some lessons from Amboseli'. In D. Anderson and R. Grove (eds). *Conservation in Africa: People, Policies and Practice,* pp. 150–67, Cambridge: Cambridge University Press.

Lutz, E. and Caldecott, J. 1996. *Decentralization and Biodiversity Conservation.* Washington DC: World Bank.

Manning, R. 1989. 'The Nature of America: Visions and revisions of wilderness'. In *Natural Resources Journal,* 29: 25–40.

Margoluis, R., Margoluis, C., Brandon, K. and Salafsky, N. 2000. *In Good Company: Effective Alliances for Conservation.* Washington DC: Biodiversity Support Program.

McCabe, J.T., Perkin, S. and Schofield, C. 1992. 'Can Conservation and Development be Coupled among Pastoral People? An examination of the Maasai of the Ngorongoro Conservation area, Tanzania'. In *Human Organization,* 51(4): 353–66.

McCracken, J. 1987. 'Conservation Priorities and Local Communities'. In D. Anderson and R. Grove (eds). *Conservation in Africa: People, Policies and Practice,* pp 63–78. Cambridge: Cambridge University Press.

Morrison, J. 1993. *Protected Areas and Aboriginal Interests in Canada.* Toronto: WWF – Canada Discussion Paper.

MRG. 1999. *Forests and Indigenous Peoples of Asia.* London: Minority Rights Group International.

Nabhan, G., House, D., Humberto, S., Hodgson, W., Luis, H., and Guadalupe, M. 1991. 'Conservation and Use of Rare Plants by Traditional Cultures of the US/Mexico Borderlands'. In M. Oldfield and J. Alcorn (eds). *Biodiversity: Culture, Conservation and Ecodevelopment,* pp. 127–46. Boulder: Westview.

Novellino, D. 1998. 'Sacrificing Peoples for the Trees. The Cultural Cost of Forest Conservation on Palawan Island'. In *Indigenous Affairs,* 4: 4–14.

Oldfield, M. and Alcorn, J. (eds). 1991. *Biodiversity: Culture, Conservation and Ecodevelopment.* Boulder: Westview Press.

Pimbert, M. and Pretty, J. 1995. *Parks, People and Professionals: Putting Participation into Protected Area Management.* Geneva: United Nations Research Institute for Social Development (UNRISD). Discussion Paper 57.

Posey, D. 1999. *Cultural and Spiritual Values of Biodiversity.* London: Intermediate Technology Publications and United Nations Environment Programme.

Pretty, J., Guijt, I., Scoones, I., and Thompson, J. 1994. *A Trainer's Guide to Participatory Learning and Interaction.* IIED Training Series no. 2. London: IIED.

PRIA. 1993. Doon Declaration on People and Parks. Resolution of the National Workshop on Declining Access to and Control over Natural Resources in National Parks and Sanctuaries. Forest Research Institute, Dehradun 28–30 October 1993 (Society for Participatory Research in Asia).

Pritchard, S. (ed.). 1998. *Indigenous Peoples, the United Nations and Human Rights.* London: Zed Books.

Redford, K.H. and Mansour, J.A. 1996. *Traditional Peoples and Biodiversity Conservation in Large Tropical Landscapes.* America Verde Publications, The Nature Conservancy, Arlington.

Redford, K.H. and Sanderson, S.E. 2000. 'Extracting Humans from Nature'. *Conservation Biology* 14(5): 1362–4.

Robinson, J. and Bennett, E. (eds). 2000. *Hunting for Sustainability in Tropical Forests.* New York: Columbia University Press.

Roe, D., Mayers, J., Grieg-Gran, M., Kothari, A., Fabricius, C. and Hughes, R. 2000. *Evaluating Eden: Exploring the Myths and Realities of Community-based Wildlife Management.* London: International Institute for Environment and Development.

Roulet, F. 1999. *Human Rights and Indigenous Peoples: A Handbook on the UN System.* Copenhagen: International Work Group for Indigenous Affairs.

Sanford, S. 1983. *Management of Pastoral Development in the Third World.* London: John Wiley and Sons.

Schwartzman, S., Moreira, A. and Nepstad, D. 2000. 'Rethinking Tropical Forest Conservation: Perils in Parks'. *Conservation Biology,* 14(5): 1351–7.

Scoones, I. et al. 1992. *The Hidden Harvest: Wild Foods and Agricultural Systems.* London: IIED; Geneva: WWF; Stockholm: SIDA.

Simpson, T. 1997. *Indigenous Heritage and Self-Determination: The Cultural and Intellectual Property Rights of Indigenous Peoples.* Copenhagen: International Work Group for Indigenous Affairs and Forest Peoples Programme.

Stevens, S. (ed.). 1997. *Conservation through Cultural Survival. Indigenous Peoples and Protected Areas.* Washington DC: Island Press.

Taber, A., Navarro, G. and Arribas, M. 1997. 'A New Park in the Bolivian Gran Chaco – An advance in tropical dry forest conservation and community-based management'. In *Oryx,* 31(3): 189–98.

Terborgh, J. 1999. *Requiem for Nature.* Washington DC: Island Press.

Turton, D. 1987. 'The Mursi and National Park Development in the lower Omo Valley'. In D. Anderson and R. Grove (eds). *Conservation in Africa: People, Policies and Practice,* pp. 169–86. Cambridge: Cambridge University Press.

Veber, H., Dahl, J., Wilson, F. and Waehle, E. (eds). 1993. '... *Never Drink from the Same Cup'. Proceedings of the conference on Indigenous Peoples in Africa.* Copenhagen: International Work Group for Indigenous Affairs and the Centre for Development Research.

Verolme, H. and Moussa, J. (eds). 1999. *Addressing the Underlying Causes of Deforestation and Forest Degradation: Case Studies, Analysis and Policy Recommendations.* Washington DC: Biodiversity Action Network.

Vickers, G. 1981. 'Some Implications of Systems Thinking'. In *Systems Behaviour Education by Open Systems Group.* London: Harper and Row and Open University Press.

WCD. 2000. *Dams and Development: A new Framework for Decision-making. Report of the World Commission on Dams.* London: Earthscan.

WCPA 1999. *Principles and Guidelines on Indigenous and Traditional Peoples and Protected Areas*. Gland: WCPA, IUCN, WWF (International).

Wearne, P. 1996. *Return of the Indian: Conquest and Revival in the Americas*. London: Cassell.

Weber, R., Butler, J. and Larson, P. 2000. *Indigenous Peoples and Conservation Organisations: Experiences in Collaboration*. Washington DC: World Wildlife Fund (USA).

Wells, M. and Brandon, K. 1992. *Peoples and Parks: Linking Protected Area Management with Local Communities*. Washington DC: World Bank/USAID/WWF.

West, P.C. and Brechin, S.R. (eds). 1991. *Resident Peoples and National Parks: Social Dilemmas and Strategies in International Conservation*. Tucson: University of Arizona Press.

Western, D. and Wright, R.M. (eds). 1994. *Natural Connections. Perspectives in Community-based Conservation*. Washington DC: Island Press.

Wood, A., Stedman-Edwards, P. and Mang, J. 2000. *The Root Causes of Biodiversity Loss*. London: Earthscan.

World Bank. 1991. *Indigenous Peoples*. Operational Directive 4.20. Washington DC.

—— 1994. *Resettlement and Development: The Bankwide Review of Projects Involving Involuntary Resettlement 1986–1993*. Washington DC: World Bank Environment Department.

—— 2001. *World Bank Operational Manual: Operational Policy 4.12 'Involuntary Resettlement'*. Washington DC: World Bank.

WRM. 1990. *Rainforest Destruction: Causes, Effects and False Solutions*. Penang: World Rainforest Movement.

WWF. 1996. *WWF Statement of Principles: Indigenous Peoples and Conservation*. Gland: WorldWide Fund for Nature International.

2

Negotiating the Tropical Forest

COLONIZING FARMERS AND LUMBER RESOURCES IN THE
TICOPORO RESERVE

Miguel Montoya

Introduction

The last three decades have seen a worldwide surge in awareness and concern about the loss of natural forest environments, resources and wildlife.
Simultaneously, man's encroachment on unexploited or reserved areas has
increased, with factors such as population increase, unrelieved poverty, and
state agricultural policies contributing to scenarios where peasants resort to
colonizing new lands in search of a living. In contrast to the optimism of the
1950s and 1960s, when peasant farmers were often portrayed as a positive
force in opening new territories to agricultural production for the benefit of
the nation,[1] this sector has more recently been depicted as a threat to the
environment and resources of not only particular nations, but the entire
world, as deforestation and fires threaten the global climate.

This chapter discusses a farming community located in the foothills of the
Venezuelan Andes, settled in what is known as the Ticoporo Forest Reserve.[2]
Since its declaration as a reserve in 1955, the face of the Ticoporo Forest has
been radically transformed through the conflictive interaction of colonizing
farmers, lumber interests and state policies, and the forest, a natural habitat of
tropical birds, fish, and fauna, has deteriorated greatly. Perceptions guiding
current programmes to regulate the situation view landless colonists as predatory and as a threat to the area, and consequently, state programmes seek to
limit land use and commercial transactions within the reserve, with the aim of

restoring state control of the area. Occupants are encouraged to plant commercial species of trees on part of their holdings in exchange for permission to keep limited amounts of cattle on their land claims.

Here, I will argue that the practices of smallholding farmers in this region are ideally suited to land colonization and not necessarily anti-conservationist; rather, the programmes themselves and the legal and political situation compel people to undertake destructive survival strategies. Ultimately, there is a better chance of saving and augmenting lumber resources by negotiation, and by supporting existing colonizers' land rights, than by implementing policies limiting occupants' integration into national economic life.

Background: Farmers, Lumber Companies and the State in the Reserve

The Ticoporo Forest, originally 269,147 hectares of natural forest, is located at the southern base of the Venezuelan Andes, in the state of Barinas which lies in the western part of the country. It is bounded by the Andean foothills, or piedmont, in the north; and by a series of rivers that flow down from the Andes – the Anaro River in the east; the Quiu and Zapa Rivers in the west. The Suripá River forms its southern border. Other rivers that flow through the reserve from the Andes are the Bumbun, the Socopó, the Old Socopó, and the Michay.

There are few written sources available to provide information about the history of this particular area. The Andes were among the first areas of Venezuela to be colonized, because of their cool climate, which more resembled that of Europe, but the nearby plains towns of Barinas (founded 1577) and Ciudad Bolivia (earlier called Pedraza) also date back to early colonial times. These towns did not flourish early on, however, and were devastated by Indian warfare, Civil War, and tropical diseases such as malaria during the first centuries of their histories.

The Ticoporo Reserve is first mentioned in written accounts by Francisco Alvarado, who in his *Memorias* (1961) describes a trip he made in 1870 between Cúcuta, on the Colombian border, and the town of Barinas. Descending the Sierra Nevada from one of the southern villages in Mérida state, he arrived in Santa Bárbara de Barinas in two days, and there sought a guide to take him through the forest to Pedraza. He arrived in Pedraza after three days' journey through a deserted and dangerous forest, with no inhabitants except a few nomadic Indians, and many rivers and streams that were difficult to cross.[3] Alvarado describes the zone as an extensive area in which one would go hungry if one did not have provisions, although it would be possible to hunt the abundant wild pigs and turkeys (Tosta 1989, citing Alvarado 1961). Oral accounts I collected from some of the first colonizers of the zone contain similar elements. Settlers spoke of the many rivers, difficult to cross, but full of fish; the great quantities of wild game to be had in the region, which helped them survive in the period before first crops could be harvested; and the profusion of enormous trees, called *majumba*, and considered to be testimonials of the great fertility of the land. Also, parts of the reserve are marshlands that flood with the yearly rains, serving as unique ecosystems for certain species of birds and fish.

The Barinas piedmont, where the Ticoporo Reserve is located, is a somewhat special case among Latin American frontiers. Export market demand has not played a role in its colonization, nor have large-scale economic interests participated. The region was opened up to agriculture by smallholders, who primarily produced for their own needs. Cattle soon became an important part of their livelihoods for a variety of reasons: the fact that dairy farming and cattle keeping are usually part of the farming practices of their villages of origin, the suitability of the land for this activity, its profitability and flexibility, and their isolation and consequent difficulties in marketing agricultural crops. The relatively stable price for beef nationally was also an incentive for farmers to increase herds.[4]

The first settlers in the Ticoporo area were Andeans, who migrated down from the mountains, particularly from the Pregonero area of the Uribante district and from the southern villages of Mérida state. During the first decades of the 1900s, smallholders had cleared and planted land in the foothills for coffee cultivation, especially in the lower-lying parts of the Uribante, an important coffee-growing district. Although the remote Andean foothills were not viable for larger agricultural enterprises such as the coffee haciendas, for independent smallholding farmers the boom in coffee prices at this time was an incentive to colonization. When coffee prices fell irreparably with the Great Depression that began in 1929, however, the coffee farmers of the Uribante expanded their dairy farming and cattle fattening activities to supplement their declining income. In the most isolated areas with poor access to markets, cattle raising became the primary activity, which, with its demand for more land as the population grew and as herds increased, led to a successive colonization of lower lying plains areas.

Older colonizers speak of having worked while young herding cattle along the trade route that was used to bring wild bulls from deep within the plains states to the Uribante district for fattening. Through this activity they came into contact with the unsettled lands of the piedmont, and many of them eventually colonized land there. By the 1950s, this wave of colonizers had reached the Ticoporo forest, where the town of Socopó, just outside the reserve, was founded in the early 1960s.

Venezuelan dictator Pérez Jiménez (deposed in 1958), who invested a great deal in the modernization of the nation, declared the Ticoporo Forest a national reserve in 1955, with the aim of conserving its lumber resources for future exploitation. By then it was already inhabited, in the early stages of colonization, and with the decree the settlers in the forest suddenly found themselves transformed into illegal occupants of national lands. The new reserve regulations caused difficulties for colonists; it was no longer possible to register landholdings, as these were now prohibited within the reserve area, and the National Guard, who began to patrol the zone, tried to keep

colonists from deforesting or establishing housing. Harassment may have dis-
suaded some families from staying, but was still largely ineffective against the
influx of new colonists to the area, many of whom did not know about, or
understand its changed legal status. The construction of a road into the
reserve for the purpose of oil exploration in the late 1950s also provided easier access into the restricted territory.

The immense extensions of fertile, unoccupied land continued to attract
both land-poor and landless farmers from the mountain regions, and they
were aided in their colonization efforts by Colombian labourers from nearby
border areas, who came to work for the higher wages being paid in
Venezuela. The merchants and larger landowners of the growing town of
Socopó, just outside the reserve, also fuelled the pace of deforestation. Their
commercial activities provided them with the means to employ labour to
deforest larger claims than those that the average peasant family generally
undertook to clear. In this way, then, the inflow of occupants and deforestation of the reserve continued.

Faced with the fact that considerable portions of the reserve were occupied, had been cleared and were being farmed, the state legalized the colonization of two large extensions. After the fall of Pérez Jiménez in 1958, pressure
for land resulted in 40,000 hectares near the town of Capitanejo being
removed from the reserve and transformed into the Paiva-Capitanejo settlement, where land was parcelled out. In 1972, the National Agrarian Institute
distributed another 43,000 hectares outside Socopó to occupants.[5]

Socopó expanded further when several lumber companies set up their
headquarters there. As of 1970, the remaining 170,000 hectares of the reserve
were divided into four units which were contracted to three different lumber
companies, Contaca, Emallca, and Emifoca,[6] for exploitation, with the
responsibility for reforestation (Table 2.1). The Ministry of Agriculture
received the fourth unit, earmarked for experimentation in tropical forestry,
and this was eventually turned over to the Ministry of the Environment
(MARNR),[7] which loaned it to the Department of Forestry of the University
of the Andes (ULA), for experimental use (Figure 2.1).

The division of the reserve into concessions sparked off a series of conflicts
between the lumber companies and the occupants, who were determined to
keep and cultivate their land claims. In the mid-1970s, a census was made of all
the occupants, and they were compensated by the government for the value of
their holdings or *bienhechurias*,[8] and expelled to put an end to the conflicts. For
a number of years, there was stricter control of the borders, but with time, controls were relaxed, resulting in a reoccupation of many parts of the reserve.[9]
The concession holders took different approaches towards occupation: while
privately-owned companies managed to keep their holdings free through careful vigilance, state-operated entities were unable to do so for political reasons,
and eventually began to undertake programmes designed to foment co-exist-
ence between lumbering interests and the farmers. By the end of the 1980s
nearly all of ULA's and Emifoca's concessions were reoccupied, and Emallca
had recently won a legal struggle against an organized occupation.

Table 2.1 Concessions Granted within the Ticoporo Forest Reserve

Unit	Company	Year	Hectares
1	Emifoca	1982	45,750
2	Contaca	1970	40,775
3	Emallca	1972	60,300
4	Experimental	1970	24,000

It should be pointed out that government policy has contributed substantially to the deforestation of the reserve. While on the one hand supporting the land use ordinances of the zone which declare it unconditionally closed to habitation and agriculture, on the other hand, the government frequently fails to enforce these ordinances. Neither the *Guardia Nacional* in charge of patrolling the zone, nor MARNR officials stationed there have adequate resources to keep insistent peasants out. However, peasants' accounts of how they pay off the guards so that they will allow them to bring in cattle, motor saws, and housing materials reveal another aspect of the local reality.

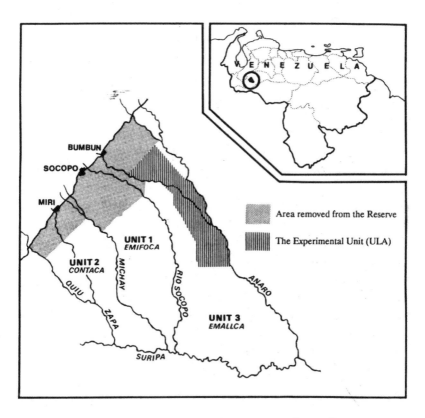

Figure 2.1 Lumbering Concessions in the Ticoporo Forest Reserve
Source: ULA Facultad de Ciencias Forestales. Proyecto CT-02.

Although large areas of the reserve have been colonized spontaneously by migrating peasants, organized occupations have also played a role in the occupation of the reserve, particularly since the end of the 1970s. Peasant organizations with links to important national political parties, such as the *Federación Campesina* and the *Liga Agraria*, have been agents in such occupations. Particularly in the year prior to national elections, such organizations attempt to distribute land to peasants, and in this context, parts of the reserve have periodically been demarcated and lots of 20 to 40 hectares allocated to peasant families, while the National Guard and Ministry of the Environment turn a blind eye.[10] Should the concession-holder undertake to evict occupants, they are faced with a time-consuming juridical process. Each individual family must be cited and proceeded against before they have occupied a plot during one year, after which it becomes much more difficult, if not impossible to legally oust them. Frequently, occupying families consist of a young couple and several school-age or infant children, with the mother and children living in a shanty on the plot while the husband works as a day-labourer to earn income to invest in the colonization venture. The eviction of women and their small children creates particularly bad press during election years, making government-backed concession-holders less inclined to take action: a fact of which organizers and occupants are both well aware.

Hence the Ticoporo Reserve can be seen to serve several opposed functions. To some extent it remains a reserve for the nation's lumber needs; it has provided, and continues to provide wood for national industry.[11] But additionally, it is a source of land that can be distributed to peasants as an incentive to party loyalty prior to national elections; providing a way for organizations with political agendas to maintain power. The national government serves its own needs by putting forth one policy publicly and allowing another to be followed surreptitiously; ignoring forest invasions while nonetheless maintaining its rhetoric about the need for conservation of forest resources and biodiversity.

The Peasant Economics of Land Colonization

The process of land colonization underway in the Ticoporo Forest has changed during the years because of the shifting nature of its incoming occupants. In the piedmont and plains during the first half of the 1900s, colonization was undertaken by groups of men[12] who deforested new farms during certain months of the year, departing from a fixed land base or income in another zone. An area attractive for settlement would be discovered or sought on hunting or fishing expeditions to the hinterlands, for example, and a stake marked off. During the following dry season a small group of men would clear an area, and, supporting themselves by hunting and fishing and with provisions they brought, they planted foods such as *yuca*[13] and a grove of bananas. A hut might also be constructed, and clearings fenced with barbed wire. This signalled occupation to others who might have their eye on the area, and provided food and shelter for a larger undertaking the following year, when several hectares would be cleared for corn cultivation. At the end of the autumn rainy season deforestation would begin, continuing until the clearing was burned and corn planted just before the spring rains. Corn, the main staple of many Latin American peasants, is, as noted by Ortiz, the crop that civilizes the wilderness (Ortiz 1980: 200). Often improved pasture seed was also spread as the corn was planted, so it would grow in the clearing after the corn harvest, producing the first pastures for cattle. When the area was considered safe from predatory animals and a measure of comfort existed – a rudimentary dwelling, an easily accessible water source, staple foods – a man would bring his family, if possible, to make colonization faster, easier and more pleasant. With the family came domestic animals – cows, pigs and chickens, all of which fell within the wife's domain.

In Ticoporo, the original occupants followed a similar pattern as they settled the lands outside of Socopó; departing from a base in the mountains or in a different area of the plains. Generally it was not the poorest families that undertook colonization ventures, because of the necessity of maintaining the family during the first years. The colonizer (or his wife) needs to have some income, savings, or another piece of land that can provide for them during the initial stages of the process. It should be noted, however, that many colonizers do not actually stake a claim, but buy a claim that already has a small clearing, stable crops and a hut, thereby shortening the process. Also, areas undergoing settlement in this way differentiate socially. While the original colonists may be of about the same means, inevitably some of them are unable to continue or decide to sell their holdings, and financially better off colonists take their place. Some of these may employ local labour, thus providing a niche for the poorer to enter as wage-labourers or caretakers.

As years pass and *bienhechurias* grow, these lands increase in value. A common practice, documented by Delgado de Bravo (1985, 1986) is to sell the original landholding and move further into the frontier, repeating the process; the sale of the land providing capital for seed, fencing, and most importantly, more cattle.[14] The economic nature of colonization can hardly be sufficiently emphasized. In opening and 'civilising' new lands, peasant families put their major resource – their labour – to use in a very efficient way. They both produce crops for their own sustenance and for sale, and as they work, improve the land, enhancing its value. If land titles can be registered, prices rise further, and in colonization zones the state eventually invests in roads, schools, and other infrastructure such as electricity, further heightening the land values. Consequently, although farming in the forest is difficult and dangerous, it often pays off richly.

Gudeman and Rivera (1990) have discussed in detail the economy of peasant farmers similar to those now inhabiting the reserve, and describe how practices of thrift and keeping resources 'within the house' allow these marginal family farming enterprises to flourish. Virgin land is particularly attractive because agricultural yields are higher, and on a frontier there is plenty of space to maintain expanding herds of cattle, as well as smaller domestic animals such as pigs and chickens. Capitalization through animals and land value is important for peasant families, because it is often the only way in which they can accumulate enough savings to provide them with a measure of economic security, and obtain a base on which the next generation can reproduce itself.

In the Ticoporo Reserve, it is plain that providing a base for future generations is an important motivation in colonization. As one drives along the Kimil, the now-paved road that leads from Socopó to the present entrance of the reserve, one passes about forty sizeable family farms, with few exceptions still owned by the original colonizing families, or by long-time secondary settlers. Of these colonizers, the majority also hold land claims deeper within the reserve, often overseen or worked in cooperation with offspring.

The time and manner of colonization makes a difference in chances for success. Having arrived at an auspicious moment, staked large claims, and having profited when their holdings were removed from the reserve regimen in 1972, many of the original colonizers on the Kimil were able to completely clear their original farms, make them produce, and repeat the process further in the reserve without ever selling any land. Subsequent settlers have seldom been as fortunate. Later claims, made when the influx to the area accelerated, are smaller, and still in the 'illegal' part of the reserve. Those who have obtained land through land distributions seem to fare the worst of all for a number of reasons. Such lots are smaller, 40 or sometimes only 20 hectares; insufficient to support a family well. Also, as the land is distributed to party supporters, neighbours are usually strangers, making the cooperation needed in colonization efforts more difficult. Settlers in these areas tend to be poorer, and are often without the family labour resources needed to make the venture work, and sometimes even without sufficient knowledge and experience in agriculture. Such colonization is misery indeed, with isolated nuclear households barely subsisting, in the most primitive conditions, forced to take local wage labour to survive, but without the hope of ever capitalizing sufficiently to become independent producing units.

Settlers and Lumber Companies: the Limits of Cooperation

Since the lumber companies obtained concessions within the reserve, there has been continuous animosity between occupants and the companies. The *Guardia Nacional* stationed at the borders of the reserve has the difficult duty of keeping new occupants out, and of making it impossible for those already within the reserve to continue deforestation as they would like. Since cattle-ranching is colonists' preferred activity, cattle are forbidden within the reserve, and must be smuggled in and out by the occupants, along with permanent building materials such as cement blocks, and other materials and tools of colonization. Occupants have, over time, discovered ways of overcoming such difficulties; and a system of clandestine paths, contacts, and networks exists which allows families to overcome the restrictions to which they are formally subject. Still, the prohibitions make farming and ranching within the reserve time-consuming and costly. Many colonists reason that since the lumber companies and the guards want, and are out to protect, the trees, then the sooner they get rid of the trees on their claims, the sooner they will be left to farm in peace. Naturally, such beliefs increase the pace of deforestation, with colonists sometimes even setting fires in the unoccupied parts of the reserve with the aim of pushing the companies out of the area. Rather than protecting the reserve, the effect of the vigilance and restrictions is to encourage needless destruction.

By the late 1980s, both Emifoca and the ULA's experimental unit were nearly completely occupied. A census of the settlers in these areas was made, and Emifoca administrators and representatives from the ULA were conducting dialogues with occupants in the hope of gaining support for silvio-pastoral programmes. The basic content of these programmes was to offer families concessions in allowing limited cattle-keeping and building in the reserve in exchange for colonists' commitment to planting trees on 20 per cent of their holdings.[15] Families were also given assurances that they would be allowed to stay on their claims, but were refused the right to sell them. If they left the area, the holdings would immediately revert to the concession.

Occupants were less than eager to accept these programmes. Legal title to their land was a fundamental goal for all of them, including the right to sell. Most were managing to enter and leave the reserve, and to bring necessities in and out by negotiating with the *Guardia*; and thus the programmes offered them rights which they already had acquired in practice, if not on paper. Few were willing to plant trees on 20 per cent of their holdings. The majority had small claims, and asked themselves, on which part of my holdings? The areas that flood in the rainy season, or the drier areas, where I built my house and plant my crops? And how will my family and I benefit from these trees, which take 20 years to mature, if we cannot sell the land? Only occupants with larger holdings, or who already owned land outside the reserve, could see any merit in these programmes. Altogether, only a fraction of the occupants agreed to take part, and most of these only paid lip service to the regulations of the schemes. The rest ignored them, until they came to an end, as Emifoca left its concession in the early 1990s, and as the ULA increasingly withdrew personnel from its experimental station.

Having worked hard to design the programmes, and dependent on their success in order to replant and manage their concessions, the ULA and Emifoca felt that the settlers were obstinate and ungrateful. They were, after all, illegal occupants, and were being offered a way to legalize their situation and to participate in a lucrative agricultural scheme. Tree cultivation, the forestry engineers calculated, would bring in several times as much income in the long run as raising cattle on the same area. The seedlings would be very inexpensive, and technical assistance would be provided.[16] It was completely unreasonable for occupants to demand land ownership, they felt, for what would remain of the reserve, then? The devastation wrought by uncontrolled colonization was already more than evident.

The Colonization Process, Capitalization, and Land Consolidation: New Alternatives?

Indeed, cultural and political factors and a variety of actors have combined to make the Ticoporo Reserve an excellent example of how *not* to manage a forest reserve.[17] But given the current situation, with Emallca's and Contaca's concessions also having been affected by spontaneous and organized invasions during the late 1990s, how might the situation possibly be improved?[18]

The Ticoporo Forest has been seriously damaged as a natural reserve, and only minor portions of it remain relatively untouched. The rest is in varying stages of colonization and agricultural production. Certain areas can only be described as farmland. In others, now being gradually deforested, the lumber companies long ago took out the largest, most valuable trees, leaving smaller trunks to grow, and other species untouched. Colonists say that they leave the valuable species when they deforest their plots, but the considerable amount of wood being taken out of the reserve under cover of night and sold as contraband at a fraction of the market price attests to the fact that a good deal of this wood is being exploited. Additionally, MARNR authorities complain that the areas once reforested by the lumber companies, hitherto respected and untouched by occupants, are now also under attack.

'A ésta reserva, la llamaría la reserva de los negocios,' (I would call this reserve a reserve for doing business) a long-time area resident said one Sunday afternoon, as we sat discussing the reserve on an occupant family's front porch. Indeed, he had a good point. The reserve has provided many actors with a good income of one kind or another. Occupants, ranchers, guards, lumber *contrabandistas*, local carpenters, peasant leaders, regional politicians – all have profited illicitly from the reserve; while the state (via taxation of concession holders), the lumber companies, and their workers have made legal profits. Out of the reserve flows a steady stream of agricultural products: cattle, milk and cheese, considerable harvests of corn and sorghum, *yuca* and plantains, all of which are technically illicit but nonetheless commercialized, reaching the national markets. What would happen if everyone were allowed to do their business legally in the reserve?

To legitimize and register occupants' holdings as their private property would, I believe, have a beneficial impact on the current situation, if occupants were to progressively pay for the holdings, and if such a scheme were to go hand in hand with agricultural incentives for silviculture, and investment in local industries in the wood-working sector. The special status of the area as a reserve of lumber for the nation should be maintained, but with the occupants integrated into a programme encouraging and supporting lumber production, and local industrialization, operating through market mechanisms. In other areas of Latin America, such as in Chiapas, peasants are planting trees to trap carbon in planned schemes in which polluting companies in the developed countries offset their emissions by funding pollution-reducing projects in other countries. Such new incentives in the growing carbon emissions trade may also be of interest to the peasants of the Ticoporo region.[19]

Peasants should come to view the trees on their lands not as a threat to be done away with, but as a possible source of income and a fund for their futures. The trees on a landholding should constitute a part of its commercial value, which would gradually encourage reforestation; and forest products could be sold at market prices, profiting peasants more than is the case today. The growth of local woodworking industries – already underway, as many small carpentry shops have sprung up in Socopó in recent years – could serve as a source of much needed employment alternatives. Such a strategy would eliminate lumber smuggling overnight, and enable the *Guardia Nacional* and MARNR to concentrate their energies on keeping vigil over the much smaller area of the reserve which has not yet been disturbed, in the hope of keeping it as a true reserve. The cost to the state would be reduced; and the money could instead be invested in encouraging reforestation and woodworking industries.

The farming families of the Socopó community located outside of the reserve do not belittle the value of the forest they have cleared from their fields. It is not uncommon for them to complain that now they must purchase even the least bit of wood they need. 'If I was to do this all over again, I'd leave a big lot uncleared in the middle of my farm,' one farmer commented. Along the Kimil, many of the legal farms have invested in 'living fences', as the teak trees that are planted along the borders between fields are called. There are also a number of hectares of teak and other tropical trees planted on some larger farms, even inside the reserve. On passing such plantations, peasants will frequently comment on their monetary value, even referring to them as being 'better than a bank account'. Respect for private property is strong in this farming culture, while the property of the state belongs to no one, and can be appropriated by those strong enough to take it.

With the legitimization of occupants' landholdings, one obviously gives up the idea of somehow restoring the area to a forested wilderness. There is no hope of that at this point: the biodiversity and fauna of the zone have suffered far beyond repair. One should recognize the fact that the state has failed in its attempt to protect the reserve, that its borders are highly permeable, and that no administration is likely to enforce the reserve regimen in practice. Making the occupants stakeholders in silvicultural schemes in which it is easy and inexpensive to plant trees, and which guarantees them the right to their land and its resources, might avoid the area becoming a purely cattle-breeding or agricultural zone, as has happened in nearby frontier areas.

A reflection on what might be called the 'natural history', or life cycle of frontier areas provides an indication of the likely outcome of granting private property rights to reserve occupants.[20] Legalizing land claims in the reserve will raise land prices, and encourage some families to sell out and move on. Their holdings are likely to be acquired by neighbours seeking to enlarge their farms, the first step in a series of land transactions leading to the gradual consolidation of small farms into larger, more economically viable agricultural enterprises. Such larger entities are more readily able to invest in silviculture than smaller farmers, who need every bit of land to provide income for immediate consumption.

Providing small farmers with viable alternatives to colonization is another way in which state authorities can help protect what remains of the reserve, as well as other forested areas. This is more difficult, but one suggestion is to establish agricultural or industrial schools in the area, where farmers' off-spring can learn a commercial trade. Colonizing new land is very hard work, and requires great sacrifices from settlers, who live isolated lives, for many years without the benefits of the more civilized world, such as roads and transportation, electricity, telephones (although cellphones have recently pro-vided alternative communication possibilities), education, medical facilities, shops, meeting places, churches. It is not a life one chooses if there are viable alternatives at hand, and the children of farmers seldom go on to colonize other areas unless they have direct financial and logistical support from their parents. Rather, the offspring of the usually large farming families tend to gravitate towards towns, and entrepreneurial or unskilled urban jobs, unless they inherit or otherwise gain access to land. Education of a practical nature can expand this second generation's possibilities, and with time, also raise the standard of living in the region as a whole, as successful farmers seek to diversify their economic activities.

Here, then, I have suggested a solution for the continuing crisis in the management of the Ticoporo Forest Reserve. Unfortunately, I do not think it is a solution that would meet with much approval from the current Chávez administration, with its paternalistic stance and distrust of the free market. But in the end, no government can care for its people, but must provide them with the means to care for themselves: clear legal rights, education, information, efficient institutions and a transparent and functioning market. The importance of the market is obvious: the reserve, as we have seen, is per-meated with business transactions of different natures. The goal of future plans should not be to limit these, but to channel them in ways that can serve the reserve's original aims.

Notes

1 In Latin America, some such schemes have been the construction of the Transamazon high-way in Brazil (Moran 1980, 1989), projects sponsored by the IDB in Peru and Bolivia in the 1960s (Schuurman 1980: 107–8) and colonization projects such as the San Julian settlement in the east-ern lowlands of Bolivia (Painter 1989). In Venezuela, the Turén agricultural colony (Llambí 1988: 72–101) is an example of a planned agricultural scheme, although undertaken on a limited scale.

2 This chapter is based on fieldwork conducted in the area of the Ticoporo Forest Reserve between 1999 and 2001, for the project 'The Tropical Forest as Frontier: Processes of Coloniz-ation and Smallholder Integration in Western Venezuela,' financed by SAREC (Swedish Agency for Research Cooperation with Developing Countries). Earlier research in the area was carried out between 1987 and 1990; see Montoya (1996).

3 At this time, then, the reserve was uninhabited. However, archeological finds indicate that the area was at least temporarily inhabited by indigenous peoples, who built the raised walkways and mounds which can still be seen within the reserve. Little is known about the purpose of these, but they may have provided a refuge and means of transport during the annual periods of flooding (Zucchi 1965–6, 1969).

4 See Llambí (1988: 168–204) for a discussion of government policy (price regulation, subsidies and imports) and the production of meat and milk in the frontier region of Perijá between 1948 and 1983.

5 These occupants were allowed to buy their land claims from the National Agrarian Institute by paying for it in quotas over a number of years.

6 Contaca is privately owned and operated in the reserve until 1998. Emallca has mixed state and private capital and was, as of January 2001, reducing personnel and restructuring its operations. Emifoca, a state company, functioned (with some interruptions) in the reserve until 1991. All companies were required to run reforestation programmes, in addition to making rational use of their concessions.

7 The Ministry of the Environment is called the MARNR, the *Ministerio del Ambiente y Recursos Non-Renovables* (Ministry of the Environment and Non-Renewable Resources).

8 *Bienhechurias* are the improvements which an occupant makes on a land claim. Since the land can not be bought or sold, it is these improvements which are negotiated in land transactions. In Ticoporo they typically consist of cleared land, fencing, semi-permanent crops (bananas, plantains, and *yuca*) a dwelling, wells, and corrals.

9 Contributing to the new inflow of settlers were families who had been compensated for the loss of farms affected by the construction of the Uribante-Caparo hydroelectric scheme in the neighbouring Andean states. Those with low indemnities could not afford to buy new farms, and turned to buying *bienhechurias* in the reserve, where prices were substantially lower. Montoya (1996) discusses the migration routes of farming families compensated by the Uribante-Caparo hydroelectric project, and the social consequences of the dam scheme.

10 Land invasions led by pseudo-leaders with political agendas have been frequent in periods prior to and just after elections in Venezuela, since the death of dictator Juan Vicente Gómez in the 1930s. See Sandoval (2000).

11 The many small carpentry shops located around the town of Socopó also make use of lumber from the reserve.

12 Colonizers are ideally groups of related men or acquaintances from a particular farming area; for example, a man might undertake to colonize with his brothers or teenage sons, or with nephews or other family members also in search of land.

13 *Yuca* is the Venezuelan term for sweet manioc, and is a food staple throughout the country.

14 Cattle are often a part of land colonization processes because of their very suitability; they provide nourishment (and company) for humans, and are flexible investments: when need arises they can be herded to market. Agricultural crops are, on the contrary, difficult to bring to markets from isolated areas, hard to sell locally where everyone cultivates, and spoil (are lost) if not sold or consumed.

15 Teak trees were the main tree crop to be planted.

16 Nugent (1991) has noted some of the limits of the type of environmental management that Emifoca and the ULA were attempting. Reacting to ideas about the environmental management of forest resources, and citing Peters (1989), Uhl and Jordan (1984) and Uhl and Buschbacher (1985), he points out that a simple demonstration of the profitability of forest exploitation alone will not in itself generate the conditions providing for the reproduction of a peasantry in the long term, and ensure the future of forest resources. More than profitability per se is needed for such programmes to function in practice – such as agrarian reform, stable markets, and a larger agro-strategy.

17 Kottak (1999) points out the importance for ecological anthropologists to study how different levels of analysis – local, regional, national and international – link and vary in time and space and impact on the environment.

18 The current Chávez administration does not sanction land invasions, in the belief that solutions to these must be negotiated; a stance that has led to increasing numbers of rural and urban land occupations throughout the nation (Regalado 1999).

19 For more information about carbon-trapping schemes, see *The Financial Times Weekend* section, 11–12 November 2000.

20 Turner's thesis on the importance of the frontier in American history contains a valuable description of frontiers undergoing settlement that also proves accurate for the Barinas pied-

mont. Quoting *Peck's New Guide to the West*, published in Boston in 1837, Turner described a distinct series of three waves of settlers to arrive in western frontiers. The first were the pioneers, who seldom owned the land, depended greatly on the natural vegetation, hunting, simple agriculture, and a few domestic animals. When these pioneers began to lack 'elbow room', they sold their holdings to a second wave, who increased the fields, built roads and bridges, better houses, schools and courthouses. The final wave were the 'men of capital and enterprise', who brought the trappings of civilization: 'orchards, gardens, colleges and churches' to the frontier. They bought out the second generation of settlers when their holdings had risen in value, enabling these to move further into the interior to become 'men of capital and enterprise in turn' (Turner 1962: 19–21). A similar succession of settlers can be observed in the Barinas piedmont: the first who make a rough clearing, sell it and move on, the second wave, which settles down and improves the farms. The third wave are the successful settlers who gradually buy out the second, and develop agricultural enterprises that are well integrated into the national economy, and with time, diversify their business in other directions.

References

Alvarado, F. 1961. *Memorias de un Tachirense del siglo XIX*. Caracas: Biblioteca de Autores y Temas Tachirenses, no. 14.

Delgado de Bravo, M.T. 1985. *Dinámica socioespacial del proceso de ocupación de tierras en la Unidad Experimental de Ticoporo*. Trabajo de Ascenso, Facultad de Ciencias Forestales, Escuela de Geografía, Universidad de Los Andes, Mérida.

Delgado de Bravo, M.T., Rojas López, J. and Valbuena Gómez, J. 1986. *Proyecto CT7–1. Estudio Socioeconómico de los Ocupantes de la Unidad Experimental de Ticoporo*. Mérida: Facultad de Ciencias Forestales, Universidad de Los Andes.

Financial Times Weekend. 2000. 'The Carbon Trappers,' 11–12 November.

Gudeman, S. and Rivera, A. 1990. *Conversations in Colombia: The Domestic Economy in Life and Text*. Cambridge: Cambridge University Press.

Kottak, C. P. 1999. 'The New Ecological Anthropology'. *American Anthropologist*, 101(1): 23–35.

Llambí, L. 1988. *La Moderna Finca Familiar*. Caracas: Fondo Editorial Acta Científica Venezolana.

Montoya, M. 1996. *Persistent Peasants. Smallholders, State Agencies, and Involuntary Migration in Western Venezuela*. Stockholm: Stockholm Studies in Social Anthropology, Almquist and Wiksell.

Moran, E. 1980. 'Mobility and Resource Use in Amazonia'. In F. Barbira-Scazzocchio (ed.). *Land, People and Planning in Contemporary Amazonia*. Cambridge University: Centre of Latin American Studies, Occasional Publication no. 3.

—— 1989. 'Adaptation and Maladaptation in Newly Settled Areas'. In D.A. Schumann and W.L. Partridge (eds). *The Human Ecology of Tropical Land Settlement in Latin America*. Boulder, Colorado: Westview Press.

Nugent, S. 1991. 'The Limitations of Environmental Management: Forest utilization in the Lower Amazon'. In D. Goodman and M. Redclift (eds). *Environment and Development in Latin America: The politics of sustainability*. Manchester: Manchester University Press.

Ortiz, S. 1980. 'The Transformation of Guaviare in Colombia: Immigrating peasants and their struggles'. In F. Barbira-Scazzocchio (ed.). *Land, People and Planning in Contemporary Amazonia*. Cambridge University: Centre of Latin American Studies, Occasional Publication no. 3.

Painter, M. 1989. 'Unequal Exchange: The dynamics of settler impoverishment and environmental destruction in Lowland Bolivia'. In D.A. Schumann and W.L. Partridge (eds). *The Human Ecology of Tropical Land Settlement in Latin America.* Boulder, Colorado: Westview Press.

Peters, C. et al. 1989. 'Valuation of an Amazonian Forest'. *Nature,* 339: 655–6.

Regalado, R. 1999. 'Las invasiones están volviendo ingobernable a Venezuela'. *El Nacional,* 20 March, p. E-2.

Sandoval, W. 2000. 'Invasiones ocurren en cada elección', *El Universal,* Caracas, 29 February, pp. 2–4.

Schuurman, F. 1980. 'Colonization Policy and Peasant Economy in the Amazon Basin'. In F. Barbira-Scazzocchio (ed.). *Land, People and Planning in Contemporary Amazonia.* Cambridge University: Centre of Latin American Studies, Occasional Publication no. 3.

Tosta, V. 1989. *Historia de Barinas. Tomo III. 1864–1892.* Caracas: Biblioteca de la Academia Nacional de la Historia.

Turner, F. J. 1962. *The Frontier in American History.* New York: Holt, Rinehart and Winston [First published in 1920].

Uhl, C. and Buschbacher, R. 1985. 'A Disturbing Synergism between Cattle Ranching, Burning Practices and Selective Tree Harvesting in the Eastern Amazon'. *Biotropica,* 17(4): 265–8.

Uhl, C. and Jordan, C.F. 1984. 'Succession and Nutrient Dynamics Following Forest Cutting and Burning in Amazonia'. *Ecology,* 63: 1476–90.

Zucchi, A. 1965–6. 'Informe preliminar de las excavaciones en el Yacimiento La Betania, Estado Barinas, Venezuela'. *Boletín Indigenista Venezolano,* Año IX, no. 1–4, Caracas.

—— 1968. 'Algunas hipótesis sobre la población aborígen de los Llanos Occidentales de Venezuela'. *Acta Científica Venezolana,* 19: 135–9.

3

Compatibility of Pastoralism and Conservation?

A TEST CASE USING INTEGRATED ASSESSMENT IN THE
NGORONGORO CONSERVATION AREA, TANZANIA[1]

*Kathleen A. Galvin, Jim Ellis, Randall B. Boone, Ann L.
Magennis, Nicole M. Smith, Stacy J. Lynn* and *Philip Thornton*

Introduction

A major challenge for conservation agencies and advocates is formulating workable compromises between wildlife conservation and the people who live with wildlife. This is sometimes difficult because conflicts expand as human populations expand and because each different situation has its own peculiar dimensions. Various ecological, social, political and economic factors impinge on virtually all human–wildlife interactions, but the weight of each factor varies from one case to another. Thus, despite the attractive advantages of integrating conservation with human development, i.e., community-based conservation, many obstacles remain.

Community-based conservation is a concept aimed at protecting biodiversity by engaging local people in the conservation process. It emerged as conservationists and others realized that although national parks may be effective in protecting wildlife, flora and ecosystems, this protects only limited areas, while most wildlife occur outside parks (Wells and Brandon 1992; Gadgil et al. 1993; Redford and Mansour 1996). National parks may also have shortcomings due to the fact that many parks in the developing world are poorly funded and may be unable to afford adequate protection for biodiversity. The community-based conservation concept derives from the notion that local communities, if they have a stake in conserving local resources, will help to protect biodiversity. The biosphere reserve is one model of community-based conservation that theoretically allows for local population involvement in management of the protected areas. Integrated Conservation–Development projects form another type of community-based development. These attempts to integrate people and conservation have great promise, although the long-term success of community-based conservation remains problematic (Wells and Brandon 1992).

Are pastoralism and conservation compatible in East Africa? History suggests that this is surely true and that in some sense, community-based conservation has always been an integral part of the pastoralist way of life. But modernization processes and changes in pastoral populations and land use have altered the patterns of interaction between pastoralists, wildlife, and their jointly occupied ecosystems. Whereas much work has been done to try to understand the changing relationships between pastoralists and the state, few studies assess the effects of modernization and change on wildlife and co-habiting resident pastoralists. Are pastoralism and conservation still compatible in East Africa today, as in the past?

Wildlife and Pastoral People in Ngorongoro Conservation Area

Interactions between conservation policy and human ecology have been the focus of our interdisciplinary research in the Ngorongoro Conservation Area (NCA), where Maasai pastoralists live with a diverse and concentrated wildlife population. Ngorongoro is an 8292 sq. km conservation area located in northern Tanzania, adjacent to Serengeti National Park (Figure 3.1). The NCA was established as a joint conservation and human use area in 1959, following the forced removal of Maasai pastoralists from Serengeti National Park. The NCA was to be the home of the extirpated Serengeti Maasai in perpetuity. Thus it became one of the first and longest running experiments in community-based conservation, a paradigm that has become popular in the 1990s. Today the NCA remains a very successful conservation area and the premier tourist attraction in northern Tanzania. However, the resident Maasai and their supporters have claimed for a long time that the conservation policies of the NCA Authority have detrimentally affected their land use and thereby their economic well-being, have undermined their food security and general welfare and are responsible for a downward spiral of economic deprivation (Arhem 1985; Parkipuny 1997). If these allegations are valid, then in this instance, conservation has not been easy to integrate with the advancement of human welfare over a period of forty years; the concept of community-based conservation may be questionable, based on the Ngorongoro experience.

The NCA harbours one of the most spectacular and beautiful landscapes in Africa. Volcanic peaks of the Ngorongoro highlands rise steeply to over 3000

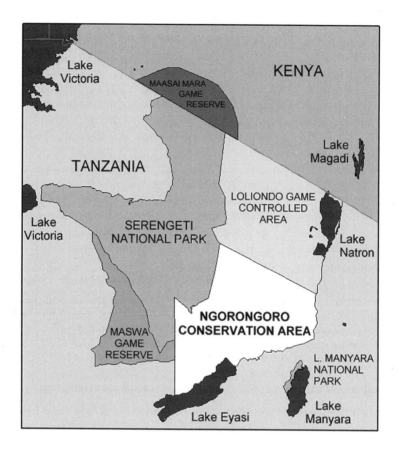

Figure 3.1 Map of the Greater Serengeti Ecosystem

metres, above Olduvai Gorge, the Rift Valley and the Serengeti plain. The landscape and vegetation diversity supported across this escarpment is startling, with tropical montane evergreen and bamboo forests at high elevations and shortgrass plains at the base levels. Between the montane forest and the plains are woodlands, bushlands and grasslands of a vast variety of forms and composition (Herlocker and Dirschl 1972). The centrepiece of the NCA is Ngorongoro Crater, the 300 sq. km caldera of an extinct volcano. The crater has several water sources and its grasslands support a year-round population of herbivores and predators, which is seasonally supplemented by migrants (Moehlman et al. 1997). Herbivores include wildebeest, buffalo, gazelle, zebra, eland, elephant, rhinoceros and others. Until very recently, the crater supported the most dense population of predators (mostly lions and hyenas) known in Africa (Kruuk 1972).

In addition to the wildlife, the NCA supports Maasai pastoralists and agro-pastoralists. In 1959, when the Maasai were removed from the Serengeti, there were approximately 10,000 Maasai resident in the NCA; over the ensuing forty years the population has expanded to about 50,000 Maasai

(Runyoro 1999). Land use has intensified and many Maasai have become agro-pastoralists, cultivating small plots of maize, beans and other products (Kajazi et al. 1997). Agriculture was prohibited between 1975 and 1991; however, since 1992 the Maasai have been permitted to conduct limited cropping (Runyoro 1999). Current conservation policies limit the amount of area that can be cultivated and outlaw grazing in some parts of the NCA. Thus as the human population expanded, land use and conflicts intensified, the Maasai sensing that their welfare and economic status were declining.

We hypothesized that conservation policy, through its limitations on land use, might have a detrimental affect on Maasai well-being as measured through household economy, health and nutritional status. In a long-term study begun in 1991, we have examined the claims of the NCA Maasai with respect to the impact of conservation policy on their land use and livestock holdings, their nutritional status and income levels, and the effects of Maasai land use on wildlife habitat. Our research established that a large percentage of the NCA Maasai cannot support themselves, but must be supported, in part by wealthier friends and relatives. Overall nutritional status was low and all NCA Maasai are in a chronic state of undernutrition (Galvin 1994, 1995, 1997; Galvin et al. 1994b). Our research supports Maasai claims of economic difficulties, but we were not able to confirm that the causes of their problems were rooted in the policies of the Ngorongoro Conservation Area Authority (NCAA). It is plausible that many of the problems experienced by the NCA Maasai are typical of pastoral populations elsewhere in the region or more generally throughout Africa (Grandin et al. 1991; Galvin 1992). This is an important point. If conservation policy is undermining the Maasai economy, then the whole concept of community-based conservation may be questioned due to the difficulties in this, one of the longest running trials of such a system. If, on the other hand, the problems are not attributable to NCAA policies but to other causes, then the concept of community-based conservation remains viable, even though this specific case is under stress. With this in mind we have been conducting a comparative study of human welfare and land use between the NCA and the adjacent Loliondo Game Controlled Area (LGCA) where conservation policies are less restrictive on Maasai land use. Cultivation is practised by most households in Loliondo and there are few restrictions on grazing and agriculture, unlike the case in the NCA.

The trade-offs among different possible conservation policies are being explored through use of an integrated modelling and assessment system (IMAS) which we developed for the NCA (Rainy et al. 1999). This chapter discusses some research results and modelling efforts currently underway, which are designed to understand implications of alternative policy and management decisions for pastoralists, wildlife, and ecosystems. We expect: 1) to determine if pastoral resource exploitation remains compatible with biological conservation, given the conditions that exist in Ngorongoro today; 2) to demonstrate those specific paths to pastoral development that are more environmentally benign than other options; and 3) to demonstrate that integrated human-ecological research, and the use of advanced technology, such as GIS, remote-sensing and ecosystem modelling, can play a useful role in the resolution of conservation/development conflicts. Because the situation in northern Tanzania is representative of conservation/development dilemmas throughout the drylands of Africa and Asia, results should test the applicability of the IMAS approach to other regions. A major challenge for biological conservation is understanding the complex interactions between biodiversity, ecology, political-economy and social dimensions of land use. These factors comprise the core components of a human-ecological system that in much of eastern Africa consists of pastoralists, wildlife, and livestock. At the centre of this complex system is land use, specifically the implications of land use for conservation, biodiversity and human well-being.

Research Design

Field research was conducted in the NCA during 1997 and was extended to include the Loliondo Game Controlled Area (LGCA) in 1998 and 1999. The research compared NCA Maasai economy, land use, health and nutrition, with that of Maasai living in the LGCA.

The LGCA lies adjacent to and north of the NCA (Figure 3.1). Climate patterns and general ecology are virtually identical to the NCA, and the LGCA is also inhabited by Maasai pastoralists and a small population of Sonjo farmers. The research design assumes that because the two regions are quite similar in terms of climate, ecology, and culture, we should be able to discern how conservation policy, which is much more restrictive in the NCA, influences Maasai land use and human welfare, by comparing these variables between regions. During the course of our study, we did find some important landscape-related differences between the regions (Lynn 2000), which seem to have an important influence on Maasai livestock management. Nevertheless, where critical aspects of land use are concerned that directly influence human welfare, we concluded that land use differences are based largely on policy-related restrictions on grazing and cultivation in the NCA; no restrictions exist in the LGCA.

Methods

Forty-nine in-depth household surveys were conducted, 23 in the LGCA and 26 in the NCA, from May through December 1998. Figure 3.2 shows the household sites as assessed through a geographic positioning unit. The number of people in the household include the head of the household, his wives and their children. Personal interviews were conducted at each household, usually with the heads of household. Following this, wives and other women in the household were interviewed separately. In these surveys, data were collected on land use patterns, household composition, income and expenditures, livestock and agricultural production, and marketing, among other things. Some results are described below.

Livestock numbers were transformed into Tropical Livestock Units (TLUs): one cow/steer = 1 TLU; one goat/sheep = 0.125 TLU; one bull = 1.25 TLU; and one calf = 0.6 TLU (Dahl and Hjort 1976; Galvin 1992). Wilcoxon W non-parametric tests were conducted to test the differences between the LGCA and NCA livestock holdings and agricultural plot size. Descriptive statistics were used to analyse the economic data. When appropriate, *t*-tests for statistical significance were run to compare data from the LGCA and the NCA.

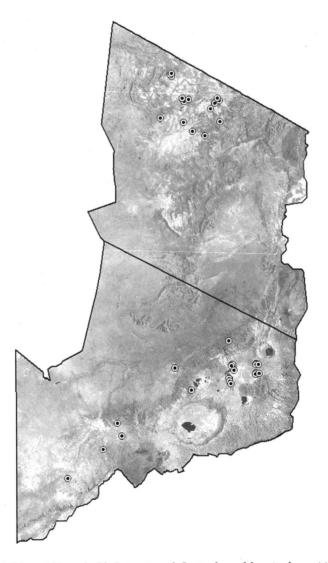

Figure 3.2 Map of Households Interviewed. Latitude and longitude positions taken with use of a GPS

Anthropometric measurements were taken on individuals in the LGCA (n = 224) in June and July 1998 and in the NCA (n = 650) in February and March of 1999. In addition, a sample of Maasai from the NCA were measured in July 1998 (n = 95). Measurements of height, weight, upper arm circumference (UAC) and triceps skinfolds (TSF) were taken on individuals, depending on their age (WHO 1995). We analysed the effect of years and regions (LGCA vs. NCA) on the anthropometry measures. ANOVA (analysis of variance) was used to detect annual differences between NCA Maasai nutrition in 1998 and 1999.

We compared nutritional status (weight, height), of children in LGCA and the NCA. Children above the age of two were grouped into three age/sex groups: 2 to 5.9 year-olds, 6 to 12.9 year-olds and 13 to 17.9 year-olds (Table 3.1). We also compared body mass index (BMI) scores (wt/ht^2, a measure of leanness), UAC and TSF measurements for adults in the two regions. Adults were grouped into four age/sex groups: 18 to 29.9 year olds, 30 to 39.9 year olds, 40 to 49.9 year olds and 50 years of age or older (Table 3.2). ANOVAs were used to assess differences in anthropometric measures. In addition, pairwise differences were determined for adult BMI scores by age group by region.

Results

Table 3.1 Sample Size of the Children's Anthropometric Surveys by Age/Sex Group and Region

| | | Age groups | |
	2–5.9	6–12.9	13–17.9
Girls			
Loliondo	14	10	5
NCA	119	142	27
Boys			
Loliondo	9	10	5
NCA	86	44	8

Table 3.2 Sample Size of the Adult Anthropometric Surveys by Age/Sex Group and Region

| | | Age groups | | |
	18–29.9	20–39.9	40–49.9	50+
Women				
Loliondo	76	37	2	6
NCA	103	33	10	18
Men				
Loliondo	12	11	2	25
NCA	18	18	16	8

Livestock Numbers and Crop Cultivation

Figure 3.3 shows livestock to human ratios for the LGCA and the NCA as measured by TLUs per person. In our sample, LGCA Maasai have more than three times as many TLUs per person (x = 10.3) as the Maasai who live in the NCA (x = 2.8) (p < 0.001) (for confirmation see Runyoro 1999). Moreover, LGCA Maasai have, on average, agricultural plots three times the size (x = .3 acres/person) of the NCA Maasai (x = .1 acres/person) (p = 0.002) (Figure 3.4). Figure 3.5 shows the livestock/human ratios arrayed against the acreage per human ratios. The majority of the NCA households are clustered together and 87 per cent of them have below the theoretical minimum of 6 TLUs per person needed for food security in pastoral populations (Brown 1971; see Galvin 1992; Homewood 1992 for further discussions on TLUs among pastoralists). The figure shows that a much lower percentage (42 per cent) of LGCA households are below this minimum.

Household Economics

Households in the NCA are significantly larger on average, with a mean of 22 people, whereas for LGCA the mean is 15 (p = 0.008) (Table 3.3). A census (NCAA/NPW 1994) of the entire NCA reported an average of 8.2 people per household. Our study shows almost three times as many people per household as in the 1994 study. Furthermore, recent estimates suggest there are about 30 people per household in NCA (Runyoro 1999). Natural population increase and emigration to the NCA since the ban on cultivation was lifted may account for this increment (McCabe et al. 1997). Also,

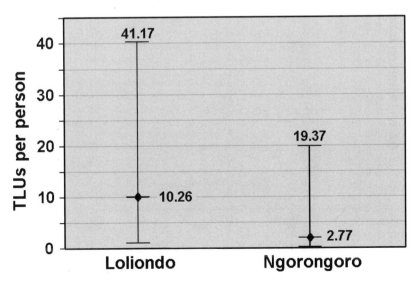

Figure 3.3 Livestock to Human Ratios (TLUs per person) for Loliondo and for Ngorongoro

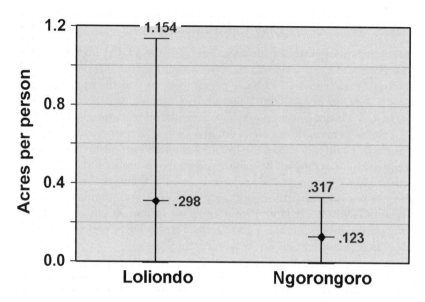

Figure 3.4 Acreage to Human Ratios (Acres Cultivated per Person) for Loliondo and for Ngorongoro

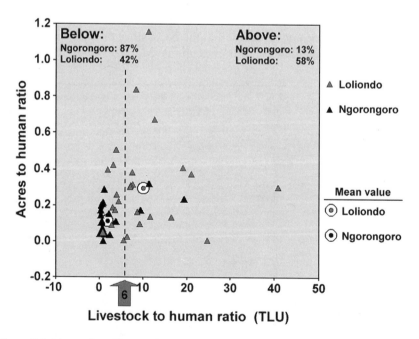

Figure 3.5 Livestock to Human Ratios Compared to Acreage to Human Ratios with an Estimate of the Minimum Number of TLU's Needed per Person for Food Security

anecdotal evidence suggests that health care has improved in accessibility and effectiveness during the last several years and may affect child morbidity and mortality (Dr. Msechu, Endulen and Dr. Said Montaghe, Wasso hospitals, personal communication).

Livestock sales were higher in the NCA than in LGCA (Table 3.4). The mean number of cattle sold in Loliondo as a percentage of the total herd size was 3.7 per cent; in the NCA it was 8.2 per cent. The same pattern exists for sheep and goat sales (Table 3.5). Households in Loliondo are generally located further from the livestock markets than those in the NCA, which may partly account for the higher sales in the NCA. In addition, discussions with pastoralists suggest that people sell diseased livestock, and households in the NCA appeared to have more diseased animals than those in Loliondo. Agricultural yields also revealed differences between the two regions. There was tremendous variability in agricultural yields (Table 3.6), but yields per person were generally about twice as high in Loliondo as those in the NCA.

Table 3.3 Demographic Statistics of Interviewed Households

	Location	*Mean*	*SD*	*P-value*	*N*
Age, head of	Loliondo	53.7	11.0	NS	26
household (years)	NCA	57.7	9.9	NS	15
People in the	Loliondo	15.0	7.3	<0.01	25
household	NCA	22.0	9.5	<0.01	20
Wives	Loliondo	2.7	1.5	<0.02	27
	NCA	4.0	2.0	<0.02	23
Children	Loliondo	10.6	6.3	<0.01	26
	NCA	17.2	8.1	<0.01	20
Houses in	Loliondo	7.3	4.6	NS	26
homestead	NCA	8.3	3.4	NS	20

Table 3.4 Cattle Sold per Household as a Percentage of the Total Cattle Herd

Location	*Mean (%)*	*SD*	*P-value*	*N*
Loliondo	3.7	2.3	NS	14
NCA	8.2	7.5	NS	15

Table 3.5 Goats and Sheep Sold per Household as a Percentage of the Total Goat and Sheep Herd

Location	*Mean (%)*	*SD*	*P-value*	*N*
Loliondo	3.8	11.1	NS	16
NCA	7.4	9.2	NS	21

The effects of livestock production and agricultural production influence household income in the two regions (Table 3.7) (Smith 1999). In both locations, livestock sales were the most important source of income, followed by crop sales. However NCA households report crop sales as being almost three times as important as for households in Loliondo.

Nutritional Comparisons

The results of comparing nutritional indices among NCA children in 1998 and 1999 showed that there was no effect of year on height of children but weight differences were significant (p = 0.02; x = 25.9 kg for children in 1998 and x = 23.4 kg for children in 1999). Adult male and female BMI scores, UACs and TSFs for NCA Maasai in 1998 were not different than those in 1999.

Figure 3.6 shows that, in general, girls and boys in Loliondo tended to weigh more than their NCA counterparts, but the differences were not significant. Among 2 to 5 year-old boys the difference in mean weight was in the order of 15 per cent; among the 6 to 13 year olds the difference was 17 per cent and among the adolescents it was 4 per cent.

The mean BMI score for all adult women from Loliondo is 19.4 whereas among women in Ngorongoro it is 18.5. Loliondo mean BMI score for men is 19.7 versus 18.7 for men from the NCA. Figure 3.7 shows adult female and adult male BMI scores for specific age groups. BMI scores were significantly different among the 18 to 29.9 year old women (p<0.01). The other age-specific values were not significantly different. Loliondo men's BMI scores for specific age groups also tended to be greater than those of men from Ngorongoro but the differences are non-significant. Adult TSFs were significantly different by region for women (p< 0.001) and for men (p< 0.001) (Figure 3.8).

Summary and Implications

Table 3.6 Kilos of Maize and Beans Harvested in 1997/1998

	Location	Mean (%)	SD	P-value	N
Per Acre	Loliondo	484.8	422.3	NS	15
	NCA	476.3	365.0	NS	5
Per Person	Loliondo	190.8	277.4	NS	12
	NCA	86.2	113.5	NS	4

Table 3.7 Sources of Income and their Relative Contribution to the Household Economy

a. Loliondo	Rank order of importance				Total value	Importance (%)
	1 (1.0)	2 (0.5)	3 (0.25)	4 (0.125)		
Livestock sales	24	8			28.0	88
Crop sales		2	1		1.4	5
Animal medicine	1	1			1.0	3
Plow Labor				1	0.5	2
Milk Sales			2	1	0.4	1
Beadwork			1		0.3	1
Livestock skins				1	0.1	0
Total	25	11	4	3	31.7	100

b. NCA	Rank order of importance				Total value	Importance (%)
	1 (1.0)	2 (0.5)	3 (0.25)	4 (0.125)		
Livestock sales	14	8	2		18.5	83
Crop sales	1	4		1	3.1	14
Honey sales			2		0.5	3
Milk sales				1	0.1	0
Total	15	12	4	2	22.2	100

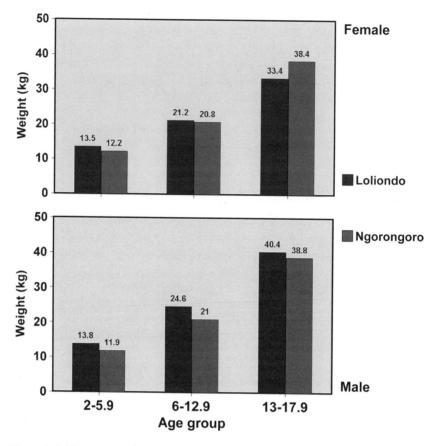

Figure 3.6 Children's Mean Weights by Age/Sex Group and Region

- The Maasai of Loliondo clearly have more resources available to them than do the Maasai in the NCA, as measured by livestock holdings and agricultural plot size. The reason for the differences in area cultivated is a direct result of NCAA limitations on agricultural plot size in the NCA (McCabe et al. 1997). The issue is more complicated where livestock are concerned. There are policy restrictions on grazing in the NCA, but these restrictions are not so severe as to account for the vast differences in livestock holdings. However, in the NCA the wildebeest migration excludes Maasai livestock from important wet-season forage resources, preventing the traditional transhumant migration of livestock; i.e., moving into the highlands during the dry season and using the plains during the wet season. This is because wildebeest calves transmit malignant catarrhal fever to cattle, a fatal disease. In the past, the Maasai apparently harassed the wildebeest away from plains areas grazed by cattle, or fenced water holes, denying water to the wildebeest (McCabe et al. 1997). This is no longer

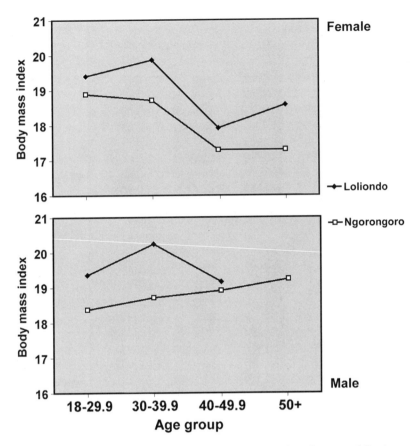

Figure 3.7 Adult Mean Body Mass Index Scores by Age/Sex Group and Region

possible within the NCA, although such actions could be used in the LGCA. Thus livestock nutrition in the NCA is constrained by lack of access to the wet season range, and other disease problems (such as East Coast fever) are exacerbated because cattle are confined to the highlands during the wet season. Reduced nutrition and increased disease incidence combine to limit production, reproduction and early survival of NCA cattle (Machange 1997).

- Households in the NCA are larger than in the LGCA. More livestock and agricultural produce are sold in the NCA than in LGCA, even though production is lower in the former. It is likely that livestock and crops are used to make up for food shortfalls; we know that most expenditures go to purchasing food (Smith 1999).
- The Maasai children from Loliondo tend to have higher anthropometric measures than children from the NCA.

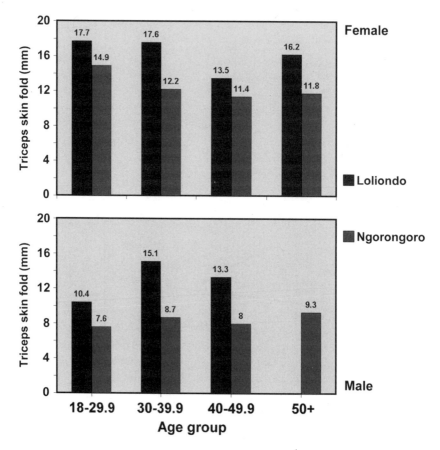

Figure 3.8 Adult Mean Triceps Skinfold Values by Age/Sex Group and Region

- Adults of Loliondo also tend to show higher nutritional status (BMI scores). TSF measurements were also significantly higher among Loliondo women and men than among adults in Ngorongoro.
- These results suggest that conservation policy affects resources available to the NCA Maasai and this may influence nutritional status of the population, especially adults. Children in Ngorongoro tend to be better buffered from nutritional stress than are adults, a pattern common among pastoral populations (Galvin 1992; Galvin et al. 1994a).
- These initial comparisons demonstrate that the Maasai of Loliondo are better off than the NCA Maasai. Some of these differences (e.g., crop acreage) are clearly attributable to conservation policy. However, population density is greater in the NCA (6.0/km²) than in the LGCA (3.9/km² with the use of population numbers from the 1988 Tanzanian census) which also affects access to resources. Other factors such as landscape variation and distance to markets may also contribute to variations in human welfare and resource access.

Integrated Modelling and Assessment System

How could the situation in the NCA be modified to improve human welfare without compromising conservation value? Data from the above research projects have been combined with other information in an Integrated Modelling and Assessment System (IMAS) based on spatial-dynamic computer modelling, geographic information systems, remote sensing and field studies, to address this question.

The IMAS is enabling alternative policy and management strategies to be objectively explored, debated, implemented, and reassessed. Figure 3.9 shows a conceptualization of the problem. Land use interacts with ecosystem structure and dynamics through such processes as vegetation and herbivore production, which are in turn driven by climate. Development and conservation policies influence land use, with subsequent impacts on pastoral welfare, livestock production, wildlife conservation efforts, and ecosystem integrity. Culture influences land use. However, land management, modernization processes and increasing interaction with national economies are also affecting land use, with subsequent impacts on other variables.

At the core of the IMAS is SAVANNA, a spatial-dynamic ecosystem model that was developed for pastoral ecological research in Turkana District, Kenya where we worked for over ten years. SAVANNA simulates plant growth; soil water budgets; herbivore foraging, energetics, weight, population dynamics and spatial distributions; pastoral herd management; and energy flows to pastoralists.

We are in the process of developing a socioeconomics submodel for SAVANNA. The results of ecological changes or policy initiatives are felt pri-

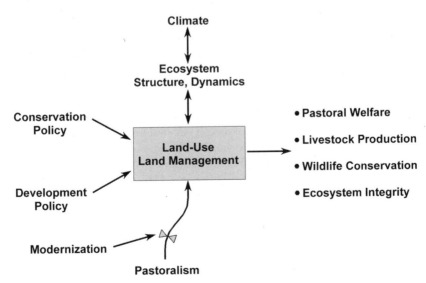

Figure 3.9 Conceptual Model of Pastoral Ecological Interactions

marily and most directly at the household level, with changes in income, food security and nutritional status. There is a simple rule-based framework for the socioeconomics submodel. First, the household has to meet its food requirements (the subsistence submodel); however, land use policy, population size and distribution and development policy all affect the ability of people to meet their basic subsistence requirement. If there is a shortfall, this is made up by recourse to various options, such as livestock sales or purchasing maize. Second, the household is assumed to manage for investment and disinvestment of livestock through purchases and sales. Third, is discretionary consumption with consequent impacts on cash reserves. Then there are cropping decisions. The outcome results in a particular level of food security as determined by cash availability and food energy. The SAVANNA model, when integrated with the socioeconomics submodel, will be able to objectively quantify the impacts of different management scenarios on factors of livestock production, pastoral welfare, wildlife, and ecosystem integrity (Galvin et al. 2000).

Some of the most important factors affecting land use in the NCA and the LGCA are increasing human populations, conservation policies and wildlife-livestock disease interactions (Runyoro 1999). Land use by NCA pastoralists is intensifying due to human population growth, while livestock grazing is greatly restricted by transmission of diseases from wildlife to livestock. For example, cattle are restricted to the midlands and highlands of NCA during the wet season because of the risk of the transmission of malignant catarrhal fever from wildebeest calves inhabiting the lowland plains (McCabe 1995; Rwambo et al. 1999), limiting the carrying capacity of the system. Goats and sheep have no such restriction. Using SAVANNA, we ran a simulation with the number of livestock increasing by 50 per cent. Cattle numbers declined dramatically during a dry period in the early 1980s (Figure 3.10a), whereas goat and sheep populations declined marginally. An example of the spatial output produced by SAVANNA is shown (Figure 3.10b), demonstrating a decline in total vegetation offtake following the collapse of the cattle population.

It is unlikely that Maasai herders would allow livestock populations to collapse. It is more likely that livestock populations would remain constant through stocking guidelines or outside inputs. Therefore, another example is where we set the populations of livestock constant and conducted another SAVANNA analysis. Signs of overstocking of the system were present, such as an increase in unpalatable herbaceous vegetation over time (Figure 3.11a). Wildlife populations generally declined with increased livestock populations (resident grazing antelope and zebra), unless not in competition with livestock, such as resident wildebeest (Figure 3.11b) which primarily inhabit Ngorongoro Crater, where livestock are not allowed to graze.

Why use an IMAS to sort out these processes? First, use of a simulation model provides a quantitative profile of the pastoral system as a whole, with its important components. Too often policy decisions are made on piecemeal information or no information at all. Policy and management decisions have an impact on all components of the system and this is one way to assess those integrated impacts. Second, use of an IMAS is relatively objective. Various stakeholders may have differing goals, but each goal can be set up as a scenario in the model and impacts on livestock, wildlife, people and the ecosystem can be determined. Management decisions are often made on the basis of narrow goals without an understanding of the impact of those goals. This assessment system provides a way to get around those limitations.

There are constraints on the use of an IMAS, however. First, the results are only as good as the assumptions and data that are entered into the system.

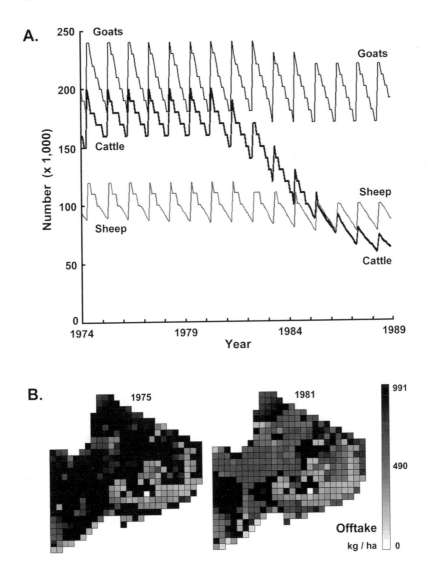

Figure 3.10 SAVANNA Ecosystem Model Results with Livestock Increased by 50% Compared to Current Conditions. Changes in livestock populations (A) and the spatial distribution of the total off take of vegetation (B) are shown.

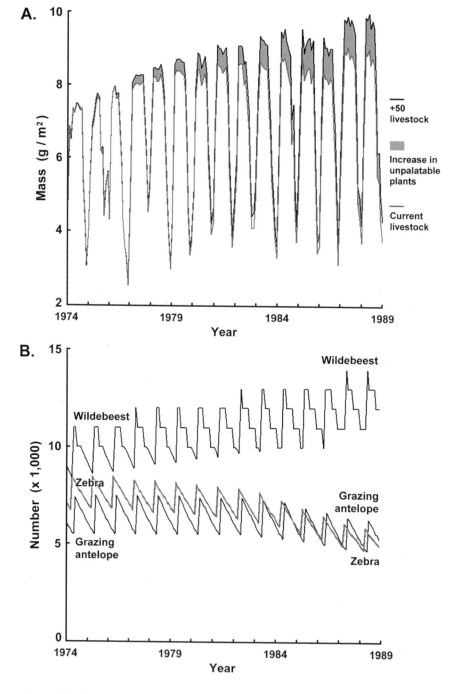

Figure 3.11 When livestock populations are held constant at 50% of their current levels, unpalatable herbaceous biomass increases over time (A) (the shaded area shows the increase through time) and most wildlife populations decline (B).

Second, what good is an IMAS if no one uses it? We have been developing the IMAS in collaboration with African University faculty and students, government agencies and NGOs representing pastoralists (e.g., African Wildlife Foundation and others) and wildlife interests (e.g., World Wildlife Fund). We have conducted demonstrations of the IMAS at several venues in Tanzania. We intend to enhance the ability of these groups to assess food security and environmental situations in livestock-based systems, including the NCA and Loliondo regions and other areas with conservation-development conflicts. As part of this effort we are developing a user-friendly interface for the model. The interface, called SavView, provides a series of windows allowing users to set values for use in models that address management questions (Figure 3.12a is an example) and a series of windows that allow users to change the geographic layers used in modelling (Figure 3.12b). For example, a user might select a map that had additional water sources added. The interface allows the user to run the model, and display results as charts showing changes over time (Figure 3.13a) and maps showing spatial distributions over time (Figure 3.13b). SavView allows users to conduct analyses that include countless variations of seven themes representing changes in: 1) rainfall amounts over time; 2) livestock and wildlife populations; 3) livestock mortality and productivity; 4) the areas in which livestock and wildlife may graze; 5) water supplies; 6) agriculture; and 7) human populations. The results of analyses reflecting sixteen potential management questions were included on a CD-ROM that was distributed to interested parties.

Training is an integral component of this endeavour. We are training Tanzanian graduate students and national park staff, and we have provided training to interested organizations. Twenty-six scientists and managers, mostly East Africans, have received in-depth training in IMAS techniques during two workshops. The model is available for use in the Community Services Centre in Arusha (AWF), and at the University of Dar es Salaam. Establishment and use of the IMAS is an iterative process and results are not expected overnight. We believe that this approach, though in a rudimentary stage, has a promising future in adaptive management for human-environment issues of biodiversity and human development.

Conclusions

Ngorongoro Conservation Area does not qualify as a community-based conservation programme in the strict sense. The resident Maasai do not receive direct monetary benefits from the presence of, or proceeds derived from, wildlife. There is no revenue sharing in the NCA. On the other hand the NCA does provide some direct benefits to residents, such as grain in times of food shortage; greater access to the goods and services available in the conservation area; limited employment opportunities and limited opportunities to sell goods to tourists. However, living in a conservation area also carries direct costs for the Maasai. The research reviewed here shows that the Maasai living in NCA

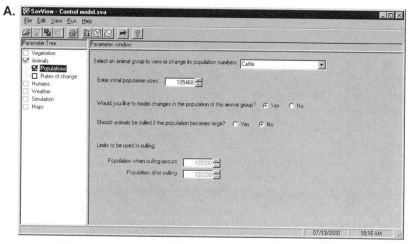

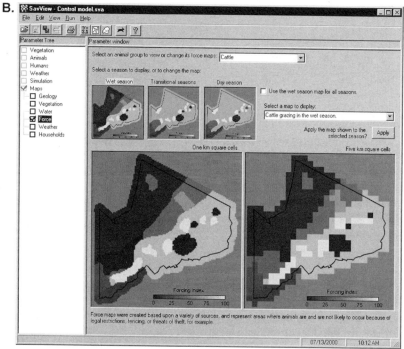

Figure 3.12 A computer interface called SavView alows users to enter values (A, as an example), and change maps (B) used in modelling

do indeed exist in a state of some economic deprivation as compared to their neighbours in the Loliondo Game Controlled Area. It seems reasonable to presume that many of the regional differences in material wealth and nutritional status are attributable to constraints on land use, resulting from NCAA conservation policy. The conclusion that we draw from this study is that the costs of

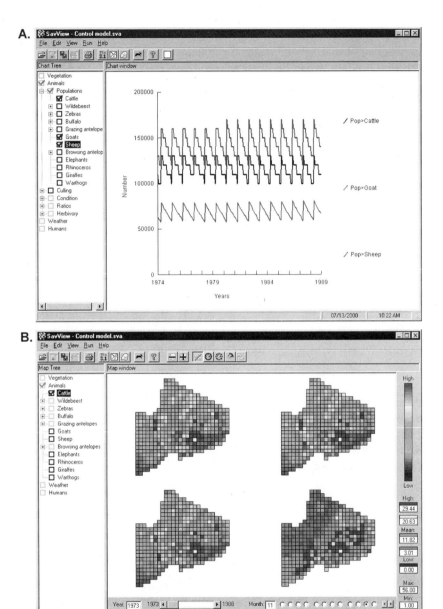

Figure 3.13 The Computer Interface will Produce Charts of Changes in the Ecosystem Over Time (A) and Changes in Spatial Distributions Over Time (B)

wildlife conservation are partially borne by the residents of conservation areas. Thus area residents will have to be somehow compensated for these costs, if their welfare is not to be compromised by conservation programmes.

Unfortunately, the situation at NCA has reached the point where there are no easy avenues for optimizing conservation and human development. The

NCAA has several options for improving the welfare of residents, but each option carries a significant cost to conservation.

Human welfare would be likely to improve if:

- *Livestock herds were larger.* Programmes aimed at reducing disease incidence would be beneficial. The costs of these programmes would be significant because diseases are many and virulent. In addition, increased livestock numbers would have implications for native herbivores. If the goal of a livestock programme were to increase herds to the point that there were 6 TLUs per person, this would require a total of about 300,000 TLUs versus the current 140,000 livestock TLUs (Runyoro 1999) now resident in NCA. The area might support a herd of this size, but it would almost certainly drastically reduce the forage and space available for native herbivores, and reduce the conservation value of the NCA.
- *Cultivation areas were expanded.* Our current estimate suggests that there is about 0.04 ha of land cultivated per person (Lynn 2000) in the NCA, or a total of about 20 sq. km of cultivated land within the conservation area. If the amount of land per person was increased to that found in the LGCA (0.122 ha/person) this would raise the total amount of land cultivated to about 60 sq. km, still less than 1 per cent of the total area of the NCA. However, some question whether *any* cultivation should be permitted inside a premier conservation area.

Other even more extreme and politically more difficult options include reducing the size of the resident Maasai population, or de-gazetting part of the current reserve. Each of these choices will have costs, benefits, supporters and detractors. The IMAS is designed to provide insight into the costs and benefits of these sorts of difficult policy options, when the time is past for easy solutions and 'no-cost' alternatives.

Note

1 Research supported by the US National Science Foundation Grants (SBR 9100132 and SBR 9709762). This publication was also made possible through support provided by the Office of Agriculture and Food Security, Global Bureau, United States Agency for International Development, under Grant No. PCE-G-98–00036–00. The opinions expressed herein are those of the authors and do not necessarily reflect the views of the US Agency for International Development.

References

Arhem, K. 1985. 'Pastoral Man in the Garden of Eden. The Maasai of Ngorongoro Conservation Area, Tanzania'. *Uppsala Research Report in Cultural Anthropology.* Uppsala.

Brown, L.H. 1971. 'The Biology of Pastoral Man as a Factor in Conservation'. *Biological Conservation,* 3: 93–100.

Dahl G, and Hjort, A. 1976. 'Having Herds. Pastoral herd growth and household economy'. *Stockholm Studies in Social Anthropology* No. 2. University of Stockholm.

Gadgil, M., Berkes, F. and Folke, C. 1993. 'Indigenous Knowledge for Biodiversity Conservation'. *Ambio*, 22: 151–6.

Galvin, K.A. 1992. 'Nutritional Ecology of Pastoralists in Dry Tropical Africa'. *American Journal of Human Biology*, 4: 209–21.

—— 1994. Food choice during drought among Maasai women pastoralists of northern Tanzania. Invited poster presented at the session on Famine Foods: Food Choice under Conditions of Scarcity, at the annual meetings of the American Anthropological Association, Atlanta, 30 Nov.–4 Dec.

—— 1995. 'Conservation Policy and Human Nutrition in Ngorongoro Conservation Area, Tanzania'. Invited paper presented at the session on Natural Resource Management in Eastern and Southern Africa: Issues of Sustainability and Conservation at the annual meetings of the Society for Applied Anthropology, Albuquerque, 29 March–2 April.

—— 1997. Biological conservation and human nutrition by geographical location in the Ngorongoro Conservation Area, Tanzania. Poster presented at the Human Biology Association meetings. St. Louis, 1–2 April.

Galvin, K.A., Coppock, D.L. and Leslie, P.W., 1994a. 'Diet, Nutrition and the Pastoral Strategy'. In E. Fratkin, K.A. Galvin and E.A. Roth (eds). *African Pastoralist Systems: An Integrated Approach*, pp. 113–32. Boulder: Lynne Rienner.

Galvin, K.A., Ellis, J.E., McCabe, J.T. and Moehlman, P. 1994b. 'Wealth and Nutrition among Maasai Pastoralists in a Conservation Area, Tanzania'. Poster presented at the annual meetings of the American Association of Physical Anthropologists, Denver, April. *American Journal of Physical Anthropology Supplement* 18: 91–2.

Galvin, K.A., Thornton, P. and Mbogoh, S. 2000. *Integrated Modeling and Assessment for Balancing Food Security, Conservation and Ecosystem Integrity in East Africa.* Final Report, Socio-Economic Modeling Component, 1997–2000 to GL-CRSP, U.S. AID.

Grandin, B.E., Bekure, S. and Nestel, P. 1991. 'Livestock Transactions, Food Consumption and Household Budgets'. In P. deLeeuw, B. Grandin and P.J.H. Neate (eds). *Maasai Herding: An Analysis of the Livestock Production System of Maasai Pastoralists in Eastern Kajiado District, Kenya*, ILCA Systems Study 4. Addis Ababa: ILCA.

Herlocker, D.J. and Dirschl, H.J. 1972. 'Vegetation of the Ngorongoro Conservation Area, Tanzania'. *Canadian Wildlife Service Report Series*, no 19. Ottawa: Queens Printer.

Homewood, K.M. 1992. 'Development and the Ecology of Maasai Pastoralist Food and Nutrition'. *Ecology of Food and Nutrition*, 29: 61–80.

Kajazi, A., Mkumbo, S. and Thompson, D. 1997. 'Human Livestock Populations Trends'. In D.M. Thompson (ed.). *Multiple Land-Use: The Experience of the Ngorongoro Conservation Area, Tanzania*, pp. 169–80. Gland, Switzerland and Cambridge, UK: IUCN.

Kruuk, H. 1972. *The Spotted Hyena: A Study of Predation and Social Behaviour.* Chicago: University of Chicago Press.

Lynn, S. 2000. 'The Effects of Conservation Policy and Ecology on Pastoral Land Use Patterns: A case study of Maasai land use in northern Tanzania'. M.S. Thesis, Department of Range and Ecosystem Science, Colorado State University.

Machange, J. 1997. 'Livestock and Wildlife Interactions'. In D.M. Thompson (ed.). *Multiple Land-Use: The Experience of the Ngorongoro Conservation Area, Tanzania*, pp. 127–42. Gland, Switzerland and Cambridge, UK: IUCN.

McCabe, J.T. 1995. 'Wildebeest Maasai Interactions in the Ngorongoro Conservation Area of Tanzania'. Final report submitted to the National Geographic Society.

McCabe, J.T., Mollel, N. and Tumainai, A. 1997. 'Food Security and the Role of Cultivation'. In D.M. Thompson (ed.). *Multiple Land-Use: The Experience of the Ngorongoro Conservation Area, Tanzania*, pp. 397–416. Gland, Switzerland and Cambridge, UK: IUCN.

Moehlman, P., Runyoro, V.A., and Hofer, H.. 1997. 'Wildlife Population Trends in the Ngorongoro Crater'. In D.M. Thompson (ed.). *Multiple Land-Use: The Experience of the Ngorongoro Conservation Area, Tanzania*, pp. 59–70. Gland, Switzerland and Cambridge, UK: IUCN.

NCAA (Ngorongoro Conservation Area Authority)/NPW (Natural Peoples World). 1994. *Census Results*. NCAA, Tanzania.

Parkipuny, M.S. 1997. 'Pastoralism, Conservation and Development in the Greater Serengeti Region'. In D.M. Thompson (ed.). *Multiple Land-Use: The Experience of the Ngorongoro Conservation Area, Tanzania*, pp. 143–68. Gland, Switzerland and Cambridge, UK: IUCN.

Rainy, J., Rainy, M., and Harris, E. 1999. 'Integrated modeling, assessment, and management of regional wildlife-livestock ecosystems in east Africa'. Report from a workshop held under the Regional Economic Development Services Office for east and southern Africa, U.S. Agency for International Development. 6–8 July 1999. Nairobi.

Redford, K.H. and Mansour, J.A. 1996. *Traditional Peoples and Biodiversity Conservation in Large Tropical Landscapes*. Arlington, VA: America Verde Publications. The Nature Conservancy, Latin America and Caribbean Division,

Runyoro, V.A. 1999. '1998 Aerial boma count, 1999 People and livestock census, and Human population trend between 1954 and 1999 in the NCA'. Report from the Research and Planning Unit, Ngorongoro Conservation Area Authority, Ngorongoro Crater.

Rwambo, P., Grootenhuis, J.G., DeMartini, J., and Mkumbo, S. 1999. 'Animal disease risk in the wildlife/livestock interface in the Ngorongoro Conservation Area of Tanzania'. Report prepared under the USAID Global Livestock Collaborative Research Support Program.

Smith, N.M. 1999. 'Maasai Household Economy: A comparison between the Loliondo Game Controlled Area and the Ngorongoro Conservation Area, northern Tanzania'. M.A. Thesis. Department of Anthropology. Colorado State University, Fort Collins.

Wells, M. and Brandon, K. 1992. *People and Parks: Linking Protected Area Management with Local Communities*. Washington DC: International Bank for Reconstruction and Development.

WHO. 1995. WHO Expert Committee on Physical Status: The use and interpretation of anthropometry. Who Technical Report Series 854. Geneva: WHO.

4

Giving Conservation
a Human Face?

LESSONS FROM FORTY YEARS OF COMBINING
CONSERVATION AND DEVELOPMENT IN THE
NGORONGORO CONSERVATION AREA, TANZANIA

J. Terrence McCabe

Introduction

As the human population of the earth grows there is an increased emphasis on the preservation of what remains of the planet's special places and
important natural resources. The number of protected areas and national
parks has increased dramatically over the past twenty years, especially in
the developing world. New models of conservation have also been introduced, many that emphasize the incorporation of indigenous peoples into
the conservation process. However, despite the importance of linking
conservation and human development, for both the protection of natural
resources and for the economies of indigenous peoples, there have been
few examples of real success. One problem is that these Integrated Conservation and Development projects are relatively new, and that lessons
learned from failure as well as success are just beginning to be understood.
Another problem is that despite the rhetoric that advocates bringing
indigenous peoples into the conservation process, often there seems to be
little common ground or even communication between those who advocate

for indigenous rights and human development and those who advocate for conservation of natural resources, especially wildlife. Of course there are exceptions, the attempts to bring the Aboriginal peoples into the management of National Parks in Australia being one example; the Campfire Programme in Zimbabwe (Communal Areas Management Programme for Indigenous Resources) being another. However, recent books by Ghimire and Pimbert (1997), Stevens (1997), Neumann (1998) and Honey (1999) illustrate how difficult this task has been. Indeed other chapters in this volume will attest to the fact that the overall record has not been encouraging, especially with respect to protected areas.

One area where the attempt at incorporating indigenous rights, development, and conservation has a long history is in the Ngorongoro Conservation Area (NCA) in Tanzania. For the last eleven years I have been conducting research there, and much of this work has been at the intersection of conservation and human development efforts. In this chapter I examine the history of this multiple use strategy and try to draw some lessons based on this forty-year-old experiment in multiple land use.

During the last decade I have observed that the relationship between the Maasai and the Ngorongoro Conservation Area Authority has often been contentious and characterized by mistrust on both sides. The Maasai have been supported by local and international organizations advocating human development and indigenous rights. The conservation effort is supported by national institutions including the Conservation Authority itself, and a number of very active wildlife oriented NGOs, such as the Frankfurt Zoological Society. The rapidly changing Maasai economy and the growth of the human population have posed serious challenges for conservation. Restrictions on land use, especially the former prohibition on cultivation, have seriously impacted the livestock based economy of the Maasai residents of the NCA. Both constituencies have held large meetings and conferences to discuss problems, but very rarely do the two groups meet together. This point was emphasized in January 2000 in the town of Arusha. During a four-day period about fifty people, including Maasai living in the NCA, development workers, and land rights activists met in one hotel. At the same time the board of directors for the Ngorongoro Conservation Area were meeting in another hotel. There was no communication between the two groups. Opinions and perceptions were reinforced but no progress was made in bringing the two sides together.

I believe that what I have observed in the NCA is characteristic of many places where protected areas are surrounded by, or incorporate, indigenous people. The lessons from Ngorongoro have relevance far beyond the borders of the Conservation Area. However, before specifically addressing the Ngorongoro case, I think that it would be useful to discuss why conservation issues should be important to social scientists; and to examine some of the underlying perceptions of the land and the people upon which conservation in Africa has been based, especially as that pertains to the pastoral regions of the continent.

Why Should Conservation be Important to Social Scientists?

The earth's natural resources and biodiversity are limited and precious resources. This fact lies at the heart of projects and programmes that emphasize the importance of 'sustainable development'. Despite the confusion over what 'sustainability' actually means or how to operationalize the concept, it is well recognized that we need to conserve our natural resource base for succeeding generations. This issue cross-cuts disciplinary boundaries and is as important to social scientists as it is to natural scientists.

The need to conserve natural resources and to protect biodiversity is reflected in the increase in the number of protected areas and the land area set aside for the purpose of conservation. In this process millions of people have been and will be negatively impacted through the loss of their land and restrictions on their livelihood activities. From 1900 to 1950, approximately 600 protected areas were established worldwide (Ghimire 1994); five years ago there were almost 10,000 protected areas encompassing approximately 5 per cent of the earth's surface (Stevens 1997). Although this may seem like a lot of land, conservationists hope to double this in the near future so that 10 per cent of the earth's land area is in some type of protected area status (McNeely 1993 in Stevens).

Much of the increase in land under protected status has occurred in the developing world. Table 4.1 lists those countries in the developing world where 10 per cent or more of the land is classified as 'protected'. Although the areas listed in Table 4.1 are classified as 'protected', they are not necessarily national parks. The first area to be designated a 'National Park' was Yellowstone National Park in the United States, and most national parks throughout the world are based on this model. Before 1992 the International Union for the Conservation of Nature (IUCN) definition of a national park, based on the Yellowstone model was:

> A large area where 1) one of several ecosystems are *not materially altered by human exploitation and occupation,* where plant and animal species, geomorphological sites and habitats are of special scientific value, educational and recreative interest or which contains a natural landscape of great beauty; and *2) where the highest competent authority of the country has taken steps to prevent or eliminate, as soon as possible, exploitation or occupation in the whole area* and to enforce effectively the respect of ecological, geomorphological or aesthetic features that have led to its establishment (West and Brechin 1991: xvii emphasis theirs).

In other words local peoples were excluded, evicted, or relocated.

This definition was somewhat softened in 1992 in recognition of the fact that excluding all indigenous peoples can result in antagonism between local populations and the national park, and that 'there are few areas in the world that have not been shaped by interaction with humanity, to some extent' (Phillips and Harrison 1999: 262). The revised IUCN definition of a national park is:

> protected area managed mainly for ecosystem protection and recreation – natural area of land and/or sea designated to (a) protect the ecological integrity of

Table 4.1 Developing Countries where 10 per cent or more of the Land is Classified as 'Protected'

Asia	**Latin America**
Bhutan	Chile
Brunei	Costa Rica
Nepal	Cuba
Pakistan	Dominican Republic
Sri Lanka	Guatemala
Thailand	Honduras
Indonesia (just under 10%)	Nicaragua
	Panama
Africa	Venezuela
Benin	
Botswana	
Central African Republic	
Kenya	
Madagascar	
Malawi	
Rwanda	
Senegal	
Tanzania (25%)	
Zimbabwe	

Source: Ghimire (1994).

one or more ecosystems for present and future generations, (b) exclude exploitation or occupation inimical to the purposes of designation of the area and (c) provide a foundation for spiritual, scientific, educational, recreational and visitor opportunities, all of which must be environmentally and culturally compatible (Quoted in Stolton and Dudley 1999).

Although this allows for some occupation of national parks by indigenous communities, evictions and the denial of rights are not uncommon. For a more detailed account of recent evictions from Mkomazi national park see Brockington 1999.

Alternative Paradigms to the Yellowstone Model

In 1980 IUCN published the World Conservation Strategy which challenged the traditional national park model for conservation. The WCS advocated incorporating local indigenous peoples into the conservation process. For the first time conservationists were called to take the needs of resident and local peoples into conservation plans. This was followed in the mid 1980s by the World Bank which began funding a programme called Integrated Conservation and Development projects in which local peoples were to be incorporated into development projects and should benefit economically from them. Today there are a multitude of alternative approaches to conservation which range from including people in national parks (as found in Canada and Britain) to co-management.

Clearly the displacement of indigenous peoples is a great concern to human rights activists and social scientists. As social scientists we should question the need to adhere to the Yellowstone model. We should also be proactive in trying to make our understanding of how local peoples use natural resources and how they view 'conservation' useful to those who design and implement conservation projects. I believe that the protectionist stand where areas are alienated from local people and maintained by paramilitary forces will have utility in only a small number of 'protected' areas. The future for conservation will depend on how well or how poorly local communities are integrated into conservation projects and programmes. From this perspective, social scientists and natural scientists will be equally important in this conservation paradigm.

Wildlife Conservation in Africa

An examination of the history of wildlife conservation in Africa, especially in eastern Africa, reveals the importance of two images or myths which have been central to the design of conservation policy. The first relates to the Western perception of the African environment, and the second concerns lingering colonial perceptions of indigenous peoples, in particular pastoralists.

Jonathan Adams and Thomas McShane in their book *The Myth of Wild Africa* trace the formation of an image of Africa as a wilderness to the stories told by explorers such as Mungo Park, Richard Burton, and David Livingstone; and to the representation of these images by painters such as Henri Rousseau.

> The heroic figures from the golden age of African exploration searched for the sublime and found it; here was a refuge from industrial, despoiled Europe. To an eager audience steeped in romanticism, and to generations that followed, the tales of the explorers created an Africa that was both a paradise and a wilderness, a place of spectacular but savage beauty (Adams and McShane 1992: xii).

The implications of this image of Africa with respect to land use policy began to become manifest during the colonial period. Anderson and Grove have noted that:

> The colonial relationship … allowed Europeans to impose their image of Africa upon the reality of the African landscape. Much of the emotional, as distinct from the economic, investment which Europe made in Africa has manifested itself in a wish to protect the natural environment as a special kind of Eden, for the purposes of the European psyche rather than as a complex and changing environment in which people actually had to live. (Anderson and Grove 1987: 4).

The second image was that of the pastoral people who inhabited the arid and semi-arid rangelands of eastern Africa: the land also inhabited by the large herds of migratory ungulates and other animals which were so attractive to conservationists. The most well-known of these people were the Maasai who lived in what is now Kenya and Tanzania. The view of the Maasai around the turn of the century is aptly summed up in the following quotation by Lord Lugard:

> The Masai country has at present the disadvantage that its inhabitants are purely pastoral, and hence there is no food or cultivation in the country, though the soil is rich and the country fairly well watered. The warlike instincts of the Masai, moreover, render them at present an obstacle to peaceful development, and a terror to the more industrious and agricultural tribes around them (Lugard 1893, vol. 1: 147, quoted in Collett 1987: 136).

The administrative goals became focused on settlement and agricultural expansion. In 1905 the Maasai of Kenya were removed to two reserves and in 1911 they were all settled on the southern reserve. During this time colonial settlement was progressing rapidly and European farms needed to be protected from wildlife. It was also recognized that income could be generated from hunting. In the early 1900s the Maasai Reserve was also gazetted as a game reserve.

During the 1920s the colonial administration in East Africa began to view all pastoral production systems as mismanaged and the land overstocked and over-grazed. This view was given academic credence by the views of Melville Herskovits with his paper on the 'Cattle Complex in East Africa' (Herskovits 1926). The Dust Bowl catastrophe in the United States raised concern about soil erosion and conservation in the arid and semi-arid lands throughout the world. All of this led to a view that wildlife conservation was the most sound way of managing the East African rangelands.

During this time a more humanistic view of African people began to emerge: one in which the indigenous people were seen as part of culture, not part of nature. This separation of nature and culture lies at the heart of Western concepts of environmental preservation. It was accepted that for wildlife to be preserved, special areas free from human habitation must be created. The national park model, developed in the United States, was adopted in East Africa as the way to conserve wildlife and preserve the environment. The first national park in which the local pastoral people were evicted was created near Lake Manyara in the 1930s.

Following the Second World War a number of other parks were created and pastoralists evicted. This trend continued throughout the 1950s and 60s and even accelerated in the post independence years (for more detail see Neumann 1998). The one exception to the Yellowstone model was that of the Ngorongoro Conservation Area; but its history has been chequered and the multiple use strategy has gone through periods of severe stress.

Ngorongoro

The Ngorongoro Conservation Area encompasses 8300 sq. km in Tanzania. Topographically, it is made up of several volcanic peaks and adjacent plateaus with elevations of approximately 4000 metres; the highlands give way to the vast plains of the Serengeti which form approximately 50 per cent of the NCA in the western section (see Figure 4.1). The highlands receive substantially more rainfall than the plains with annual precipitation averaging 800–1200 mm for the higher elevations while the lower elevations receive only 300–400 mm.

Recent archaeological investigations strongly suggest that pastoralists have occupied this area for at least 2500 years; and that the Maasai have lived in the Ngorongoro region since the middle of the nineteenth century (Collett 1987; Borgerhoff-Mulder et al. 1989). The Maasai pastoral economy is based on the keeping of cattle, supplemented by the raising of small stock and small-scale cultivation. Typically, three or more Maasai families and their animals live together within a thorn bush enclosure, with each family utilizing a particular gate; this social and spatial unit is usually referred to as an *enkang* or *boma*. Enkangs are relatively sedentary, changing locations only once every few years. Within the Conservation Area, livestock follow a seasonal round in which people and livestock congregate at the semi-permanent settlements in the dry season and disperse into temporary camps in the wet season.

The Ngorongoro/Serengeti ecosystem was gazetted as a national park in 1940, but at that time allowed both pastoralists and cultivators to continue with their traditional way of life. During the early 1950s conflict between the pastoralists, cultivators and park authorities resulted in the division of the park into what is now known as Serengeti National Park, and the NCA. Serengeti National Park was to be administered as a typical national park, with no human habitation outside of the park authorities. The NCA, however, was to be administered with a dual mandate; to protect the interests of conservation on the one hand, and to protect the interests of the resident Maasai on the other hand. Although the Maasai were allowed to reside within the NCA, they were subject to a series of policy changes that had significant impacts on their pastoral livelihoods. Some areas were closed to pastoralists; in other areas, such as the Ngorongoro and Empakaai Craters, cattle could be grazed but no settlements were allowed (prohibition of settlements in the craters occurred sometime around 1974). However, the most important restriction resulted from an amendment passed in 1975 which prohibited all cultivation within the Conservation Area. Because the Maasai living within the Ngorongoro area had traditionally depended upon local cultivators for grain, this was perceived as a great hardship. In 1992 this restriction was lifted by an order from the local member of parliament, but without consultation with the Conservation Area Authority. In 1998, the President announced that small-scale cultivation would be allowed. However, some conservationists have recently advocated that the World Heritage designation for the NCA be rescinded because of cultivation, again causing the Authority to question the multiple use strategy.

Pastoralism and Change in Ngorongoro

The most serious challenge for conservationists committed to the multiple use concept in the NCA has been the relatively rapid social and economic

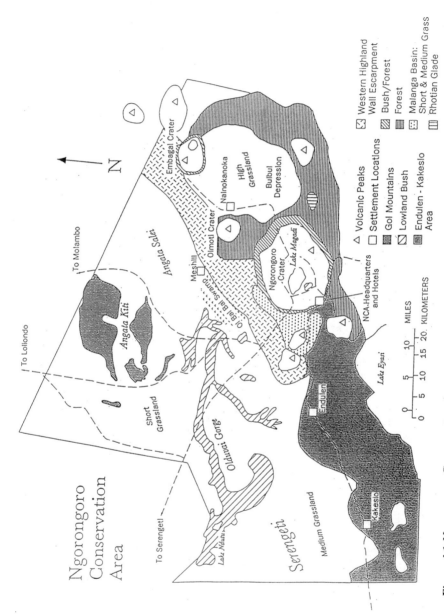

Figure 4.1 Ngorongoro Conservation Area

change that has been characteristic of most pastoral peoples in East Africa for the last twenty to thirty years. The Maasai in the NCA have been subject to similar social and economic influences as their pastoral neighbours; however, their ability to cope with new stresses has been limited by conservation policy. In addition, conservation policy itself has created stresses that their pastoral neighbours have not had to contend with.

In many cases policy seems to have been predicated on misperceptions or myths concerning the livelihood strategies of the Maasai. Despite the evidence that the Maasai of the NCA were highly dependent on grain, and that people were more sedentary than their ancestors, when I first conducted an examination of the Maasai economy in 1989 I was told by the Conservation Area Authority that the Maasai depended upon milk, meat and blood for their subsistence and that their calls for increased access to grain were really an attempt to cause trouble. At this time the Authority had access to the numerous reports and papers published by Homewood, Rodgers and Arhem who had conducted extensive research in the NCA in the early 1980s (Homewood et al. 1987; Homewood and Rodgers 1991). This lack of attention to the human side of the dual mandate has resulted in local residents feeling that they have borne most of the price for conservation, yet it was their ability and desire to co-exist with wildlife that has preserved the vast herds of wildlife still found in the NCA. In the section below I discuss major stresses and policy changes that have affected Maasai pastoralism from the early 1960s to the present. I have divided the time periods by significant changes in conservation policy in the NCA.

1957–1974

The eviction of all Maasai pastoralists from the Serengeti National Park had serious repercussions outside the Park boundaries. There were two major impacts for residents of the NCA. Many Maasai (but certainly not all) regularly used the Serengeti plains for grazing during the wet season. The area around the Moru kopies was especially important in this regard. The restrictions on grazing meant that the lowlands in the NCA would be subject to more grazing pressure and areas free from wildebeest calves more difficult to find (see below). Secondly, many of the Maasai who were residents of the Serengeti were relocated into the NCA, especially the area in between Endulen and Kakesio. This both increased grazing pressure and intensified the use of other resources (water and areas for cultivation).

Perhaps the biggest challenge to the Maasai livelihoods during this time, however, was the dramatic increase in the wildebeest population that occurred from the early 1960s to the early 1970s. During this period wildebeest numbers expanded from approximately 240,000 animals to 1,600,000. The cause for this increase is probably related to the eradication of rinderpest in the livestock population, which served as a reservoir for the disease. The impacts on the Maasai were twofold. First, wildebeest calves carry a virus that transmits the disease malignant catarrhal fever; this is benign to the wildebeest but fatal to cattle. The virus is shed in the ocular and nasal secretions of the wildebeest calves until they are about three months old. The wildebeest migration comes into the NCA during the months from January to April (the rainy season) and it is during this time that the wildebeest give birth. The only way for the Maasai to ensure that their cattle do not contract the disease is to avoid areas where wildebeest calves have been grazing. The traditional grazing pattern involved moving to the lowlands during the rainy period so that the cattle could utilize the highly nutritious grasses that grow there, and recover from the stresses of the previous dry season. With the irruption of the wildebeest population, the entire lowlands became dangerous places for the Maasai to graze their cattle. They have responded by adjusting their migratory cycle so the cattle leave the plains when the wildebeest begin to calve, and return when the wildebeest calves are not infectious (Maasai say this is when the calves turn from brown to black), or when the wildebeest are thought to have left the area. One of the net results of this adjusted grazing pattern is that cattle enter the dry season in poorer condition than previously. An old Maasai man in Nainookanooka told me in 1995: 'The wildebeest come and take all our grass with them, but they leave their disease behind.'

Another result is that there is less grazing available for the Maasai cattle in the dry season, and they respond by taking their animals into the protected Northern Highland Forest as conditions become dry. The Conservation Area Authority and many conservationists have not seemed to appreciate the systemic relationship between the wildebeest migration and Maasai grazing patterns, and the penetration of livestock in the Highland Forest is deemed to be evidence of improper livestock management strategies.

1974–1992

During this time period the Maasai were subject to four severe stresses. The first was the imposition of a ban on cultivation, the second was restrictions on the use of the highland craters, the third was the spread and increase in livestock disease, and the fourth was a dramatic increase in the human population. Although the ban on cultivation during the 1950s precipitated the crisis that resulted in the splitting of the NCA from Serengeti National Park, by the early 1970s conservationists again felt that cultivation and wildlife conservation were incompatible. In 1974 new rules were enacted that made any cultivation within the NCA illegal. The local residents protested, but felt that they did not have any ability to influence decisions that were already made. Some people continued to cultivate illegally, and when caught, were subjected to fines and sometimes imprisonment. Very little that had happened throughout the history of the NCA had more negative impacts on the relationship of the local residents to the Authority than the ban on cultivation.

The same laws that banned cultivation also contained restrictions on the use of Ngorongoro, Empakaii and Olmoti Craters. Many Maasai families were permanent or temporary residents of these areas, and based on interviews that I have conducted it is clear that Ngorongoro Crater was a very important resource for people and livestock living throughout the highland areas. Highly nutritious grasses grew on the Crater floor, and there were a number of sites of permanent water. A few large bomas were located within the Crater itself, but many more people and livestock depended upon the resources there at different times during the year. It is also clear that the Maasai living in the Crater were not consulted, or even well informed about their eviction. The eviction process was described by Henry Fosbrooke, the first Conservator of the NCA.

> Early one morning in March 1974 three Land Rovers entered the Crater, one going to each boma. They carried the personnel of the paramilitary Field Force Unit. Without explanation and without notice they ordered the immediate eviction of the inhabitants and their cattle. Their possessions were carried out by transport of the Conservation Area Authority and dumped on the roadside at Lairobi. No explanation was given and no arrangements made for the re-settlement of the evacuees (Fosbrooke 1990, quoted in Neumann 1998: 147).

Clearly this was a great hardship for those affected, but combined with the ban on cultivation, it sent a very clear and disturbing message to the Maasai throughout the Conservation Area.

During this time there was also a dramatic increase in tick borne diseases. Some attribute the severity of new outbreaks to the fact that acaricides were available at times in the past, but were no longer available by the 1980s. Nevertheless it is clear that East Coast Fever, and 'Ormilo' (bovine cerebral theriolossis) had become very important diseases in the NCA by the 1980s. There was not only high mortality from these diseases, but cattle that survived often gave less milk because of the disease load.

In 1989, I led a team of researchers in an assessment of the state of the Maasai economy sponsored by the IUCN. The results of this research have been published elsewhere (McCabe et al. 1992; McCabe et al. 1997), and indicated that the Maasai economy was in a state of serious decline. The main conclusions of the study are:

- Grain accounted for a significant part of the Maasai dietary intake (similar to the results of research conducted by Katharine Homewood).
- Grain was obtained by selling livestock and buying grain in local shops.
- More livestock needed to be sold to meet grain requirements than could be replaced through natural reproduction.
- The road system was adequate for tourists, but in many areas where there were large concentrations of Maasai, the roads were impassable once it began to rain.
- Many of the local residents were involved in a downward economic spiral, and people saw little hope for their long-term survival.
- There was significant malnutrition and undernutrition among the children of many Maasai households.
- An increasing human population and a stable or fluctuating livestock population meant that each generation had less access to livestock than the generations that preceded them.

- The human and livestock population trends observed in the NCA were characteristic of many pastoral peoples in East Africa at the time of the study.
- Other pastoral peoples were responding by diversifying their economies, and especially through the adoption of agriculture.

1992–present

I continued to work in the NCA throughout the 1990s on a number of projects, and in 1995 I had the opportunity to restudy the Maasai economy. This was especially important because it had been three years since the ban on cultivation was lifted, and the results of this study would be directly relevant to the design of a new management plan that was taking place at the time. Some of the more important results of this 1995 study are:

- Approximately 85 per cent of families in the NCA had adopted cultivation since the ban was lifted.
- Families from all wealth categories were participating in agriculture, although a higher percentage of poor families cultivated than wealthy families.
- Plots were small, averaging slightly less than 1 acre per household.
- Yields were low, averaging between 7.5 bags of maize per acre to 2.2 bags of maize per acre (one bag = 90–100kg.)
- Crops grown by individual families contributed approximately one-half of total grain consumed by the family.
- The percentage of the family herds that needed to be sold dropped dramatically, as did the number of reproductive livestock.
- The percentage of children under 5 classified as undernourished dropped slightly.
- The percentage of children under 5 classified as malnourished dropped significantly.
- People in the Conservation Area felt that they had regained control over their lives, and could plan for the future.
- The relationship between the Authority and the local residents improved substantially.

Since 1995 a number of positive developments, from the Maasai perspective, have taken place in the NCA. The most important of these has been the implementation of a large restocking and veterinary programme funded by the Danish International Development Agency (DANIDA). The goal of this programme is to restock 3000 families classified as destitute, and who have fallen well below the number of livestock units necessary for a mixed agro/pastoral livelihood. In addition to the restocking programme, DANIDA is giving the initial funds and infrastructure necessary to provide privatized veterinary services to the areas of the NCA where they are most needed. One of the major problems facing the residents has been very restricted access to veterinary care and drugs. This programme could have a dramatic impact on the health and reproductive capacity of the livestock kept by the Maasai; and thus on their pastoral livelihoods.

Recent Developments

The improvements in the economy have proved to be a double-edged sword. On the one hand, things are better for the Maasai. On the other, the spread of cultivation, coupled with the increase in the human population have been a cause of great concern to conservationists. Again there is talk of re-evaluating the entire multiple use strategy. Some conservationists have written to UNESCO advocating that the World Heritage designation be rescinded because of the extent of cultivation in the NCA. Based on conversations that I recently had with the current Conservator, this is being taken quite seriously. There is also some immigration as cultivators from outside the NCA have been moving in. The escalating human population is considered a very real threat and there does not seem to be any obvious solution to the problems caused by increased and intensified use of the natural resources of the area. One critique of the Conservation Area Authority has been that no attempt has been made to bring the Maasai into the decision-making process and to give them a real and tangible vested interest in conservation. The traditional age set leaders, the *ilaigwenak,* provide a structure for discussion between the NCAA and the Maasai, but this has not been utilized. On the other hand a new organization 'the Pastoral Council' was put in place following the 1995 management plan. It originally included a number of NCAA managers and this undermined the decision-making process of the Council. The new Pastoral Council includes six traditional elders, six ward representatives, six women, six youth representatives, fourteen village chairmen, and the Conservator (Danida 1997). Although this is a step in the right direction there remains a high degree of scepticism regarding the influence of the Pastoral Council in management decisions.

At the same time, there has been a vigorous attempt to preserve and defend the rights of the Maasai living in the NCA. This has been spearheaded by Professors Shivji and Kapinga, both professors of law at the University of Dar es Salaam, who have published a book *Maasai Land Rights in Ngorongoro, Tanzania* (1998). The plight of the Maasai living in the NCA has also caught the eye of the national and international press, but often the reporting has been rather sensational, and not always accurate. A story that appeared on 25 February 2000 in the Tanzanian newspaper, the *Business Times,* begins: 'The human and civil rights of over 40,000 Maasai pastoralists in the Ngorongoro Conservation Area in the north-east of Tanzania are being gravely violated.' The caption under a picture of Maasai men dancing stated: 'Some of them, when they go back home, at the Ngorongoro Conservation Area, will be prohibited to enter the area, just as their kith and kin are barred.' Although Maasai rights may well have been violated, no residents of the NCA have been refused entry into the NCA, at least that I am aware of.

Those advocating for the preservation of Maasai rights should be supported, but in this context there is no attention being paid to the concerns of the conservationists or any recognition that Ngorongoro is truly a very special place. Instead of talking to each other both sides are becoming more entrenched in their position, and are marshalling national and international constituencies to support their side of the argument. It is my strong opinion that this will not work.

Lessons Learned from the NCA Experience

I believe that the most important lesson to be learned from the Ngorongoro experience is that there needs to be communication and trust between conservationists, social scientists, the indigenous people, and the Authority that is administering the area. In the NCA this has been a major, if not *the* major, obstacle to bringing the people into the conservation process. There is even a physical separation between those advocating for conservation goals and those advocating for indigenous rights. The conservation experts have traditionally lived and worked out of the headquarters area on the rim of Ngorongoro Crater. Social scientists, development workers, and human rights advocates have traditionally lived and worked out of the town of Endulen, 38 kilometres away. Both groups have mistrusted each other and both have been guilty of interpreting the results of research to support their ideological position.

In a broader sense it seems obvious that conservationists need to be aware of the impact that conservation policy has on indigenous peoples. However, if we as social scientists are to have an impact within the conservation community then we need to be equally receptive to the arguments that promote conservation. I have often heard conservationists demonized by social scientists, characterized as neo-colonialists at best. We have to appreciate the dedication and commitment that most professional conservationists have invested in their work, and we need to present the results of our work to that audience.

The second lesson is that combining the interests of conservation with economic development is a very difficult task. Too often rather simplistic solutions are proposed without any understanding of the complexities of human social organization, changing norms and values, or rapidly changing social and economic conditions. It is also not enough to argue for empowerment of the local people or incorporating the indigenous people into the conservation process. These may be critical for any integrated conservation and development project to work, but they will not satisfy conservation or development goals.

The third lesson is that although the issues may revolve around environmental conservation, the process by which decisions are made is political. The political arena is one in which differential power relations are critical, and some people and organizations are far more sophisticated than others in working within this context. In the case of the NCA, the international conservation organizations, and in particular Frankfurt Zoological, have had far more influence than any development organization or NGO. This is beginning to change, but in countries like Tanzania when the goals of conservation come into conflict with development goals, the outcome is almost always in favour of conservation.

The fourth and final lesson is that any project or programme that tries to bring together conservation and human development must be based on in-depth understanding of the human community and be flexible enough to cope with changing social and economic conditions. In the NCA, much of the policy that impacted the Maasai was based on a myth of who they were and how they lived. The Maasai, just like other pastoralists across the conti-nent, are diversifying their economies and becoming more sedentary. Human populations are increasing while livestock populations are not. No one fixed policy can cope with these changes.

The lessons learned from the Ngorongoro experience have relevance to human communities and protected areas throughout the world. In my mind, without real communication and trust among the 'stakeholders' the process of bringing local people into the conservation process is doomed to failure.

References

Adams, J. and McShane, T. 1992. *The Myth of Wild Africa: Conservation without Illusion.* New York: Norton.

Anderson, D. and Grove, R. (eds). 1987. *Conservation in Africa: People, Policies and Prac-tice.* New York: Cambridge University Press.

Borgerhoff-Mulder, M. et al. 1989. 'Disturbed Ancestors: Dataoga History in the Ngorongoro Crater'. *Swara*, 12(2).

Brockington, D. 1999. 'Conservation, Displacement, and Livelihoods: The Conse-quences of Eviction for Pastoralists Moved from the Mkomazi Game Reserve, Tanzania'. *Nomadic Peoples*, 3(2): 74–96.

Collett, D. 1987. 'Pastoralists and Wildlife: Image and Reality in Kenya Maasailand'. In D. Anderson and R. Grove (eds). *Conservation in Africa: People, Policies and Practice*, pp. 129–48. New York: Cambridge University Press.

Danida, Ngorongoro Pastoralist Project. 1997. Project Document. Copenhagen.

Ghimire K.B. 1994. 'Parks and People: Livelihood Issues in National Parks Manage-ment in Thailand and Madagascar'. *Development and Change*, 25: 195–229.

Ghimire, K.B. and Pimbert, M.P. (eds). 1997. *Social Change and Conservation.* London: Earthscan.

Herskovits, M. (ed) 1926. 'The Cattle Complex in East Africa'. *American Anthropolo-gist*, 28: 231–72, 633–64.

Homewood, K. M. and Rodgers, W. A. 1984. 'Pastoralism and Conservation'. *Human Ecology*, 12(4): 431–42.

Homewood, K.M. and Rodgers, W.A. 1991. *Maasailand Ecology: Pastoral Development and Wildlife Conservation in Ngorongoro, Tanzania.* Cambridge: Cambridge Uni-versity Press.

Homewood, K.M., Rodgers, W.A. and Arhem, K. 1987. 'Ecology of Pastoralism in Ngorongoro Conservation Area, Tanzania'. *Journal of Agricultural Science, Camb.*, 108: 47–72.

Honey, M. 1999. *Ecotourism and Sustainable Development: Who Owns Paradise.* Washing-ton DC: Island Press.

McCabe, J.T., Mollel, N. and Tumainai, A. 1997. 'Food Security and the Role of Cultivation'. In D.M. Thompson (ed.) *Multiple Land-Use: The Experience of the Ngorongoro Conservation Area, Tanzania.* Gland: IUCN.

McCabe, J.T., Perkin, S. and Schofield, C. 1992. 'Can Conservation and Development be Coupled among a Pastoral People: The Maasai of the Ngorongoro Conservation Area, Tanzania'. *Human Organization,* 51(4): 353–66.

Neumann, R.P. 1998. *Imposing Wilderness. Struggles over Livelihood and Nature Preservation in Africa.* Berkeley: University of California Press.

Phillips, A. and Harrison, J. 1999. 'Appendix 1: Categories of Protected Areas'. In S. Stolton and N. Dudley (eds). *Partnerships for Protection: New Strategies for Planning and Management for Protected Areas,* pp. 262–8. London: Earthscan.

Shivji, I.G. and Kapinga, W.B. 1998. *Maasai Land Rights in Ngorongoro, Tanzania.* Dar es Salaam: The Lands Rights and Resources Institute.

Stevens, S. 1997. *Conservation through Cultural Survival: Indigenous Peoples and Protected Areas.* Washington DC: Island Press.

Stolton, S. and Dudley, N. (eds). 1999. *Partnerships for Protection: New Strategies for Planning and Management for Protected Areas.* London: Earthscan.

West, P.C. and Brechin, S.R. (eds). 1991. *Resident Peoples and National Parks: Social Dilemmas and Strategies in International Conservation.* Tucson: University of Arizona Press.

5

National Parks and Human Ecosystems

THE CHALLENGE TO COMMUNITY CONSERVATION. A CASE
STUDY FROM SIMANJIRO, TANZANIA

Jim Igoe

Introduction

Community conservation initiatives in Tanzania claim to give rural Tanzanians direct control of natural resources, thereby creating incentives for sustainable resource management at the community level. In practice, however, the agendas of international conservation organizations, private tour companies, and state elites dominate these programmes. The primary objective of Tanzanian community conservation is currently to enrol local people in the protection of national parks. Ironically, the institutional legacy of national parks plays a central role in the very problems that proponents of community conservation are trying to solve. As colonial institutions, national parks in East Africa were gazetted without regard for local resource management systems, or even the seasonal migration of resident wildlife. This chapter considers the ecological, economic, and social problems that national parks have caused throughout East Africa, taking into account structural adjustment programmes which facilitate the wholesale alienation of natural resources from local users.

Community Conservation and the Institutional Legacy of National Parks

It is nearly impossible to address the issue of conservation in East Africa without reference to a paradigm of resource management known as community conservation. Stated most simply, this paradigm envisions a synthesis of conservation and development. To quote Barret and Arcese (1995: 1073), community conservation 'assume(s) that human and non-human systems are interdependent and, therefore, that the challenges of conservation and the challenges of development are inextricable'. African community conservation takes its model from Zimbabwe's CAMPFIRE programme, which began with the premise that giving local people a stake in wildlife would increase their incentive to conserve it.[1] Wildlife would become an important engine of local economic development (Kiss 1990 and Murphree 1996).

By the early 1990s the idea of community conservation had spread throughout Africa. In Tanzania, the rhetoric of community conservation has come to pervade the country's resource management and economic development policies. Current enthusiasm for community conservation resonates with the wave of economic and political liberalization that has swept Africa since the fall of the Berlin wall (Barkan 1994 and Kelsall 1998). This liberalization has been driven by the policies of international lending institutions, which see African states as inefficient and assume that 'market economies and political democracies naturally undergird each other' (Harbeson 1994: 7). In addition to free markets and multi-partyism, African countries require a new form of civil society, which will increase the capacity of citizens to participate in the governance of their countries. NGOs (Non-governmental organizations) are lauded as the 'backbone' of this newly emerging civil society (World Bank 1997: 10 and Hyden 1995: 44).

In keeping with this imperative of economic and political liberalization, community conservation in Tanzania is driven by international NGOs like the AWF (African Wildlife Fund) and the WWF (World Wildlife Fund). These organizations and others have worked closely with Tanzanian officials to formulate policies and to institutionalize community conservation in the country. Expatriate owners of private tour companies, who run community conservation projects in the villages where they operate, have assisted them in their efforts. The rhetoric of these organizations and companies advocates empowering local people to work as 'partners' with government agencies to develop sustainable resource management strategies that will bring direct benefits to their communities. In reality, community leaders are usually brought on board after the planning process is complete. Their primary role appears to be convincing local people to accept conservation projects that were designed without their input, by a small group of expatriate experts and Tanzanian elites.[2]

Prevailing relationships between planners and target communities reveal a fundamental discrepancy between the rhetoric of empowering local people and the practice of community conservation. Too often, the desires of local communities are overshadowed by the agendas of international conservation organizations and the Tanzanian wildlife authorities. An explicit objective of community conservation in Tanzania is to keep protected areas viable by enrolling neighbouring communities in their preservation. With parks as the essential units of conservation, its primary goal becomes the protection of 'critical land units essential to the integrity of the park, and the migratory routes of its wildlife' (TANAPA 1994: 1).[3]

Ironically, there is growing evidence that national parks themselves are contributing to the very problems that advocates of community conservation are trying to solve. British administrators, who assumed that African resource management systems were environmentally destructive, initially imposed the national park model in East Africa. The establishment of national parks and game reserves displaced thousands of rural people, most of whom continue to live on the margins of the protected areas from which they were evicted. The loss of natural resources to indigenous resource management systems that these evictions entailed frequently forced local people to mine natural resources in the areas to which they were restricted.

In spite of the negative impacts of protected areas on local resource management systems, they were never redefined to suit an African context in the years following independence. Newly independent governments 'accepted wholesale the imported conservation philosophy which more often than not sought to alienate local people from wildlife resources without alternatives' (Gamassa 1993: 4). The independent government of Tanzania has continued gazetting national parks to the present day. In 1988, the Department of Wildlife evicted over a thousand people from the Mkomazi Game Reserve. The paramilitary forces of the state wildlife authorities enforce the exclusion of people from protected areas. Consequently, relationships between parks and neighbouring communities have become increasingly antagonistic and occasionally violent.

Many advocates of community conservation recognize this animosity, and have focused their energies on improving park/community relations. By emphasizing community goodwill, however, these advocates frequently neglect to address the forces that represent the biggest threat to the continued viability of protected areas. This is where the link between conservation and development begins to break down. Outside of parks, development has meant putting land and other natural resources to their most profitable use regardless of ecological sustainability. World Bank/IMF structural adjustment, and its imperative that Tanzania should produce hard currency to service its national debt, has accelerated this process. Consequently, more and more land is being alienated from local producers and allocated to large-scale farms and ranching schemes. These enterprises impinge on park boundaries and obstruct wildlife migration routes. This is a process that is beyond the control of local people, who are now finding themselves squeezed between conservation and development interests with fewer and fewer livelihood options.

The problems outlined here are rarely addressed by advocates of community conservation in Tanzania. While these issues may be raised at workshops they are usually dismissed as problems that will disappear once communities are better organized, educated, and presented with viable alternatives to their current economic activities. Since community conservation in Tanzania is still in its pilot stages, however, the benefits of this expected transformation exist primarily in the rhetoric of international conservation organizations. Dialogue concerning community conservation is limited almost exclusively to the realm of ideas, realities are kept far from the public eye, and concrete case studies are rarely evoked.

In this chapter I will explore the assumptions of community conservation as it is practised in Tanzania through a detailed case study of Tarangire National Park and Maasai communities in the neighbouring Simanjiro Plain. I will demonstrate that community conservation efforts in Simanjiro do not address people's grievances concerning the loss of resources to Tarangire, nor do they evaluate the costs of these losses to local economies and resource management systems. Programmes that emphasize revenue sharing and technical development projects fail to address the cultural values and the aspirations of local people. Fixation on wildlife migration corridors ignores regional processes, which threaten both wildlife and local livelihoods. Additionally, very little consideration has been given to the social impacts of community conservation in Simanjiro. Until these fundamental problems are addressed, meaningful community conservation is unlikely to occur in Simanjiro or other pastoral areas of Tanzania.

Tarangire National Park and the Maasai of Simanjiro

Although Simanjiro District and Tarangire National Park are administered by different institutions of the Tanzanian State (the Simanjiro District Council and TANAPA respectively), conservationists and local resource users agree that they are part of a single ecosystem (see Figure 5.1). Until recently, the area that is now Simanjiro District and Tarangire National Park was part of the traditional territory of the Kisongo Maasai. Maasai pastoral systems followed a migration pattern similar to that of vast herds of wild ungulates with which they coexisted.

During the dry season (July to October) herders and wildlife concentrated around the Tarangire River (now inside Tarangire National Park) and other permanent water sources. At the onset of the short rains (October or November) they would disperse into the Simanjiro Plain to take advantage of seasonal water sources and abundant pasture (Borner 1985: 3). Wildlife began returning to the river in June, but pastoralists remained in wet-season dispersal areas as late as August, depending on conditions (Figure 5.2).

Contrary to popular stereotypes, the Maasai never lived in complete harmony with the wildlife that shared their pasture and water. Predators occasionally attacked livestock and people, and wildebeests had to be avoided during the calving season.[4] Nevertheless, herders and wildlife generally coexisted without any major problems. Additionally, important symbi-

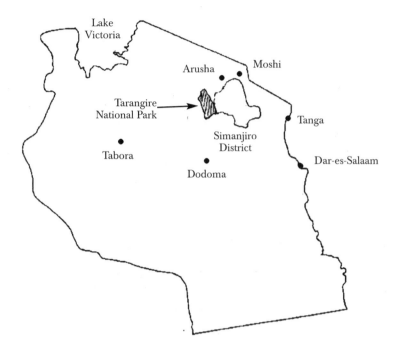

Figure 5.1 Simanjiro District and Tarangire National Park in a National Context

otic relationships did exist. Informants throughout my study area explained that controlled burning regimes encouraged new flushes of pasture that benefited wildlife as well as livestock. They also spoke of the importance of large animals in keeping back brush, which opened new grazing areas and reduced tsetse infestation.

For the Maasai of Simanjiro, the area that is now Tarangire was central to their system of transhumant pastoralism. First, the Tarangire River represents the most reliable dry-season water point in the entire ecosystem. In addition, the park contains an area known as the Silalo Swamp, which was the primary drought reserve area for the herders of the Simanjiro/Tarangire ecosystem. Finally, it contains a number of seasonal water sources, which were used as wet-season dispersal pasture during the early rainy season (see appendix to this chapter). Local herders claim that Tarangire National Park has disrupted their herding systems and contributed to the decline of Simanjiro's pastoral economy.

When the area that is now Tarangire became a game reserve in 1957, resident herders were allowed to continue living there. In 1961 there was a drought which many of my informants regard as the worst they have ever seen. During this drought, access to Tarangire and the Silalo Swamp was crucial to the survival of the herds of Simanjiro. Even today, the swamp is legendary. Informants throughout Simanjiro claim to have brought their livestock there in 1961.

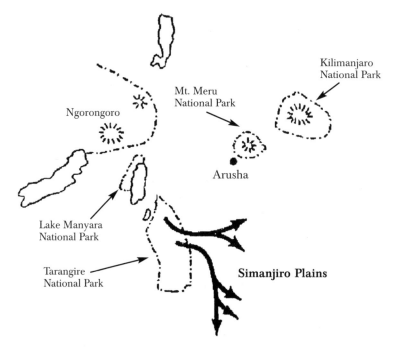

Figure 5.2 Wildlife Dispersal Patterns in the Wet Season from Tarangire to the Simanjiro Plain

In 1970 Tarangire was upgraded to a national park, and TANAPA rangers informed resident pastoralists that they would have to relocate. Since pasture and water resources to the east of the park were still relatively abundant, most herders left without being forcefully removed. The total number of people affected by the evictions is unclear. However, a household census conducted by Alan Jacobs in 1957 indicates that the Tarangire River had one of the largest concentrations of dry-season homesteads in northern Tanzania. From this census, and the accounts of local elders, it appears that at least several hundred people were displaced from Tarangire. It is important to remember that this number probably does not reflect the total concentration of herders around Silalo Swamp during periods of drought.

The loss of the swamp was profoundly felt during the drought of 1993/94. At this time, approximately thirty households (over 200 people) from Loibor Sirret (a village on the border of the park) went to rural Kondoa (south of Tarangire) in search of water and pasture. They found that conditions in this area were not substantially better than those in Loibor Sirret. The majority of these people lost most or all their livestock. Many have still not returned. Respondents in Loibor Sirret maintain that this unfortunate situation would not have occurred if they still had access to the Park and the Silalo Swamp.

Respondents in Loibor Sirret also indicated that nearby areas inside the park often get rains earlier than Loibor Sirret itself. They complained that pasture and water were nearby, and yet they were refused access to them.[5] They feel that the loss of these areas has contributed to declining livestock health and has reduced the reproductive capacity of their herds. Pastoralists forced onto the margins of the park are bitter about the tremendous loss of natural resources that the evictions entailed. They point out that the border has given wildlife an advantage over livestock. Although wildlife is still able to disperse into the Simanjiro Plain during the wet season, livestock cannot follow the wildlife back to the Tarangire River in the dry season. Herders described this arrangement as essentially unfair.

Relations between the TANAPA rangers and neighbouring communities have been consistently antagonistic. People report having their livestock confiscated by park rangers, being tied up, beaten up, sexually harassed, threatened (occasionally with guns), and arrested. Fines for being arrested and retrieving confiscated livestock run into several hundreds or thousands of dollars. Local people do not view the park as a public resource, nor do they believe that those who deny them access are interested in conservation. Respondents reported that during bad years they were forced to pay substantial bribes ($600 to $1200) for access to resources inside the park. Many were of the opinion that tourism benefited a wealthy elite, while bringing few benefits to the people who are forced to pay for national parks through the loss of their traditional natural resource base.

Community animosity toward Tarangire increased in 1989, when a group of expatriate and Tanzanian conservationists proposed the establishment of a 6000-sq. km conservation area on the eastern side of the park (see Figure 5.3). Marcus Borner, of the Frankfurt Zoological Society, outlined the details of the proposed conservation area in an article entitled, 'The Increasing Isolation of Tarangire National Park'. The article expresses concern over several large-scale farms that had impinged on park boundaries and blocked wildlife migration routes in and out of the park. It recommends a 'dual use (of the Simanjiro Plain) by Maasai pastoralists and wildlife' (Borner 1985: 95). It further recommends that agriculture be prohibited in the conservation area and that local herders be forced to destock. Borner and his colleagues showed little concern for the livelihoods of resource users living in the proposed conservation area.

Borner's recommendations exemplify a type of thinking that is overly focused on the preservation of a single national park. His perception of a protected area in crisis is based on localized observations of broader regional processes. Much of the resource pressure on the borders of Tarangire was not caused by the Maasai at all but by members of neighbouring agricultural groups who were settling throughout northern Simanjiro (see Figure 5.4). These settlers were squeezed from their homelands on Meru and Kilimanjaro, between coffee plantations on the slopes and national parks on the peaks (cf. Neumann 1992 and Spear 1997).

Landlessness on slopes of Mount Meru has now reached emergency proportions. The Presidential Land Commission reported that in 1992–3, 'Arumeru district authorities had been directed to settle landless people (about 18,000 in number) in other districts, particularly Kiteto (including

Simanjiro) and Monduli Districts' (URT 1993b: 9). The resettlement did not occur as authorities in the targeted districts were unwilling to receive such a large influx of outsiders. If it had occurred, however, it would have entailed the forced relocation of thousands of people into the very areas that the Simanjiro Conservation area was designed to protect. Similar processes in both Arumeru and Simanjiro have created large numbers of displaced people, who have been crowded onto an ever-shrinking resource base. Both the Arumeru District authorities and Marcus Borner perceived these processes as localized crises, which could be solved by either exporting the problem or by increasing centralized control of resources.

The idea of the Simanjiro Conservation area was ultimately scrapped. Many informants asserted that the lobbying efforts of influential Maasai elders defeated the plan. Although it never materialized, the thwarted conservation area has contributed significantly to the poor relations that still prevail between residents of Simanjiro and Tarangire National Park. Herders living on the borders of Tarangire still remember their evictions from the park. They are aware that the conservation area would have entailed the eviction of pastoralists living within 10 km of Tarangire, as well as the total prohibition of agriculture in the Simanjiro Plains.[6] The plan and past experiences with Tarangire rangers have made them suspicious of everything TANAPA has subsequently done in the area (cf. Kipuri and Nangoro 1996: 45).

Community Conservation in Simanjiro: Past, Present, and Future

In 1988, TANAPA introduced a programme known as Community Conservation Service (CCS). This came to Simanjiro in 1991, where it came to be known locally as 'Ujirani Mwema' (Good Neighbourliness). In recounting the history of this programme, informants often stated that communities had prevented TANAPA from expanding the boundaries of the park through force. CCS was therefore perceived as a device to trick communities into accepting the imposition of wildlife buffer zones and corridors.[7]

The programme itself is based on benefit sharing. TANAPA provides matching funds for village based development projects. An assessment of this programme written by Naomi Kipuri and Benedict Nangoro, both Maasai social scientists, gave it low ratings saying projects had been undertaken without consulting communities and that it did not give any meaningful support to the pastoral economies. In a response to the assessment, Patrick Bergin replied that it was not TANAPA's objective to provide support to the pastoral economy,[8] but simply to foster good relations between National Parks and neighbouring communities.[9]

This statement alone betrays a fundamental misunderstanding of Maasai resource management systems, and the meaning of 'Good Neighbourliness' in Maasai culture. This was summed up by a Maasai elder at a community meeting with CCS in September of 1996:

TANAPA can't teach us about Ujirani Mwema. We had Ujirani Mwema before TANAPA ever came here. If it doesn't rain here, I know that I can take my cattle to Naberera (a neighbouring village). The people at Naberera won't turn me away. They will let me stay so that my cattle will not die. They will help me, because they may need my help another year. This is Ujirani Mwema. TANAPA does not understand this. Their livestock (wildlife) come to graze in our villages, and we don't bother them. If it rains in the park we can't go there, even if our cattle are dying. If we do, we are beaten up and our cattle are taken away. This is not Ujirani Mwema. I know all about TANAPA's Ujirani Mwema. I've seen it first hand, and we don't need it here. We would all be better off if TANAPA took their Ujirani Mwema and went away.

The words of this elder were reiterated in some version by almost all of my respondents who lived within proximity of the park. Although TANAPA has been successful in implementing small technical projects in a few villages, community meetings are still dominated by similar concerns. Community suspicion of Tarangire and anything going by the label of 'conservation' has hindered community conservation efforts in the District.

Members of a local Maasai NGO, who attend and disrupt community conservation meetings in Simanjiro villages, have exacerbated this situation. CCS officers see these activists as radicals who are gratuitously disrupting what they are trying to achieve and confusing the local communities. Based on my discussions with local people, however, I am convinced that they are far from confused. Rather, they are following the lead of a group of people who they perceive as protecting their interests. The result is that in several villages no progress has been made towards community conservation. This situation is regrettable, since there is a great deal of overlap between the stated objectives of CCS and the Maasai NGOs. Additionally, the current impasse between conservationists and local NGO leaders represents a potential stumbling block to future community conservation programmes in Simanjiro.

The latest development in community conservation is the Department of Wildlife's (DoW) draft wildlife management plan. The plan includes provisions for the creation of Wildlife Management Areas (WMAs), which will allow communities to benefit from wildlife resources within their boundaries. Although these have yet to be officially legislated, the DoW has already implemented pilot WMAs and has given the green light for additional ones. The upshot of this draft legislation is that communities with high wildlife population densities can apply to their district government for status as a WMA. If this status is granted, then the community will have the right to manage wildlife within its boundaries in ways that will generate revenues for community development projects. These projects will be conceived and designed by the communities themselves; with the help of local NGOs, representatives of the DoW, and experts from the AWF.

While WMAs might bring benefits to some communities, they will be limited to those that are in relative proximity to national parks. Within these communities, benefits are likely to reach a relatively small group of people and not be the types of benefits that local people most want. Like Ujirani

Mwema, WMAs are conceived as programmes that will safeguard protected areas and their resident wildlife and do not address the real needs and desires of pastoral communities. Pastoralists are interested in the potential revenues of WMAs, but they are more concerned with the long-term viability of their herds. Their number-one stated need remains increased access to land and other natural resources; and this need cannot be met through WMAs.

An additional danger of WMAs and community conservation is that they may become ends in themselves. Such programmes involve relatively large numbers of outside actors. In addition to state and parastatal wildlife authorities, the implementation of WMAs involves international development organizations and local Maasai NGOs. These local NGOs often represent the interest of a small group of educated elite who have positioned themselves as 'gate keepers' between communities and donors. At the time of this study, two Maasai NGOs were vigorously campaigning in Simanjiro over claims to community mandates and access to donor money.[10]

As future programmes unfold there is a very real danger that local people may become commodities of, rather than participants in, community conservation. With levels of funding increasing every year, so too the number of individuals and institutions involved is proliferating. In order to be recognized as legitimate actors, these organizations and individuals must be able to claim community support.

In post-liberalization Tanzania a profusion of non-profit and private sector organizations now vie with Robert Fatton's 'predatory state' for the prize of Goran Hyden's 'uncaptured peasantry'. In this new arena, however, the productive capacity of the peasantry is less important than the peasants themselves, symbolized as rural communities. By capturing, or claiming to capture, the support of these 'communities' individuals and organizations gain access to the international funding on which they depend. This could lead to the ironic situation, where much energy is expended in the arrogation of communities while little meaningful conservation gets done.

Conclusion and Recommendations

In order to avoid the scenario described above, conservationists will need to work for true community support. To claim community consensus in the interests of expediency will in the long run create conflict, and conservationists will continue to see their efforts thwarted by the disruption of their activities. Many influential community members are convinced that conservationists are proverbial wolves in sheep's clothing. In order to gain their support, conservationists must demonstrate that this is not so.

Unfortunately, providing tangible support for direct community control of land and natural resources is politically difficult in Tanzania today. Colonial development and conservation programmes were built on a foundation of legislation that gave the government direct and centralized control of land and natural resources (see URT 1993a). To the present day these laws have remained on the books with minimal revisions. If anything, their impact has been strengthened by structural adjustment programmes and their imperative that land and natural resources be put to their most productive and profitable use. In his book *Not Yet Democracy* (1998), Tanzanian attorney and noted human rights activist, Issa Shivji outlines current efforts by Tanzanian and international activists to create legislation and institutions that are more amenable to community control of land and natural resources.[11]

Unfortunately, neither the Tanzanian government, nor the British consultant who drafted Tanzanian land law in 1998, were receptive to the ideas of these activists. A government position paper, addressing the possibility of community control over land and natural resources, states that such measures would turn the government 'into a beggar for land when required for development' and that 'the Investment Promotion Policy will be impossible when the government doesn't have a say in land matters' (quoted in Sundet 1997: 9). Shivji wryly concedes that government officials are probably correct in their observation that few rural Tanzanians would be willing to sacrifice their livelihoods for the benefit of state elites and foreign investors (1998: 82).

From this perspective, community conservation cannot simply be treated as an exercise in gaining local support for the preservation of wildlife and the protection of park boundaries. It faces a much more daunting project of addressing the ecological and social impacts of national parks on rural communities, and policies that mandate the liberalization of land and natural resources to service Tanzania's national debt. Clearly, there is no easy formula to address these types of long-standing historical problems, but at the very least it is important to acknowledge that community conservation under current conditions in Tanzania is a fundamentally political proposition.

International NGOs, as well as bilateral donors and international financial institutions, officially claim to stay away from these types of domestic political issues, since they have no authority over the internal affairs of sovereign countries. And yet there are precedents for this type of involvement in other parts of the world. In 1988, for instance, the World Wildlife Federation and the Friends of the Earth supported activists from the Kayapo ethnic group in Brazil (Turner 1993: 538). Strong support from a network of environmental and human rights NGOs allowed Kayapo activists to halt the construction of a World Bank sponsored hydroelectric dam on the Xingu River. The construction of this dam would have had devastating ecological and social impacts.

The impacts of parks and commercial farms in Simanjiro perhaps do not appear as dramatic as those caused by the construction of hydroelectric dams. Nevertheless, they have altered the ecosystem and local resource management in ways that fundamentally threaten peoples' livelihoods as well as the future of wildlife conservation in Tanzania. Enhancing local control of land and natural resources, and strengthening indigenous resource management systems, will necessarily require strategies that begin to address the impacts of these global processes and institutions. This will require tough political choices that run counter to the interests of Tanzanian elites and international investors, as well as the active dismantling of top-down attitudes and relationships that currently pervade international development and conservation circles (cf. Chambers 1996). While these types of choices are clearly unattractive, they are essential if post-cold war enthusiasm for democracy and participation is ever to be realized.

Obviously, such fundamental changes will not happen overnight. Nevertheless, important steps could be taken in the context of community conservation in Simanjiro. The threat that commercial agriculture poses to both pastoralism and wildlife provides a common ground for pastoralists and conservationists. International conservation organizations could join with international human rights NGOs (as they already do in Latin America) to lend support to community-based land rights initiatives. At the local level they could provide support to community initiatives to gain village land titles, and develop land plans that would make the alienation of village land to large-scale farms more difficult. Additionally, community conservation revenues could contribute to community efforts to evict illegal farms through Tanzanian courts. At the national level, they could provide funding to initiatives to reform Tanzanian land law. These activities would demonstrate to pastoralists that conservationists share their interests and are therefore their natural allies. This is also quite probably the only realistic strategy for protecting the integrity of Tarangire.

Another shortcoming of community conservation initiatives in Simanjiro is that they provide no tangible support for the pastoral economy and its associated resource management practices (cf. Kipuri and Nangoro 1996). While extolling the virtues of pastoralism as the only economic activity that is directly compatible with wildlife conservation, conservationists currently follow strategies that punish pastoralists and harm their economy. As pastoralists become impoverished they begin farming, an activity that conservationists perceive as a threat to protected areas and the conservation of wildlife. In fact, one of the stated objectives of Tarangire CCS is to educate people to refrain from farming in wildlife migration corridors.

Clearly, it is in the interest of both conservationists and pastoralists to have more livestock and fewer farms. This could be brought about through restocking programmes, veterinary services, and the creation of improved water sources for livestock – the things that pastoralists constantly request but which are rarely forthcoming. To implement such projects would demonstrate a true commitment to pastoralists and their livelihoods; and it is these projects that pastoralists would look for in exchange for not farming in wildlife migration corridors.

Finally, the smooth functioning of pastoral systems requires mobility and maximum access to natural resources, both of which have been greatly impeded by the creation of protected areas. Tarangire contains a drought reserve which is crucial to the entire Simanjiro pastoral system. Several authors are now asserting that the radical separation of humans and nature is a Western idea inappropriate to the African context (cf. Gammasa 1993 and Neumann 1992). Kipuri and Nangoro argue that conservationists should reconsider legislation that absolutely excludes people from national parks. They recommend that pastoralists be given limited access to strategic park resources in bad years. Clearly, occasional access to Silalo would greatly enhance the viability of pastoralism in Simanjiro.

Community sentiment on this matter is clear: if conservationists are interested in Maasai rangelands they should also be willing to make concessions for access to resources in protected areas. This would allow communities to negotiate with conservationists from a position of equals. It would also create a situation of give and take that would increase the likelihood of a positive community response to conservation agendas.

In the final analysis, true community conservation will require a comprehensive reassessment of conservation objectives. If communities are actually given control over their own natural resources, then community definitions of resource management are likely to prevail. Conservationists, who are interested in the preservation of wildlife, will need to understand indigenous ideas about conservation and search for common ground. If communities are to be partners in this process, then community agendas will need to be given equal weight. This will entail redefining development to reflect both community and conservation interest – a process that will involve compromise and a large commitment of time and resources. In such an undertaking, expedient solutions are impossible. The potential gain, however, is great: the sustainable management of resources and enhanced community well-being.

Notes

1 CAMPFIRE stands for Communal Areas Management Programme for Indigenous Resources. While the programme is widely lauded as an exemplary model for community conservation throughout Africa, empirical analysis of CAMPFIRE programmes indicates that it is beset with the same types of institutional and political problems that are described in this chapter. A study by the Indigenous Environmental Study Centre indicates that CAMPFIRE has excluded and displaced local communities, benefited private safari companies and local elites, has not contributed significantly to local incomes, and is not ecologically sustainable (Patel 1998).

2 Most of the community conservation initiatives in northern Tanzania involve revenue sharing and technical projects. The majority of these have been implemented by TANAPA's Community Conservation Service (CCS). Private tour companies, most notably Dorobo Tours and Oliver's Camp, are undertaking similar programmes.

3 This agenda is evident in the rhetoric of international conservation organizations. In a speech to Hillary Clinton, during her visit to Tanzania in 1997, Patrick Bergin (then head of the AWF's Community Conservation Service Centre) stated: 'Tanzania urgently needs to work with the communities and local government authorities in areas outside of parks and reserves, and to assist these communities by giving them the legal right, technical knowledge and the economic incentive to maintain wildlife as one form of land use in their area.'

4 Wildebeest carry a disease called Malignant Catarrhal Fever, which is fatal to domestic cattle.

5 Homewood and Rodgers' (1991) study of pastoralists in Ngorongoro indicates that the early rains are the most crucial period in the annual cycle of transhumant herding systems. Most livestock are in poor health due to the nutritional stress of the dry season that has just ended. It is crucial, therefore, that these animals be brought to wet-season pasture and water as early as possible and over the minimum possible distance.

6 In response to widespread loss of livestock and rising grain prices, most Maasai in Simanjiro now engage in small-scale agriculture. They view the possibility of a ban on agriculture as a major threat to their very well-being. Many are aware of conditions in the Ngorongoro Conser-

vation Area, where agriculture was forbidden. The Tanzanian government recently lifted the ban on agriculture in Ngorongoro.

7 Buffer zones and corridors are still mentioned *ad nauseam* in the conservation literature. They are essentially zones that are kept free of agriculture for the benefit of migrating wildlife. In the new Wildlife Management Areas proposal these are now referred to as a buffering function. Communities would be paid to keep designated areas free of agriculture based on wildlife densities within these areas.

8 In Tanzania, it is politically difficult for organs of the central state to provide direct support to traditional pastoral systems. High-ranking officials in the Tanzanian government are openly hostile to pastoral systems, which they see as an impediment to national development.

9 In an interview with Patrick Bergin, I asked what he thought about the possibility of CCS becoming actively involved in lobbying against the large-scale farms in Simanjiro. He said that this was an unlikely scenario as TANAPA officials would not want to risk their position by taking on other Tanzanian elites who are involved in land grabbing in Simanjiro.

10 For a detailed discussion of Maasai NGOs, community-level politics, and conflicts over donor money see Igoe 2000.

11 The details of the types of laws and institutions that would enhance community control over natural resources are outlined in the *Report of the Presidential Commission of Inquiry into Land Matters* (URT 1993a). Although the commission was appointed by the President of Tanzania, its report was suppressed by the Government (Sundet 1997: 9). A detailed discussion of how these laws and institutions might operate in the study area is provided in Igoe and Brockington 1999: 63–81).

References

Barkan, J. 1994. 'Divergence and Convergence in Kenya and Tanzania: Pressures for Reform'. In J. Barkan (ed.). *Beyond Capitalism vs. Socialism in Kenya and Tanzania*, pp. 1–46. London: Lynne Rienner Publishers.

Barret, C. and Arcese, P. 1995. 'Are Integrated Conservation-Development Projects Sustainable?' *World Development*, (23)7: 1073–84.

Bergin, P. 1997. 'Outline for Presentation to the First Lady of the United States'. Unpublished Speech.

Bonner, R. 1993. *At the Hand of Man: Peril and Hope for Africa's Wildlife*. New York: Alfred Knopf.

Borner, M. 1985. 'The Increasing Isolation of Tarangire National Park'. *Oryx*, 19: 91–6.

Chambers, R. 1996. 'The Primacy of the Personal'. In M. Edwards and D. Hulme (eds). *Beyond the Magic Bullet: Non-Governmental Organizations' Performance and Accountability in the Post-Modern World*. London: Earthscan Publications.

Gamassa, D-G.M. 1993. 'Marginalisation of Maasai pastoralists in Northern Tanzania'. Unpublished paper presented at Fourth Annual Common Property Conference of the International Association for the Study of Common Property, Manila, Philippines.

—— 1996. 'Community Based Wildlife Management'. Unpublished lecture presented at Sokoine University of Agriculture.

Harbeson, J. 1994. 'Civil Society and Political Renaissance in Africa', in J. Harbeson, D. Rothchild, and N. Chazan (eds). *Civil Society and the State in Africa*, pp. 1–32. London: Lynne Reinner Publishers.

Homewood, K.M. and Rodgers, W.A. 1991. *Maasailand Ecology: Pastoralist Development and Wildlife Conservation in Ngorongoro, Tanzania*. Cambridge: Cambridge University Press.

Hyden, G. 1980. *Beyond Ujamaa in Tanzania*. Berkeley: University of California Press.

—— 1983. *No Shortcuts to Progress: African Development Management in Perspective*. London: Heinemann.

—— 1995. 'Bringing Voluntarism back'. In J. Sembaja and O. Therkildsen (eds). *Service Provision under Stress in East Africa*, pp. 35–50. London: James Currey.

Igoe, J. 2000. 'Ethnicity, Civil Society, and the Tanzanian Pastoral NGO Movement: The Continuities and Discontinuities of Liberalized Development'. Doctoral Thesis. Boston University.

Igoe, J. and Brockington D. 1999. *Pastoral Land Tenure and Community Conservation: a Case Study From North-East Tanzania*. International Institute for the Environment and Development, Pastoral Land Tenure Series No. 11, London.

Kelsall, T. 1998. 'Donors, NGOs, and the State: the Creation of a Public Sphere in Tanzania'. Paper Presented at the African Globalization Conference, University of Central Lancashire.

Kipuri, N. and Nangoro B. 1996. 'Community Benefits Through Wildlife Resources: Evaluation Report for TANAPA's Community Conservation Service'. Arusha, Tanzania: mimeo.

Kiss, A. (ed.). 1990. *Living with Wildlife: Wildlife Resource Management with Local Participation*. Washington: World Bank Technical Paper No. 130, Africa Technical Department Series.

Murphree, M. 1996. 'Approaches to Community Participation'. In ODA, *African Wildlife Policy Consultation, Final Report*. London: Jay Printers.

Neumann, R.P. 1992. 'The Social Origins of Natural Resource Conflict in Arusha National Park, Tanzania'. Doctoral Thesis. University of California, Berkeley.

Patel, H. 1998. *Sustainable Utilization and African Wildlife Policy: Rhetoric or Reality?* Cambridge, MA, Indigenous Environmental Policy Center.

Shivji, I. 1998. *Not Yet Democracy: Reforming Land Tenure in Tanzania*. London: International Institute for the Environment and Development.

Spear, T. 1997. *Mountain Farmers*. Berkeley: University of California Press.

Sundet, G. 1997. 'The Politics of Land in Tanzania'. D.Phil. Dissertation, University of Oxford.

Tarangire Conservation Project. 1996. Interim Report.

TANAPA. 1994. 'Tarangire National Park CCS Strategic Action Plan'. Unpublished Report.

—— 1996. 'Statement on Government Proposal for a Mandatory Sharing of Twenty-Five Percent of Protected Area Revenues With District Councils'. Unpublished Position Paper.

Turner, T. 1993. 'The Role of Indigenous Peoples in the Environmental Crisis: The Example of the Kayapo of the Brazilian Amazon'. *Perspectives in Biology and Medicine*, 36(3): 526–45.

URT. 1993a. *Report of the Presidential Commission of Inquiry into Land Matters. Vol. I. Land Policy and Land Tenure Structure*. United Republic of Tanzania, Dar es Salaam.

—— 1993b. *Report of the Presidential Commission of Inquiry into Land Matters. Vol. II. Selected Land Disputes and Recommendations*. United Republic of Tanzania, Dar es Salaam.

World Bank. 1997. *The State in a Changing World*. New York: OUP.

Appendix: Natural Resource Management in a Transhumant Pastoral System and Natural Resource Management in a National Park

The purpose of this appendix is to illustrate graphically the incompatibility of national park style conservation and transhumant pastoralism.

Transhumant pastoralism in East Africa is a multiple use, open access system. It is multiple use because pastoralism is not practised to the exclusion of other resource use systems. Pastoralism is quite compatible with hunting, as well as commercial game viewing. Pastoralism is also compatible with small-scale agriculture under the right management conditions. In the past, symbiotic relationships existed between East African pastoralists and their agricultural neighbours. A vigorous trade in livestock for grain, as well as inter-marriage, existed between these groups. In the period following the harvest, pastoralists' livestock were grazed on the stubble of the harvested fields. The manure left by these animals served as fertilizer in the subsequent planting season.

Transhumant pastoral resource management regimes are bimodal: they follow a cyclical grazing pattern from dry to wet season. All transhumant regimes centre around a permanent source of water (the spot marked I in the model, Figure A5.1). During the dry season (June or July to October), humans and their livestock concentrate around such a water source. The area in the immediate vicinity of the water (the circle marked II in the model) is for the queuing of livestock waiting to drink. Pastoralists follow elaborate schedules for the watering of their livestock during the dry season. Systems within the queuing area are complex but generally run quite smoothly. A meeting of elder males sets the queuing schedules through consensus decisions.

The area immediately beyond the queuing area is pasture reserved for sick, immature, and lactating animals (marked III in the model). These are animals that cannot range far in search of pasture. Access to this pasture is available to anyone with a permanent homestead in proximity to the permanent water source. Access to outsiders must be negotiated, and permission will be determined by a consensus decision of a meeting of local elders. In no case are mature animals allowed to graze in this area. Intentional violation of this rule is punished by a fine of livestock. The area beyond the calf pasture (marked IV on the model) is pasture for healthy mature livestock.

Permanent homesteads (marked V in the model) are built around the perimeter of the calf pasture to ensure that everyone has equal access to it, without clustering that would lead to overgrazing. Such an arrangement also ensures that all households are equally distant from the permanent water source.

The broken arrows running from the permanent homesteads (marked Va and Vb in the model) indicate the movement of livestock during the dry season. Because of the large numbers of animals concentrated around permanent water in the dry season (often running to several thousands) pastoralists must water their animals on alternate days. On days that livestock are watered (arrow a) they are taken from the homestead in the morning to the queuing area, where they await their turn to be watered. In the afternoon,

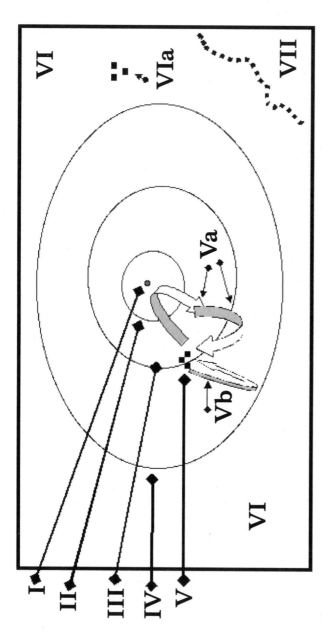

Key: I)   Permanent dry-season water source.
      II)   Queuing area for livestock waiting to drink.
      III) Pasture reserved for immature, sick, and lactating animals.
      IV) Dry-season pasture for healthy mature livestock.
      V)   Dry-season homesteads:
             Va)   livestock movement on days that animals are watered.
             Vb)   livestock movement on alternate days.

      VI)   Wet-season dispersal areas:
             VIa) wet-season camps.
      VII) Drought reserve pasture.

Figure A5.1 Generalized Model of a Transhumant Pastoral System (Used with Permission of Saruni Ndelelya)

they are moved beyond the permanent homestead into the dry-season pasture. They graze here and return home in the evening. On alternate days, livestock are herded away from the permanent water source to a different part of the dry-season pasture where they graze for the entire day.

The first rains of the wet season usually come in November or December. Once the rains begin, most of the livestock are moved to wet-season pasture (all the area beyond the concentric circles – marked by VI). During the wet season this area is replete with seasonal water sources and mineral rich pasture. Pastoralists then disperse to the water sources throughout this area, which allows dry-season pasture to recover.

Not all of the people and livestock move during the wet season. Most women, children, and elder males remain in the permanent homesteads with the sick, immature, and lactating animals. The young men take all the mature and healthy stock and move them to the wet-season pasture. During this time they will live in temporary wet-season camps (VIa in the model). The household heads may go to check on their sons and their herds periodically, then at the end of the season (usually June or July) the young men return with the livestock to the permanent homesteads and the whole process begins again.

One more feature, which is essential to transhumant pastoralism, is drought reserve areas (marked VII in the model). These are areas with pasture and water that never dry up (usually a swamp or pasture on the slopes of a mountain) and are not used during normal years. Pastoralists know that in a semi-arid environment rains are likely to fail every seven to ten years. When this happens, all livestock are taken to the drought reserve and thereby saved. Drought reserves ensure long-term viability to a system of resource management that operates in an extremely uncertain environment.

In contrast to pastoralism, the type of resource management associated with national parks is single use and closed access. No activities other than the preservation of wildlife are permitted, and human habitation is strictly forbidden. This system depends on the use of paramilitary force to ensure continued human exclusion. National parks in East Africa are usually built around permanent water sources that guarantee spectacular concentrations of wildlife. In Tanzania parks were gazetted without consideration of how they would affect pastoral resource management systems. A cursory examination of a map of Tarangire (Figure A5.2) serves to illustrate the extent of disruption caused to pastoral systems by the creation of a national park. This park expropriated the very heart of the Simanjiro pastoral system. It contains three natural resource features that were crucial to pastoral resource management: the Tarangire River (1), which was a permanent water source; the Silalo Swamp (2), which was an important drought reserve area; and an area of highland pasture (3), which was wet-season dispersal. Addressing these types of resource inequities is the central challenge of community conservation in Africa under structural adjustment.

6

The Mursi and the Elephant Question

David Turton

The call for 'community participation' in conservation projects has grown to such an extent over the past few years that it has virtually become current orthodoxy, along with similar calls for participation and 'bottom-up' planning and management in rural development projects (IIED 1994; Pimbert and Pretty 1995; and numerous references therein). The reasons for this turning away from a 'preservationist' approach, which sees local people as an obstacle to effective natural resource management, are as much biological and economic as they are moral and political. Firstly, since virtually all existing ecosystems are a function of human use and disturbance, artificially to exclude such disturbance runs the risk of reducing biodiversity rather than preserving it (Hobbs and Huenneke 1992: 324, cited by Pimbert and Pretty 1995: 21). Secondly, not only are the technical and logistical costs of attempting to exclude human activity from protected areas very high but such efforts are almost certain to fail. They will alienate the local population from conservation objectives and thus require an ever-increasing and, in the long run, unsustainable level of investment in policing activities.

I shall take the correctness of these arguments for granted, partly because, being neither a biologist nor an economist, I am not qualified to subject them to close analysis and partly because I imagine few would wish to disagree with them. But there is, of course, a huge potential here for well-intentioned rhetoric to take the place of action, or to provide a 'donor-friendly' screen behind which the same old 'preservationist' and ultimately unsuccessful policies are put into practice. The latest plan for the development of the Omo,

Mago and Nechisar National Parks, in Southern Ethiopia, is a case in point. The feasibility study for this project, which is now known as the 'Southern National Parks Rehabilitation Project' pays frequent lip service to the need to involve the local people and 'increase the tangible economic benefits' they gain from conservation (Agriconsulting 1993: 61) but, six years later, there has still been no serious effort to achieve either of these objectives.

The issue I shall address in this chapter, therefore, is not whether local participation in conservation is in principle 'a good thing', but whether it is feasible and how it might be achieved in the case of the Mursi. This will mean, firstly, giving some baseline information about Mursi natural resource management, without which it is impossible to know to what extent, if at all, present human activity in the area is detrimental to the sustainable use of its renewable resources. Secondly, I shall discuss a number of documents in which foreign advisers and consultants have presented 'top down', or 'preservationist' proposals for conservation in the lower Omo Valley. Thirdly, I shall make some recommendations for a radically different approach, based on the now conventional wisdom of 'conservation with a human face' (Bell 1987).

Mursi Natural Resource Management

The Mursi live in an oblong territory of about 2000 sq. km, bounded to the west and south by the River Omo, to the east by the River Mago and to the north by the River Mara, a seasonal tributary of the Omo (Figure 6.1). They depend on three main subsistence activities, each of which is insufficient and/or precarious in itself but, when taken together with the other two, makes a vital contribution to the economy: flood-retreat cultivation at the Omo, rain-fed cultivation in the bushbelt and cattle herding in the wooded grasslands above the 500m contour line. Cultivation is primarily the responsibility of women, and cattle herding of men. The main crop is sorghum, though some maize is also grown, together with cow peas, beans and squash. In spanning their three main natural resources (floodland, bushland and grassland), the Mursi have developed a form of transhumance which, although it takes place over a relatively small area, does not permit fixed residence, in a single locality, for any section of the population. Rights to subsistence resources are allocated in a way that reflects the physical and ecological character of the resource and maximizes the contribution it makes to the overall viability of the economy.

Floodland is a scarce resource which makes a critical contribution to the economic viability of households but the extent of which varies unpredictably from one year to the next. It must therefore be allocated in such a way that short-term adjustments can be made between the amount of land available for cultivation in any one year and the number of potential cultivators. Each Omo cultivation site is associated with a particular clan but it would be very misleading to speak of clans 'owning' land. For a clan is not an organized group but a patrilineal category of the population. Clan names are

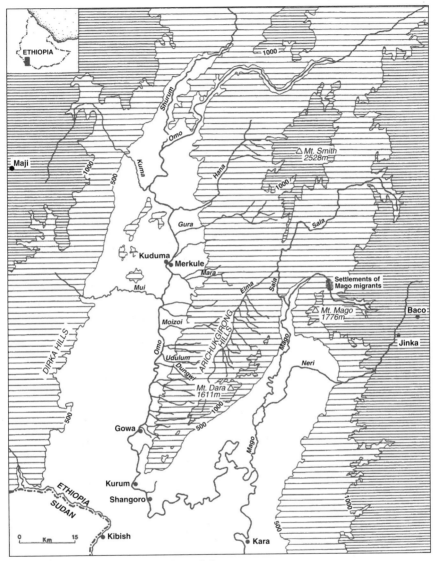

Figure 6.1 Mursiland: topography and drainage

merely labels, recording the fact that particular stretches of riverbank were first occupied by members of particular clans whose descendants now have prior rights to its use. The effective owners of riverbank land are small groups of close patrilineal kin – descendants of the same grandfather or great grandfather – who allocate land to more distant kin and affines, normally for one or two years at a time. Riverbank land, then, is collectively owned by small groups of kin, but many others may have potential or 'diffused' rights in it.

The advantage of this system, which clearly depends on obligations to kin being seen as inescapable, is that it maintains a balance between supply and demand where great flexibility is needed to ensure that the maximum benefit is gained from flood cultivation in any one year by the maximum number of individuals. It is not just that flood levels vary from one year to the next but also that the effect of a poor flood will not be uniformly felt at all cultivation sites. Security for individuals and families in these circumstances means having 'dormant' rights in riverbank land at various points along the Omo which can be activated at short notice, and this is what the moral imperatives of kinship and affinity make possible.

Rights to grazing land, on the other hand, are vested in local groups, not kin groups. Here it is necessary to explain that the Mursi are divided into five territorially based groups, or *buranyoga* (sing. *buran*), which are named, from north to south, Baruba (formerly known as Mara), Mugjo (formerly known as Mako), Biogolokare, Arioli and Gongulobibi (Figure 6.2). Each of these divisions spans the full range of natural resources, from floodland in the east to dry season grazing land in the west. The fact that they make ecological sense presumably accounts for their size and boundedness and for the strong sense of moral obligation which their members feel towards each other (cf. Spencer 1990: 215–16). They are, in short, miniature replicas and potential equivalents of the Mursi *buran* as a whole.

Each buran is associated with a particular territory within which its members have 'primary user rights' (Potkanski 1994: 17), but members of other buranyoga are granted temporary rights in the same territory at times of hardship, crisis or emergency. Collective ownership of a resource implies, by definition, that there are rules and conventions determining who shall have access to it, for how long and under what circumstances. As far as grazing land is concerned, these rules apply at the level of the buran. There is a sense in which all Mursi have a right to graze their animals anywhere in Mursiland, but the sense is this: they have a right to be granted access to areas outside their own buran in times of crisis and on a temporary basis. This applies particularly to access to dry-season grazing areas in the Elma Valley, the key constraint here, of course, being the availability of permanent water points in the Elma Valley. Since rainfall is highly variable, herders have to be alert to changing conditions on a daily basis and be ready to move their animals at fairly short notice in order to match the available water and grazing to animal numbers in a particular place.

It is often said that pastoralists own grazing land 'collectively' and livestock 'individually' but this distinction, which lies behind Hardin's vastly influential 'Tragedy of the Commons' argument (1968, 1988), is a gross over-simplification. We have just seen that the collective ownership of grazing land amongst the Mursi is compatible with controlled access at the level of the buran. As far as cattle are concerned, it is certainly always possible to identify an individual owner for any particular animal, but so many other people are likely to have actual and potential rights in the same animal that to describe this as individual ownership would be highly misleading because it would imply

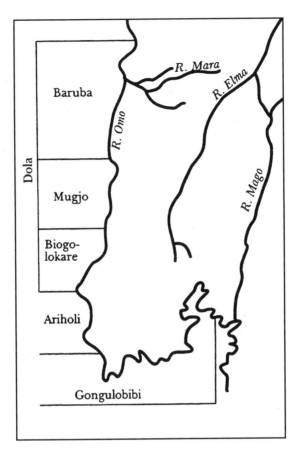

Figure 6.2 Distribution of Territorial Groups (*buranyoga*) in Relation to the River Omo. Reprinted, with permission, from David Turton, *Pastoral Livelihoods in Danger: Cattle Disease, Drought and Wildlife Conservation in Mursiland, South-Western Ethiopia,* Oxfam Research Paper No 12, 1995.

that the 'owner' could use and dispose of the animal entirely as he or she saw fit. It would be more accurate to describe cattle as owned collectively, small groups of patrilineally related men – essentially groups of brothers – having primary 'user rights' in them. This does not mean that brothers always live together and herd their cattle as a single unit. On the contrary, a man is more likely to be found sharing a settlement with his affines than with his patrilineal kin. It does mean that brothers have potential or 'dormant' rights in each other's cattle which they can activate at any time, but especially in extreme circumstances. The diffusion of rights in livestock to a wide variety of kin, affines, and 'stock associates' is a feature of pastoral resource management which has been fully described in the anthropological literature and which is, of course, an effective means of spreading risk and hedging against environmental and other uncertainties. (A particularly notable means of achieving this objective amongst the Mursi is their method of collecting and distributing bridewealth cattle (Turton 1980)). It is worth pointing out that the system for allocating rights in riverbank land, another scarce, critical and highly variable resource, has more in common with that for allocating rights in cattle than it has with that for allocating rights in grazing land.

Most East African protected areas and national parks have been created in areas used by pastoralists. One of the main justifications for this has been the 'institutional fact' (Thompson et al. 1986, cited by Warren and Agnew 1988) that pastoralists do not know how to manage the environment in a sustainable way. In particular, their combination of communal ownership of land and individual ownership of cattle locks them into a relentless drive to build up their herds until they exceed the carrying capacity of the range, thus bringing about irreversible environmental degradation – the so-called 'Tragedy of the Commons'. Being based on the abstract assumptions of games theory and the concept of the economically rational individual, the argument is elegant and convincing – until one looks at the real world.

Firstly, and as I have already demonstrated briefly for the Mursi, communal access to grazing land does not necessarily equal 'open access'. Or, to put it otherwise, a communal system can control, restrict and coordinate the behaviour of individuals through rules and conventions which they recognize it is in their own best interests to observe (Runge 1984 and 1986). Secondly, the Tragedy of the Commons argument is based on an 'economic' definition of carrying capacity (the optimal stocking density for commercial ranching) which is considerably lower than the 'ecological' carrying capacity of subsistence herding (Behnke and Scoones 1993: 3–8). Thirdly, in the arid and semi-arid grazing areas of East Africa, a stable equilibrium between animal and plant populations may never be reached because rainfall and temperature fluctuate so widely that 'it is likely that these non-biological variables will have a greater impact on plant growth than marginal changes in grazing pressure caused by different stocking densities' (Behnke and Scoones 1993: 8). And finally, the one certain conclusion to emerge, over the past few years, from the so-called 'overgrazing controversy' (Homewood and Rodgers 1987) is that the subject is so beset with conceptual confusion and so much in need of more objective methods of assessment and evaluation that great care should be taken before making any assertions about individual cases in advance of a careful study of the evidence (e.g. Warren and Agnew 1988; Homewood and Rodgers 1987 and 1991; Abel and Blaikie 1990; Tapson 1993).

I am not qualified to make such a study of the Mursi case. I can only report that there is no obvious evidence that their pastoral activities are, in the words of one recent definition of range degradation, bringing about 'an effectively permanent decline in the rate at which land yields livestock products. "Effectively" means that natural processes will not rehabilitate the land within a timescale relevant to humans, and that capital or labour invested in rehabilitation are not justified' (Abel and Blaikie 1990: 113). The same can also be said, *mutatis mutandis*, of flood retreat cultivation, although the case of rain-fed cultivation, because it depends on clearing new areas of bush every few years, is more problematic. Even if it is accepted, however, that the natural resource management system of the Mursi has the capacity to 'maintain those features of the natural environment which are essential to its continued wellbeing' (Behnke and Scoones 1993: 20), there is always the danger that the co-operative norms upon which its smooth and efficient running depends will break down under pressures brought about by economic change and state incorporation.

The greatest threat to the efficient management of natural resources by African pastoralists has come not from contradictions internal to the ecology of subsistence herding, as the 'Tragedy of the Commons' argument would have us believe, but from external pressures. Not the least important of these have resulted from well-intentioned but misguided livestock development projects (Horowitz 1986; Galaty and Bonte 1991; Dyson-Hudson 1991). The loss of key dry-season pastures, whether to agriculturalists or wildlife conservation schemes, has had a particularly disastrous impact on pastoralists. For the Mursi, the threat to these 'key pastures in wetter areas' comes from the potential development of the Omo and Mago National Parks.

Saving the Elephants

The boundaries of the Omo and Mago Parks, as they have been described since at least 1970 (they have not yet been gazetted) enclose between them the most valuable agricultural and pastoral resources of the Mursi – flood retreat land on both banks of the Omo and dry-season grazing land in the Elma Valley (Figure 6.3). It follows that, if and when these boundaries are legally established, the Mursi will be transformed overnight into illegal 'squatters' in their own territory. The area between the Elma and Omo, the only part of Mursiland not included in the parks, has been designated the 'Tama Wildlife Reserve', where 'controlled settlement and other human activity may be allowed subject to the special consent of the minister and may be phased out as required' (EWCO 1989, quoted by Sutcliffe 1992: 83). It is obvious that those who demarcated these boundaries had virtually no understanding of the human ecology of the area (Turton 1987).

In a report submitted to the Wildlife Conservation Department (as it was then called) in 1978, J. Stephenson and A. Mizuno recommended the merging of the two parks (the Mago Park literally still only existed on paper at that point) into a 'Greater' Omo/Mago National Park on the grounds that,

> The Omo and Mago will lose their value as national parks if vested human interests are permitted to exist between them. For one thing, some of the wild animals, chiefly elephant, lion and zebra will interfere to an increasing extent with the rights of the people of the Tama wedge and conversely the people will interfere to an increasing degree with the wildlife of the Tama and the two neighbouring parks' (Stephenson and Mizuno 1978: 41).

The novel suggestion of Stephenson and Mizuno for protecting the 'rights' of the inhabitants of the 'Tama wedge' (whom they estimate to number no more than 1750 individuals) is to forcibly resettle them outside the proposed

Figure 6.3 The Boundaries of the Omo and Mago National Parks

park boundaries, an exercise the 'onus' of which 'falls fairly and squarely on the Administration and not on the Wildlife Conservation Department' (1978: 49). Since they do not indicate what kind of human 'interference' would be avoided by resettling the Mursi, it is worth asking in what ways their continued presence could be detrimental to the welfare of wild animals. There are, presumably, two main possibilities: they might kill them directly or be in competition with them for the same natural resources.

The Mursi certainly have the pragmatic, unromantic, view of nature which is characteristic of those who directly gain their livelihood from it and live in daily contact with it. They would share the view that Wordsworth sadly attributed to the majority of his contemporaries, namely that 'a rich meadow, with fat cattle grazing upon it, or the sight of... a heavy crop of corn, is worth all... the Alps and Pyrenees in their utmost grandeur' (1835: 151, quoted by Thomas 1984: 257). But the corollary of this is that the Mursi do not share the urban, modernist assumption that to explore, exploit, understand, paint, photograph, document or in other ways control and dominate nature is to fulfill our potential as human beings. The result is that they do not kill animals – any more than they climb mountains – merely 'because they are there'. The main use they make of them is as a source of food at times of severe hunger, the species they most frequently hunt being the buffalo. Buffalo hides are also exchanged at such times with highland agriculturalists (who, among other things, bury their dead in them) for money and grain. They also kill elephants for their ivory, which can be used, as it has been for the last 100 years in this area to buy rifles and cattle from highland traders. In short, the Mursi kill animals to obtain economically useful products and, when necessary, to protect their cattle (this applies mainly to hyenas), but otherwise their disposition is, as Evans-Pritchard wrote of the Nuer, 'to live and let live' (1956: 267).

Since livestock use resources upon which wild animals also depend, there is obvious potential for competition between them. On the other hand, there seems to be more scope for coexistence between wild and domestic animals under a subsistence herding regime than under either commercial ranching or sedentary agriculture: subsistence herders are mobile, do not monopolize water points and are relatively sparsely settled (Homewood and Rodgers 1991: 191–2; Hillman 1993: 11–12). Although they depend heavily on cultivation, the Mursi have no permanent settlements and, apart from the Omo itself where cattle cannot be kept because of the high tsetse challenge and lack of grazing, there are no permanent water sources in their territory. It is true that the presence of relatively large numbers of people at flood cultivation sites on the Omo during the dry season must have some impact on the behaviour of wild animals. This can hardly prevent them, however, from using countless watering points along uninhabited stretches of the river, while for half the year (March to September), there is virtually no human settlement at all along the Omo. As for settlements in the grazing areas, these are always situated well away from water points, with the result that the use of these points by cattle does not exclude their use by wild animals at other times of the day and night.

Stephenson's and Mizuno's main concern is with the *protection* of wild animals – especially those of most interest to tourists – in an environment they describe as having 'retained its primeval character from ages past' (1978: 2). Thus, virtually all their recommendations have to do with the need for technical, administrative and security improvements – more roads, buildings, vehicles, game guards and guard posts. Although they describe the area as 'the country's last unspoilt wilderness' (p. 1) they consider it to be in such imminent danger from human activity that all the local people (Mursi, Bodi and '300' Chai living in the Omo Park) must be resettled as a matter of urgency, after which 'the integrity of the boundaries must be rigidly preserved' (p. 49). It would be difficult to find a set of recommendations more unrealistic in their expectations nor more calculated to stir up the bitter opposition of local people. This is an extreme statement of the 'preservationist' approach to conservation, an approach which owes more to European myths – about 'wild' Africa and about the essentially apolitical objectives of conservation (Anderson and Grove 1987) – than it does to African realities.

A more realistic and enlightened proposal for the development of the Omo and Mago parks is presented by Sutcliffe (1992), who criticizes the conservation categories used by the Ethiopian Wildlife Conservation Organization (EWCO) (as it is now known), such as 'national park' and 'wildlife reserve', on the grounds that they fail to give proper consideration to the 'basic needs of the local population' (p. 86). His proposal would divide the area, inhabited by the Mursi into three 'categories of conservation management' (pp. 87–91): the Omo and Mago National Parks, with reduced areas; the 'Mago Resource Reserve', where 'relatively low intensity human land use would be allowed to continue'; and the 'Omo River Anthropological Reserve' where 'the subsistence economy of the indigenous population' would be maintained. He also recommends that 'planning and demarcation of the new land use zones should ... be negotiated with the peoples' and that 'sharing of revenue from visitors to the area should also be catered for' (p. 92).

These proposals are a notable advance on Stephenson's and Mizuno's, for three reasons. Firstly they show a greater understanding of local subsistence systems (though it must be said this would not have been difficult); secondly they include a number of measures specifically designed to protect the interests of the local population, even to the extent of altering the park boundaries; and thirdly, they recognize the need to ensure that local people gain tangible benefits from conservation development. Unfortunately, however, enthusiasm for the proposals must be tempered with some scepticism because of two fundamental assumptions they share with earlier approaches: that it is only the needs – and the 'basic needs' at that, meaning basic subsistence needs – of the local population that have to be taken into account and that natural resource management is, by definition, an activity that can only be managed effectively 'from above'. These assumptions are, of course, connected since, if all that matters is the identification of a people's 'basic needs' (and not, for example, their knowledge, capacities, attitudes and aspirations) then it should be possible in theory (it hardly ever is in practice) for these to be identified and catered for by outside experts on flying visits from the national capital.

It is presumably because he does not question these assumptions that Sutcliffe is led, from the highest of motives, to suggest that an 'Anthropological Reserve' is established along both banks of the Omo. There are a number of objections to this proposal, some practical and some ethical. Firstly, and as is clear from what I have written above, such an area could not possibly support the 'subsistence economy of the indigenous population'. Secondly, the proposal ignores the rights and aspirations of those who would, in effect, be confined to this corridor of land along the Omo to improve their living conditions, quality of life and economic security. Thirdly, it would therefore create exactly the kind of local opposition to the conservation plans of the EWCO that would ensure their ultimate failure. And fourthly, it would be a short step from here to regard the Mursi as little better than another form of wildlife: not an endangered species, to be sure, but no more than an aesthetic enhancement of the 'national park experience' for the tourist.

By far the most ambitious and costly plans to date for developing the Omo and Mago Parks are set out in the report of a feasibility study for the 'Southern Ethiopia Wildlife Conservation Project', which has since been renamed the 'Southern National Parks Rehabilitation Project' (SNPRP). This was planned as a five-year project, focusing on the Nechisar (Arba Minch), Mago and Omo National Parks, to be financed by the European Development Fund to the tune of approximately ECU 16 million. A consultancy team, consisting of a wildlife biologist, civil engineer and economist, made helicopter and ground visits to the Mago and Omo between 17 and 23 March 1993. In their final report they note that 'It is almost certainly in the socio-cultural area that the greatest long term threats to project sustainability lie' (Agriconsulting 1993: 60). What they mean by this is that, without the 'goodwill and cooperation' (loc. cit.) of the local people, the project will not succeed. They therefore propose to 'increase the tangible economic benefits that rural people get from land used for wildlife conservation' by, among other means, 'introducing revenue sharing with rural communities' and 'giving priority to local people in opportunities for employment' (Agriconsulting 1993: 61).

It is envisaged that revenue sharing will operate through the financing of 'priority rural development projects', identified with the help of 'socio-anthropologists' and ultimately decided upon by the EWCO 'since this should relate to wildlife conservation' (p. 62). Denial or loss of such benefits would be used as a sanction to induce 'respect for the laws and rules relating to wildlife conservation and park management' (p. 62). For it is realized that it would be a hopeless task to attempt to force such respect on 'well armed, unruly tribesmen at home in a vast wilderness' (p. 62). In line with this realization it is proposed to appoint at least one assistant warden in each park to have responsibility for 'extension work' and to keep the number of game guards and outposts relatively small. This recognition that the key to the success of the project lies in the attitude towards it of the local people rather than in its infrastructural and policing capacity, is greatly to be welcomed. Unfortunately, however, it is clear from the general tone and content of the report that, like the documents discussed earlier, it is firmly based on 'top-down' and 'preservationist' assumptions.

Despite its emphasis on the importance of 'socio-cultural' factors, the report goes into detail only about the technical and infrastructural arrangements required by the project and has nothing to say about the knowledge, attitudes, customs and beliefs of the local people. The specification of relevant 'socio cultural' factors, one must conclude, is considered irrelevant to the feasibility stage of the project (unlike the specification of roads, bridges and buildings) and can therefore be taken care of after it has begun. In particular, there is no mention of the resource management skills of the people which must, by the report's own evidence, be considerable, for the Mago and Omo parks are each described as an 'impressive wilderness' (pp. 125 and 137). This of course is the language of the tourist brochure, not of ecological science, and it is aimed, presumably, at political decision-makers. The political usefulness of the wilderness myth is that it implies (a) that there are very few people currently living in or using the area – to use the report's own words, that there is little or no 'encroachment' (pp. 125 and 137) – and (b) that those who are living in and using it are a threat to its 'wilderness' character.

Revenue sharing is therefore intended to buy the 'goodwill' of local residents while denying them a decision-making role in the management of their own resources. Quite apart from the questionable morality of this strategy, it will almost certainly fail because the 'community developments' to be financed under the scheme will be allocated, on the stick and carrot principle, at the behest of the conservation authorities. The local people will thus become 'passive beneficiaries' (IIED 1994: 21), with no final say in how the benefits are distributed – a reliable recipe for failure in any development project.

There is a further serious problem here which is mentioned in the report but the implications of which are ignored: since there will be no 'excess revenue' generated by the parks in the near future and since it will be important to make some investment in 'community development' at an early stage in order to gain the 'goodwill' of the people, this investment will have to come out of project funds. Even when 'excess revenue' (presumably from tourism) does become available, it will be 'inadequate to finance capital intensive inputs … (schools etc.) and will more likely be adequate for the relatively modest running costs of such facilities' (p. 62). Once donor funds are no longer available, in other words, the revenue-sharing scheme will probably collapse.

The most significant weaknesses of the report, which force one to conclude that its emphasis on the 'socio-cultural area' is largely rhetorical, are that it contains no information about the natural resource management strategies of the local people and no evidence that any of them were informed, let alone consulted, about the project during the six-day field visit that the study team made to the Omo and Mago Parks. The willingness this shows to spend huge amounts of money on such a complex and far reaching environmental project, apparently in full recognition of the crucial importance of the 'socio-cultural area' to 'project sustainability' and yet without taking virtually any notice of the human ecology, environmental knowledge, capacities and rights of the local population is staggering.

Following the feasibility study it was decided to initiate a two-year 'preliminary phase' of the project, the aims and objectives of which are set out in an undated document entitled 'National Park Rehabilitation in Southern Ethiopia: Preliminary Phase'. This describes a number of urgent objectives for the preliminary phase 'in view of the rapid degradation of the parks' (p. 11). These include 'the gazettment of the selected focal areas, Nechisar, Omo and Mago National Parks, according to their presently recognized boundaries' and various steps to strengthen the legal, institutional and infrastructural capacities of the EWCO. What is most striking about this document, however, is its uncompromisingly negative attitude to human activity in and around protected areas, its totally unrealistic expectations of what can be achieved by military-style protection of such areas and its use of 'institutional' rather than objective facts to justify far-reaching policy proposals. The author(s) would have us believe that there is no time to lose in gazetting the Parks because degradation is proceeding apace within them. It is proposed to prepare 'a comprehensive and coherent land and resource use plan' for the Omo and Mago Parks which 'will aim to address the very serious environmental degradation currently taking place as a result of the indiscriminate build-up of livestock herds ... combined with uncontrolled exploitation of the natural resources of the national parks' (pp. 12–13).

Since no evidence is presented in support of the statement that 'degradation' is taking place, the author(s) clearly believe(s) this is a self-evident truth – despite the notorious complexity of the issues involved in identifying and measuring rangeland degradation, and despite the fact that the report of the feasibility study contains no evidence of environmental degradation in the Omo and Mago parks. There could be no better illustration of the way policy decisions, having momentous long-term implications for the well-being of people and the environment, can be based on assessments and assumptions which bear hardly any relation to the world as it really is.

While the report of the feasibility study avoids the question of resettling people living within the park boundaries simply by pretending they do not exist, this document takes the bull by the horns, stating that it will be one of the objectives of the preliminary phase 'to plan the resettlement of people who have settled in the parks' (p. 7). Note the implication of the phrase 'who have settled in the parks': one is led to believe by this that those currently living in the parks represent a recent human incursion into a previously uninhabited 'wilderness'. These so-called 'squatters' will be resettled 'in collaboration with the local authorities ... in an orderly fashion following the provision of assistance by way of materials and food for a period of six months, and provision of social infrastructure in their new villages (4 boreholes, 2 schools and 2 clinics)' (p. 9) The number of those to be moved is put at 480 in Nechisar and 1200 in the Omo (loc.cit.) although, as is so frequently the case with forced resettlement, there is ambiguity about the actual numbers. In a previous document, upon which this one is clearly based, the same figures were quoted but they referred to 'families', rather than individuals. This would be more realistic, since 1200 families would more than account for the entire Mursi population of around 6000. If we really are expected to believe that only 1200 individuals would need to be removed from the Omo Park, it would be interesting to know how this particular figure was arrived at. Those who are not to be resettled will be 'sensibilised ... in order to minimise conflictual or unsustainable resource use' (p. 9). At this point one begins to wonder whether these proposals are intended to have any connection with the 'real world' at all. The terms of reference for the preliminary phase make no mention of the need to arrive at workable proposals to ensure that local people gain tangible and realistic benefits from wildlife conservation and tourism, despite the insistence of the feasibility study team that unless this can be achieved the project will fail in the long run.

The terms of reference do, however, include the requirement for a 'sociologist or socio economist', whose task will be to undertake a 'socio-economic survey ... concerning families to be resettled, and local populations living in or near the national parks'. One can only conclude that those responsible for designing and implementing the project are not only prepared to see local people bear the main burden of its cost, even to the extent of being forced off their land with six months' food aid 'where necessary', but that they simply do not appreciate the need for local involvement, purely on grounds of efficiency.

Involving the People

But what kind of 'involvement'? The word is open to as much rhetorical abuse as that other development buzz-word, 'participation' which can include getting people to provide labour for a project or merely asking them what they would like a project to do for them ('needs assessment'). I assume that the only kind of 'participation' that is likely to lead to long-term success in any development project is 'interactive participation' (Pimbert and Pretty 1995: 26), the essential feature of which is that local people are involved, from the start in design and implementation. It must surely be accepted by now that, unless people have real power to influence the way a project is designed and managed, they will not feel that it is 'theirs', whatever (often temporary) benefits they derive from it.

Espousing as I do this notion of participation, it would clearly be contradictory if I were to recommend a management structure and set of objectives that would allow effective local participation in the SNPRP, for these must themselves be worked out with local involvement. But I can make some general observations and recommendations, based on my knowledge of the Mursi over the past 25 years, which I hope might prove helpful to the EWCO if, as I hope, it decides to revise its present approach to this project.

First, there is no rush. Talk of 'serious environmental degradation' in the Omo and Mago parks has more to do with getting donors to release funds than with the actual situation in those areas. I recognize the pressures there must be on government departments concerned with conservation, and their advisers, to emphasize the image of 'Africa in crisis' in their competition for scarce development funds. The same pressures have been felt by development NGOs, who have only recently recognized that to use heart-rending images of starving Africans in their fund-raising literature is counter-productive. By spreading a false image of the passivity and helplessness of rural Africans, such advertising promotes inappropriate aid which prolongs or increases the levels of poverty. Similarly, projects which are designed to appeal to (and therefore which confirm) the European image of Africa on the brink of ecological disaster are more likely to help bring that disaster about than prevent or mitigate it.

Second, a new feasibility study should be undertaken with the sole aim of assessing the prospects for effective local participation in the development of the Omo and Mago Valleys as a conservation area. The first objective of the study would be to analyse the natural resource management strategies of the Mursi as well as their northern neighbours, the Bodi, and to assess the environmental impact of these strategies, both positive and negative, over the past few decades. This would require a lengthy period of fieldwork by a team that should include a range ecologist and an anthropologist with specialist knowledge of pastoralist ecology in East Africa. The second objective would be to initiate a process of debate and discussion within local communities, aimed at formulating an effective management structure for the project in which these communities would have a decisive decision-making role. The third objective would be to make a thorough study of other community participation conservation schemes in Africa, especially among pastoralists. (The study team should visit some of these, especially perhaps the Ngorongoro Conservation Area in Tanzania.)

Third, the EWCO should take steps to ensure that its staff are well informed both of the need for a community-centred approach to wildlife conservation and of the history of such projects in other African countries. As the Agriconsulting report suggests, at least some national park staff should be given specific responsibility for 'extension work' and it would be enormously helpful if such staff received training in so-called 'Rapid' or 'Participatory' methods of 'Rural Appraisal'.

Fourth, it should not be assumed that it would be necessary, in order to involve the Mursi in the project at a decision-making level, to create new organizational structures. A key difficulty in the management of community conservation projects – and indeed of other kinds of 'participatory' development projects (Hogg 1992) – has been how to define the 'community' as an empirical entity. The five-fold territorial division of the Mursi population into *buranyoga* would provide decision-making units of the required nature. The relative economic homogeneity of the Mursi, their egalitarian ethos and their strong tradition of public debate and oratory make one confident that they would have no difficulty in adapting their existing methods of public decision making to the demands of a *genuinely* participatory conservation project.

Fifth, conservation development should not be seen as incompatible with pastoral development – that is with helping the Mursi to improve their food security through such means as the extension of veterinary services and the construction of water points in dry-season grazing areas. Pastoral development of this kind – aimed, that is, at improving the productivity of subsistence herding rather than at commercial offtake – is, as Homewood and Rodgers forcefully argue for the Maasai of the Ngorongoro Conservation Area (NCA), 'entirely compatible with conservation' (1991: 248).

Sixth, and in view of the last point, a decision should be taken now to alter the boundary of the Mago Park. This would be to take account, not only of flood-retreat cultivation along the banks of the Omo as suggested by Sutcliffe, but also of the vital importance of the Elma Valley to the pastoral activities of the Mursi and therefore to the viability of their entire economy. This would mean re-tracing the western boundary of the park so that it follows the top of the Omo-Mago watershed, thereby enclosing the whole of the Mago Valley but excluding the whole of the Elma valley.

Finally, and as part of a radical re-assessment of the negative role of local people in conservation development which is implicit in the documents discussed earlier, a decision should be made to drop all plans for the resettlement of families living within park boundaries. This is not to say that resettlement of local people can never be justified, under any circumstances. My argument here is that there are no grounds for believing that the long-term objectives of wildlife conservation would be served by excluding local people from the area and strong grounds for believing that such action would have the opposite effect.

Conclusion

The SNPRP, with money provided by the European tax-payer, has the potential to create a humanitarian outrage if the re-assessment recommended here is not made. Such outrages have been perpetrated many times in the past, and almost always on people who, like the Mursi, have no power to fight for their own rights and interests. It therefore falls to outsiders to attempt to fill this political vacuum, speaking up on behalf of those who cannot speak for themselves and gaining in the process a reputation for being 'trouble makers' amongst those whose income and career prospects are linked to the completion of such scandalously ill-conceived projects as the one I have outlined in this chapter.

In this case, however, we can base our appeal on grounds of efficiency as well as equity. Failure to put the interests and well-being of the local human population at the centre of the project – to approach the 'elephant question', in other words, from the point of view of the Mursi – will almost certainly ensure that it does not achieve its conservation objectives. There have been some imaginative attempts made in other African countries to involve local people, to varying degrees and with varying degrees of success, in conservation projects (e.g., Lindsay 1987; Skinner 1989; Newby 1990; and Murphree 1991; McCabe et al. 1992). It is disappointing, therefore, that a more concerted effort has not been made to benefit from this experience in the planning of the SNPRP. Given the substantial funding available, the size and significance of the areas to be 'rehabilitated' and the advantage that comes from being able to learn from other people's mistakes rather than one's own, there is an opportunity here for Ethiopia to create one of the most exciting and influential wildlife conservation projects in Africa. For the sake of us all, and especially for the sake of the Mursi and the elephants, let us hope that it is not too late for this challenge to be accepted.

The Mursi and the Elephant Question: Postscript

This chapter originated in a paper for a workshop on participatory wildlife management, which was organised by Farm Africa and the Save the Children Fund (US) and held in Addis Ababa in 1995. The purpose of the workshop was to enable Ethiopia to benefit from the experience of other African countries in promoting community-centred approaches to conservation and wildlife management. The purpose of the paper was to urge the Ethiopian Wildlife Conservation Organisation (EWCO) and the European Union's Addis Ababa office to give more serious and systematic attention to the need to make community participation an integral part of the Southern National Parks Rehabilitation Project (SNPRP), which was then about to get underway. I made no bones about the fact that my main concern was for the welfare of the affected human populations, specifically the Mursi. (This was implied by my title, which was based upon a joke about a man whose passion was elephants and who, when asked to write an essay about the 'Polish Question', chose as his title 'The Elephant and the Polish Question'.) But I based my hope of persuading the responsible authorities to make changes to the project on an appeal to pragmatism and efficiency rather than to ethics: this was a case, I thought, where what was good for the Mursi would also be good for the elephants.

My appeal fell on deaf ears. The project went ahead without major revision and those responsible for funding and implementing it appear to have seen my intervention as an attempt to undermine their conservation objectives, for the sake of the narrow self-interest of a few thousand Mursi, rather than to make them more realisable. In this postscript I shall fill in what details I can concerning the subsequent history of the project. My account will be sketchy, because of the paucity of information available to me. This comes from two project documents, an 'Inception Report' for May–July 1995 (DHV Consultants BV, 1995) and a 'Final Report' covering the period May 1995 to April 1998 (DHV Consultants BV, 1998) and from a meeting I had with various officials of the EWCO during a visit to Addis Ababa in April 1999.

The 'preliminary phase' of the project (henceforth referred to simply as 'the project') began in May 1995, when the Project Manager (from DHV Consultants, The Netherlands) took up his post. It was expected to run until May 1997 but was later extended to 30 April 1998. A budget of 2 million ecu was allocated to the project, with a further 740,000 ecu to be provided by the Ethiopian Government from 'existing counterpart funds'. Over half of this latter amount was to cover resettlement costs, although the inception report devotes only four lines to this topic, stating that 'This objective will not be further discussed at this stage and actions to achieve it are unlikely to begin before year two' (p. 11). Three 'issues' are identified in the inception report as constituting 'project objectives', the first and most important being the establishment of a 'legal framework' for conservation. This is to include the formal adoption by the government of a Wildlife Conservation Policy, the legal gazettement of the three parks and the achievement of financial and managerial autonomy for the EWCO.

A wildlife policy and wildlife law were duly drafted and forwarded to the government in April 1997 but no official response had been received by the time the project ended in April 1998. As for gazettement, the final report states that this cannot sensibly proceed until the wildlife law has been adopted and promulgated, since 'under existing law gazettement would ... lead to immediate problems relating to the presence of people in the parks. The new law provides for zoning which would avoid conflictual situations ... ' (p. 9). Nor had financial autonomy for the EWCO been achieved by the end of the project, although this goal is described in the report as 'potentially achievable'. The significance of such a step, however, has been diminished by the formal transfer of responsibility for Ethiopia's national parks from the EWCO to the respective regional authorities, which occurred soon after the project began.

The second issue identified as a project objective was 'the strengthening of conservation efforts' in the three parks. This was to include improving the capacity and morale of park staff, rehabilitating park infrastructure and planning the resettlement of local people living in the parks. Efforts to achieve the first of these objectives appear to have been relatively successful in Nechisar, where regular patrols were instituted and where 'Hostile clashes between park staff and intruders is fortunately not a feature' (p. 11). In the Mago and Omo Parks, however, such clashes continued to occur, leading to the death of a game scout in January 1998. 'The net result is that the promising start to the institution of surveillance by regular patrols has effectively come to nothing and Mago and Omo are once again freely accessible to anyone wishing to hunt' (p. 11). The rehabilitation of park infrastructure has included repairs to roads and tracks, the repair of the Omo ferry, the digging of wells and the rehabilitation of existing park buildings and the construction of new ones. Plans to resettle people living in the Omo and Mago parks were shelved, no doubt because it was soon realised that this was an impossible task, given the lack of a detailed resettlement plan, the potential opposition of the well-armed local population and the totally unrealistic budget available for this purpose – 471,000 ecu for the project as a whole. The final report states that 'No resettlement is foreseen in Omo and Mago' (p. 14). The resettlement of families from the Nechisar Park (estimated in the project documents to number 480, which is the equivalent of at least 3,000 individuals) remained an integral part of the project, though it had not been achieved by the time the final report was written. The report states that 'There is now a high degree of probability that this vital action can be successfully concluded, although not during the life of this phase'.

The third 'issue' to be addressed by the project was the 'sustainable exploitation' of the three parks. This was seen to involve the preparation of 'development strategies' for each park, the encouragement of tourism, a review of the existing ban on safari hunting, censuses of wildlife and, in a revealing turn of phrase, an effort 'To inform the local communities about the project and their role within it'. Development strategies for the parks have been included in their management plans, wildlife censuses have been carried out in the three parks and the ban on safari hunting was lifted in 1996. Apart from the production of a 'promotional booklet' and park brochures, there appear to have been no other practical steps taken to encourage tourism. As for the involvement of local people in the project, this was left to the park wardens to organise, since no project funds were available for this purpose. The final report merely states that,

> A vigorous and very productive dialogue with the communities around the parks had already been initiated by the wardens before the project began. Active and frequent discussions continue to be held. However, there is limited progress in this area because there is still nothing specific to discuss. (p. 15)

The report concludes with a recommendation to extend the first, or preliminary, phase of the project for a further year by means of a 20 per cent increase in the original grant from the European Development Fund. The purpose of the extension would be to help the Government create 'workable institutional arrangements for conservation' and

> ... to fund the resettlement of the people in Nechisar National Park. The recent efforts of the Southern and Oromia Regions to achieve this *absolutely vital goal* must be given every assistance. To stop now is to lose the initiative for ever. (p. 19, emphasis added)

On a visit to Addis Ababa in April 1999 I was able to meet several officials of the EWCO, including the General Manager, the Director of National Parks and the member of staff who had taken over as Project Leader of the SNPRP, after the departure of the ex-patriot Project Manager in May 1998. They told me that the preliminary phase of the project had indeed been extended for a fourth year, up to June 1999. It was still intended to go ahead with the resettlement of people living in the Nechisar Park, but this had been held up because of 'administrative problems'. The people had been consulted, through a workshop at which their representatives were present, and they were ready to move. Some of the building materials needed for the resettlement – including corrugated sheeting – had already been purchased. The resettlement would therefore go ahead as soon as the 'administrative problems' had been solved. It was not explained what these problems were, but I gathered from other sources that they had to do with an argument between the Southern Nations, Nationalities and Peoples Region, within which the Park falls, and the neighbouring Oromia Region, about a resettlement site for the main population living in the Park, the Guji, who are Oromo speakers. I also learned that, while there had been no forced removals, the Southern Region authorities were putting strong pressure on the Guji to leave the Park 'voluntarily', by denying them veterinary, health and educational services.

This brief account of the subsequent history of the SNPRP illustrates at least two general issues. First, there is the contrast between good policy and poor implementation. The feasibility study for the SNPRP was full of wise words and noble aspirations on the subject of community participation, but it provided none of the detailed information about local resource management and decision-making systems that would have been necessary in order to plan the project in such a way that it could have lived up to these aspirations. The failure systematically to consult local people, either before or during the course of the project, to involve them in decision-making and to ensure that they gained long-term benefits from the project was at variance with the Ethiopian Government's own policies on environmental protection, which state that policies in this area should 'promote the involvement of local communities' and 'ensure that park, forest and wildlife conservation and management programmes ... allow for a major part of any economic benefits deriving therefrom to be channelled to local communities affected by such programmes.' (Environmental Protection Authority, 1997, pp. 9–10). Second, there is the failure of donor institutions to ensure that the projects they fund, particularly those involving resettlement, are carried out in accordance with internationally agreed guidelines, designed to protect the rights and well-being of affected populations. In this case, the relevant guidelines were those of the OECD on Involuntary Displacement and Resettlement. These require, amongst other things, that, when resettlement is unavoidable, a detailed resettlement plan, with timetable and budget, be drawn up, aimed at improving or at least restoring the economic base of the resettlers. There is no evidence that such a plan was drawn up in this case, nor that the European Commission made any effort to ensure that the project complied with the OECD guidelines.

Who, then, have been the beneficiaries of this project? I shall leave that for the reader to decide but, whoever they are, they have not included the Mursi and, still less, the elephants.

References

Abel, N.O. J. and Blaikie, P.M. 1990. *Land Degradation, Stocking Rates and Conservation Policies in the Communal Rangelands of Botswana and Zimbabwe.* Pastoral Development Network, Paper 29a. London: Overseas Development Institute.

Agriconsulting. 1993. *Feasibility Study for a Wildlife Conservation Project in Southern Ethiopia: Final Report.* Rome: Agriconsulting SpA.

Anderson, D. and Grove, R. (eds). 1987. *Conservation in Africa: People, Policies and Practice.* Cambridge: Cambridge University Press.

Bell, R. 1987. 'Conservation with a Human Face: Conflict and reconciliation in African land use planning'. In D. Anderson and R. Grove (eds). *Conservation in Africa: People, Policies and Practice.*

Behnke, R.H. and Scoones, I. 1993. 'Rethinking Range Ecology: Implications for rangeland management in Africa'. In R.H. Behnke, I. Scoones and C. Kerven (eds) *Range Ecology at Disequilibrium: New Models of Natural Variability and Pastoral Adaptation in African Savannas.* London: Overseas Development Institute.

DHV Consultants BV (1995) *Inception Report, 25 May–31 July,* National Parks Rehabilitation in Southern Ethiopia Project, EU Project No 7 ACP ET 068, Ministry of Natural Resources, Development and Environmental Protection, Addis Ababa.

DHV Consultants BV (1998) *Final Report, May 1995–April 1998,* National Parks Rehabilitation in Southern Ethiopia Project, EU Project No 7 ACP ET 068, Ethiopian Wildlife Conservation Organisation, Addis Ababa.

Dyson-Hudson, N. 1991 'Pastoral Production Systems and Livestock Development Projects: An East African Perspective'. In M. Cernea (ed.). *Putting People First.* Oxford: Oxford University Press for the World Bank (2nd Edition).

Environmental Protection Authority (1997) *Environmental Policy of the Federal Democratic Republic of Ethiopia*, Environmental Protection Authority in collaboration with the Ministry of Economic Development and Cooperation, Addis Ababa.

Evans-Pritchard, E.E. 1956. *Nuer Religion.* Oxford: Clarendon Press.

EWCO. 1989. *Wildlife Conservation Report to the Agro-Ecological Zonation Study.* Addis Ababa: Ethiopian Wildlife Conservation Organisation/Ministry of Agriculture.

Galaty, J. and Bonte, P. (eds). 1991. *Herders, Traders and Warriors: Pastoralism in Africa.* Boulder, Colorado: Westview Press.

Hardin, G. 1968. 'The Tragedy of the Commons'. *Science*, 162: 1243–8.

—— 1988. 'Commons Failing'. *New Scientist*, 22 October.

Hillman, J.C. 1993. *Ethiopia: Compendium of Wildlife Conservation Information. Volume I: Wildlife Conservation in Ethiopia.* The Wildlife Conservation Society International, New York Zoological Park, Bronx NY 10460, USA/Ethiopian Wildlife Conservation Organisation, P.O. Box 386, Addis Ababa, Ethiopia.

Hobbs, R.J. and Huenneke, L.F. 1992. 'Disturbance, Diversity and Invasion: Implications for conservation'. *Conservation Biology*, 6: 3240.

Hogg, R. 1992. 'NGOs, Pastoralists and the Myth of Community: Three case studies of pastoral development from East Africa'. *Nomadic Peoples*, 30: 122–46.

Homewood, K.M. and Rodgers, W.A. 1987. 'Pastoralism, Conservation and the Overgrazing Controversy'. In D. Anderson and R. Grove (eds). *Conservation in Africa: People, Policies and Practice.*

—— 1991. *Maasailand Ecology: Pastoralist Development and Wildlife Conservation in Ngorongoro, Tanzania.* Cambridge: Cambridge University Press.

Horowitz, M.M. 1986. 'Ideology, Policy and Praxis in Pastoral Livestock Development'. In M.M. Horowitz and T.M. Painter (eds). *Anthropology and Rural Development in West Africa.* Boulder, Colorado: Westview Press.

IIED. 1994. *Whose Eden?: An Overview of Community Approaches to Wildlife Management.* London: International Institute for Environment and Development.

Lindsay, W.K. 1987. 'Integrating Parks and Pastoralists: Some lessons from Amboseli'. In D. Anderson and R. Grove (eds). *Conservation in Africa: People, Policies and Practice.*

McCabe, J.T., Perkin, S. and Schofield, C. 1992. 'Can Conservation and Development be coupled among Pastoral People? An examination of the Maasai of the Ngorongoro Conservation Area, Tanzania'. *Human Organisation*, 51(4): 353–66.

Murphree, M.W. 1991. *Communities as Institutions for Resource Management.* Occasional Paper of the Centre for Applied Social Sciences. Harare: University of Zimbabwe.

Newby, J. 1990. 'The Air-Tenere National Nature Reserve'. In A. Kiss (ed.). *Living with Wildlife: Wildlife Resource Management with Local Participation in Africa.* Washington DC: World Bank.

Pimbert, M.P. and Pretty, J.N. 1995. *Parks, People and Professionals: Putting 'Participation' into Protected Area Management.* Discussion Paper 57. Geneva: United Nations Research Institute for Social Development.

Potkanski, T. 1994. *Property Concepts, Herding Patterns and Management of Natural Resources among the Ngorongoro and Salei Maasai of Tanzania.* Pastoral Land Tenure Series No. 6. Drylands Programme. London: International Institute for Environment and Development.

Runge, C.F. 1984. 'Institutions and the Free Rider: The assurance problem in collective action'. *Journal of Politics*, 46: 154–81.

—— 1986. 'Common Property and Collective Action in Economic Development'. *Proceedings of the Conference on Common Property Resource Management.* Washington DC: National Research Council, National Academy Press.

Skinner, J. 1989. *Towards Better Woodland Management in the Sahelian Mali.* London: Overseas Development Institute.

Spencer, P. 1990. 'Pastoralism and the Dynamics of Family Enterprise'. In C. Salzman and J. Galaty (eds). *Nomads in a Changing World.* Series Minor XXXIII. Naples: Dipartimento di Studi Asiatici, Istituto Universitario Orientale.

Stephenson. J. and Mizuno, A. 1978. 'Recommendations on the *Conservation of Wildlife in the Omo-Tama-Mago Rift Valley of Ethiopia*'. Report submitted to the Wildlife Conservation Department of the Provisional Military Government of Ethiopia, Addis Ababa.

Sutcliffe, J.P. 1992. *Peoples and Natural Resources in the North and South Omo and Kefa Administrative Regions of Southwestern Ethiopia: A Case Study in Strategic Natural Resource Planning.* National Conservation Strategy Secretariat, Ministry of Planning and Economic Development, Addis Ababa. (Draft)

Tapson, D. 1993. 'Biological Sustainability in Pastoral Systems: The Kwazulu case'. In R.H. Behnke, I. Scoones and C. Kerven (eds). *Range Ecology at Disequilibrium: New Models of Natural Variability and Pastoral Adaptation in African Savannas.* London: Overseas Development Institute.

Thomas, K. 1984. *Man and the Natural World: Changing Attitudes in England, 1500–1800.* Harmondsworth: Penguin Books.

Thompson, M., Mortimor, M. and Hatley, T. 1986. *Uncertainty on a Himalayan Scale.* London: Milton for Ethnographia Press.

Turton, D. 1980. 'The Economics of Mursi Bridewealth: A comparative perspective'. In J. Comaroff (ed.). *The Meaning of Marriage Payments.* New York and London: Academic Press.

—— 1987. 'The Mursi and National Park Development in the Lower Omo Valley'. In D. Anderson and R. Grove (eds). *Conservation in Africa: People, Policies and Practice.*

Warren, A. and Agnew, C. 1988. *An Assessment of Desertification and Land Degradation in Arid and Semi-Arid Areas.* Paper 2, Drylands Programme. London: International Institute for Environment and Development.

Wordsworth, W. 1835. *Wordsworth's Guide to the Lakes* (5th Edition) (Edited by E. de Selincourt, 1977).

7

Forced Resettlement, Rural Livelihoods and Wildlife Conservation along the Ugalla River in Tanzania[1]

Eleanor Fisher

Introduction

In the twentieth century, the conservation of wildlife within protected areas in East Africa involved radical change in the relationship between people, land and natural resources. Population resettlement played a part in this change; areas of land now protected under wildlife and forestry laws were once populated and people were moved – forcibly or otherwise – by colonial and post-colonial authorities.[2] However, the links between population resettlement and the gazettement of protected areas for conservation purposes are complex and have led people to re-interpret and contest both resettlement and conservation goals in many different ways.

Understanding how experiences of population displacement and resettlement in the past are given expression in the present is critical if we are to appreciate fully the nature of people's connections to many protected areas in East Africa today. These connections include productive activities, but they encompass also the meanings people give to conservation and to their interactions with representatives of the state and conservation agencies.

This chapter takes a case from western Tanzania, near the Ugalla River, which is today enclosed within a protected area, Ugalla Game Reserve. It is based on archival sources and ethnographic fieldwork undertaken in Tanzania between 1992 and 1994.[3] The discussion focuses on the transformation of Ugalla from a territory inhabited by people to a protected area conserved for wildlife. In this case, the human population was forcibly moved in the mid-1920s as a public health measure. This eviction can be seen as a 'critical event' that opened up the possibility for the territory to be used in new ways. Approximately thirty years after the human population was resettled, Ugalla was gazetted for the conservation of fauna and flora. In effect, the land was transformed from a tribal homeland into a state managed protected area, governed by conservation goals, policies, values and practices.

What makes this an interesting case is that, despite resettlement policies and conservation laws restricting access to the area, the original inhabitants and other rural people have never been totally excluded. Moreover, both colonial and post-colonial authorities have recognized that resource use in Ugalla is important to people's livelihoods. Part of the reason people have managed to maintain a presence in the area is through the way tribal identity and resettlement history have been used as a means to negotiate access to Ugalla. For, after the human population was resettled, they continued to lay claim to their past area of habitation and, in so doing, maintained a presence within it in ways that generated different historical trajectories to those envisaged in the original resettlement plan.

One way these claims were consolidated was through use of Ugalla on a seasonal basis for productive activities such as hunting, fishing, honey hunting and beekeeping. Over time this led to the development and institutionalization of beekeeping and fishing as specialized livelihood activities permitted within Ugalla Game Reserve. In this respect, forced resettlement spawned the beginning of a contradiction between administrative objectives – those that drove both resettlement actions and later conservation planning – and local people's need to make a livelihood and desire to have access to Ugalla on their own terms. Thus, in more recent years, the conservation of wildlife in a purportedly 'natural' area, has met with counter-tendencies that have generated a field of action in which it is possible to observe different interpretations of what constitutes the locality of Ugalla. These interpretations emerge in an on-going tension between state claims over Ugalla as a game reserve, and local people's claims over Ugalla as a place where they wished to carry out productive activities.

Sleeping Sickness Resettlement from the Ugalla River: A Critical Event in the Life of the People

People living near the Ugalla River were forcibly resettled as the British colonial response to a sleeping sickness epidemic that erupted in western Tanganyika during the mid-1920s.[4] It was held that by re-locating the population into concentrated settlements away from the river, tsetse fly, a carrier of the sleeping sickness trypanosome, could be kept 'at bay' and people protected from the disease.

Resettlement took place in three consecutive stages corresponding to the dry seasons of 1925, 1926, and 1927. People were moved to seven areas known as resettlement 'concentrations' each situated approximately 50 to 80 kilometres from the river.[5] It is estimated that a total of 9191 people were moved; although whether all came from near Ugalla River, and how accurate the process of enumeration was, is not known.[6]

Historical accounts by men who lived through the evacuation tell of a Doctari Makaleni (Dr Maclean) who came to tell them they had to move (Fisher, field diary 1993). They describe how huts were burnt, how people were afraid of the lorries that came to transport them, and of old people who refused to leave and who, left behind, were eaten by hyenas. After the move, people experienced a lot of hardship. One man described how:

> when we moved the life was bad and we were not blessed. At this time there was sleeping sickness … [in Ugalla] … but even so there was a lot of food … Even before we moved, when we came to sell beeswax people here would say 'no don't come' and we would reply, 'no we eat our ugali … [staple food] … of cassava and finger millet differently from you' … [i.e. prefer our own customs] (Fisher, field diary 1993).

Some older people still refer to the resettlement area as a 'counterfeit country'. The move also generated political conflict, as previously independent chiefdoms were seemingly subordinated to others in the resettlement areas.

In the 1930s, as the sleeping sickness epidemic spread, resettlement concentrations became established in other parts of the region. Agents of modernization, expressed through Christian missions, education, health care, agricultural and veterinary extension work, all gravitated towards the spaces delimited by the resettlement schemes.[7] The hopes of administrators are echoed in sentiments expressed by the Provincial Commissioner for Western Province: '[i]n their ignorance bush natives must realise the advantages of safe community life properly administered, must eventually out-weigh the joys of being left alone …'[8] This is not to say that there was no resistance to the new centralizing trajectories of colonial administrators. Once moved, some people simply ran away to live in the forest, as was the case on a number of occasions in Ugalla.[9]

As a window to assist us to examine the consequences of forced resettlement from Ugalla, we can conceive resettlement as a critical event that radically transformed the nature of people's presence in the locality. This is not to say that other events (e.g. warfare, long-distance trading, German colonization) did not have a significant impact on the local population. However, population resettlement was a situation in which individual and collective relationships to the locality were irrevocably changed in a very short period of time. The dramatic nature of this change fed into a process in which people had to encompass and internalize new experiences, social relationships, livelihoods, memories of the past, and visions of the future (c.f. Long 1997). By viewing resettlement in this way, our attention is directed not only to the shock people experienced, but also to their responses to the event, including their capacity to rebuild their lives. It also enables us to examine how the act of resettling the population was an event that precipitated change in land use in Ugalla and necessitated specific, on-going, administrative actions on the part of the British colonial government.

Identifying People According to Productive Categories: the Rise of Administrative Problems

Once Ugalla had been depopulated, a potential for the area to be exploited in new ways had been created: past inhabitants began to generate different connections to the area, and new interests started to arise. As one old man described:

> you know when someone leaves the place he has lived, he has his memories ... every year it was an obligation to return. People went by foot ... in two or three days you would arrive with nothing and make *kangara* ... [beer made with honey] and drink lots, eeeh. In the resettlement area, people said 'you should not make *kangara* ... [due to attempts to control drunkenness and make people use staple crops 'sensibly'] (field diary 1993).

Despite people's desire to return to live in Ugalla, district and regional administrative officials would not permit this to happen and, to uphold the objectives of the resettlement policy, the administrative status of the area was changed. Because of the continued threat of sleeping sickness, the region surrounding the Ugalla River became categorized as a 'quarantine zone' on public health grounds. People were only permitted to enter the quarantine zone on terms established by administrative officials, namely as part of an accepted seasonal labour migration to work as 'fishermen'. Other categories of people were actively excluded from the area, they included 'hunters', 'settlers' and 'women'. Thus began an annual struggle between officials wishing to control people's presence in the area, and people who wanted to return to Ugalla each dry season.

In the way administrators sought to control people's access to Ugalla, we see that after resettlement these rural people became categorized according to the forms of the productive resource use they engaged in. These categories are important because legitimacy was only accorded to certain types of productive activity, while other activities were banned. Thus, prior to the move, written descriptions of people living near the Ugalla River describe, for example, 'the Wagalla', 'native Africans', 'taxpayers' or 'sleeping sickness cases' (see Burton 1860; Reichard 1892; Cameron 1877; AWSID 1916; [RH.Mss 2551, 1919]; Maclean 1929a/b). However after resettlement, and in keeping with wider regional transformation, one can witness a change as people became referred to as 'fishermen', 'hunters', 'honey hunters' and 'agriculturists'. Each of these categories is linked to a productive activity, particularly those carried out by men. In effect, these economic categories fed into official perceptions of the problems confronting the administration of Ugalla, generating a perceived need for new administrative solutions.

Hand in hand with categorization of people according to productive activities, was the fact that people themselves were generating new associations with Ugalla, based on these forms of resource use. Productive categories of resource use were not simply imposed from outside; they also reflected long-term change in the nature of productive activities for the people concerned.

In the 1930s, development in the region was biased towards the promotion of agriculture, but it was recognized that this was not the only activity that men undertook (women are largely invisible in the referenced archival sources).

> Agriculture is, and I hope it always will be, the main occupation of the natives ... [but] ... it is a fact that it does not occupy anything like all their time ... [the] ... slack season ... [lasts for] ... half the year ... during which time, in this Province probably 10,000 men spend their spare time every year in collecting honey and fishing.[10]

Despite notions of non-agricultural activities being 'spare time', permitting men to fish along the Ugalla River was recognized as important for those concerned.

One difficulty associated with fishing on the River was that men continued to catch sleeping sickness. As a consequence, a native authority ordinance was instituted that covered the area around the Ugalla River (Tabora District Book 1935).[11] It stated that fishermen proceeding to the Ugalla River were to go in large parties under a leader; any sick were to be sent at once to hospital; and the whole party was to proceed for medical examination at the end of the season.

In interview, the grandson of a man who acted as a fishing leader told how: 'one day, at the time we moved, we chose one person as a leader and he was a fisherman'. He proceeded to describe how:

> it was yourself and your own nets, you said you were going to build a camp ... but to have a camp first you had to come and see the Chief ... each leader of every camp was given permission ... it was essential that you came to see the Chief to build in a certain place. He was the leader of all the river from Koga to Silongwe, he was the Chief of the area ... If a person had done something really wrong he was brought here to Siri and then chased from the riverside (Fisher, field diary 1994).

Although referred to as a 'chief' the leader was not one in an administrative sense but a skilled and respected fishing leader. A key point in this description is the fact that the fishing leader was not simply imposed by administrative officers, but held an existing and accepted role. This is important; not only do we have to understand how the colonial administration exercised its authority on the fishermen, but also how the fishermen kept and renegotiated their right to use the river and to be responsible for their own well-being.

The fact that a fishing ordinance was implemented is relevant because it reveals how attempts were made to control people's access to the area they had previously inhabited. In this sense, we can see the policy of resettlement, and subsequent actions to prevent people from catching sleeping sickness, acted as a means to discipline the local population.

Alongside fishermen, 'native hunters' were a problem for the administration of Ugalla. This was also due to the threat of sleeping sickness and the need to control people's presence near the river. However, 'the problem' of native hunting was treated very differently from that of fishing. Unlike fishing, it was illegal and led to direct confrontations with colonial administrators, many of whom were keen hunters. Ugalla River was the main sports hunting ground in the region, and native hunters were blamed for game being 'shot out'. As a result, attempts were made to register or confiscate their muskets and some men were taken to court.[12]

In addition to fishing and hunting, people tried to resettle in Ugalla as small groups returning on an ad hoc basis. Because of the continued threat of sleeping sickness, and also because new settlements represented failure for those charged with maintaining the resettlement policy, the colonial authorities kept forcibly re-evacuating settlers. For example, in 1938–9, a new rise in sleeping sickness cases was recognized and linked to fishing, honey hunting and 'a tendency to shift back into individual communities'. Some 48 'families' and 'fishmongers' were reported to be residing at the river or in nearby bush (west Ugalla). All were forcibly removed because, in the view of the Provincial Commissioner, 'there would have been 480 by this time next year'.[13]

An interesting social aspect of the dynamic of colonial attempts to prevent people from returning to live in Ugalla was a prohibition on women going to the river. This was seen as an important means to stop seasonal settlements from becoming permanent (c.f. Mblinyi 1989 on Tanganyikan colonial policies restricting female migration).[14] Indeed, successive representatives of both the colonial and post-colonial administrations have sought to exclude women from Ugalla because they were – and are – thought to generate problems; not only through the threat of establishing new settlements, but also as the cause of fighting and drunkenness within seasonal camps, thus upsetting what was – and is – considered to be social order (with implicit inferences to prostitution and lack of moral control).

From the limited texts available on Ugalla prior to the mid-1920s, and from oral historical accounts, it would appear that women did not hunt, fish or honey hunt.[15] Nonetheless, they were living in villages in the area and undertook many different productive and reproductive activities. Absence of habitation and tasks associated with women after the 1920s, suggests a very different use of the space from that time, with dry season productive activities undertaken by men in Ugalla being separated from other tasks and aspects of daily life in villages in the region.

Whether people's sense of belonging to Ugalla was erased by resettlement and the administrative categorization of people (men) according to productive activity can be disputed. What is clear is that some people approached the new situation through the reorganization of seasonal activities in such a way that a livelihood could be made in Ugalla and associations with the area maintained. This was to the exclusion of certain categories of people, which included women, children and the old or disabled, as well as others who did not carry out forest activities. In practice, fishing may not have been the only activity that men engaged in or even the main reason for going to the river. However, over time occupational activities did change and become more market oriented, with certain forms of production being institutionalized as key forms of resource use in Ugalla: for example, fishing and beekeeping – to the exclusion of others, such as agriculture and hunting.

To summarize, while the authorities recognized that they could not prevent people returning to the area outright, they sought to confine entry to 'fishermen' for a limited part of the year, asserting authority indirectly through a fishing leader, and by using the discipline of medical intervention on people's bodies. In contrast, 'native hunters' had to have their weapons registered or were taken to court; women were excluded to prevent 'unruly behaviour' and permanent settlement, while re-settlers were forcibly evicted. At one level, the resettlement policy was unsuccessful because people could not be prevented from returning to the area or from catching sleeping sickness. However, there was a dramatic change in the nature of human occupation and in the use of resources that did enable greater government control over and organization of the population.

The Enclosure of Land for Conservation: Ugalla River Game Controlled Area

The changing status of Ugalla fed into a process of re-categorizing and re-evaluating the area by a variety of social actors who represented different interests. A bridge was built across the River, lumbering took place, and Ugalla drew the attention of colonial scientists anxious to carry out botanical, geological, entomological and zoological research (Phillips 1931; Jackson 1936; Milne 1936; Potts 1937; Glover 1939).[16] In addition, it became part of a fashionable western hunting circuit for wealthy tourists seeking to shoot lion, greater kudu, black sable, roan and water fowl (Sayers 1930; Rodger 1954; Moffett 1958).[17]

The presence of these new social actors in Ugalla gradually brought about a change in the administration of the area. Whereas public health issues had been central to administration in the region through the 1920s and 1930s, from the 1940s a new administrative emphasis was placed on the need for rational forms of natural resource management and development planning (Iliffe 1979: 436).[18] As part of this process, in the 1950s extensive areas of western Tanganyika (land from which people had been forcibly resettled due to sleeping sickness) were enclosed within forest reserves and game controlled areas. This included the region surrounding the Ugalla River and meant that people's rights to use resources or to settle were severely restricted through new forestry and fauna preservation laws.

The absence of settlements or agricultural activities near the Ugalla River projected an image of the area as uninhabited. This representation provided extra cultural and social value for outsiders to whom it appeared a 'natural landscape'. Its flora and fauna could be valued according to scientific (and sporting) standards, as against the economic motives of entrepreneurs or the livelihood interests of rural people. This contributed to Ugalla being gazetted as a conservation area in which wildlife were controlled and protected. The first part of Ugalla to be gazetted was a small portion in the east, the result of entomological experiments on tsetse fly, not for wildlife preservation per se.[19] This new status of Ugalla as a conservation area was achieved largely thanks to the earlier history of population resettlement.

In the new policies on natural resource management and development planning, we see the perpetuation of administrative attempts to keep people out of Ugalla, albeit for conservation rather than sleeping sickness prevention purposes. At the same time, however, administrators familiar with the area recognized that the exclusion of rural people was impossible to administer and, importantly, that Ugalla played a significant role in their livelihoods.

The status of activities such as fishing in Ugalla was influenced by wider change as forms of seasonal production were given increased emphasis within regional development. In the 1940s and 1950s, the colonial government was actively trying to promote the beeswax and fishing 'industries' (Rodger 1954). Indeed, for the first time reference was made to 'beekeeping' in administrative documents, with the term 'honey hunters' transformed to the more progressive category of 'beekeepers'. This was in keeping with Ugalla River having become one of the main locations for extension work by the newly created Beekeeping Section (Smith 1994).[20]

In 1952, D.K. Thomas, the Game Ranger for the Western Range, promoted the need for a game controlled area in the Wala–Ugalla River area in order to control the hunting and lumbering that was taking place, and to provide a breeding sanctuary for the types of animals that inhabit Brachystegia woodland (Thomas 1961).[21] Thomas did not point to native hunters as the main culprits of game destruction; instead, extensive 'illegal and unsporting hunting' by resident British, Arabs and Greeks and by foreign tourists was held to blame, while Greek entrepreneurs were held responsible for lumbering.

Support for 'native rights' in the proposed protected area was voiced by a number of colonial officials. For example, in promoting the need to conserve Ugalla, the Provincial Commissioner argued that '[the GCA would not] ... affect the interests of Africans, as it is all fly country and the only inhabitants are fishermen'.[22] But the Game Ranger was strong in his advocacy of the interests of these fishermen: 'it is however essential that Africans be allowed to continue their fishing activities in the proposed controlled area'.[23] Similar support was given for beekeepers by the District Commissioner for Mpanda and the Beeswax Officer in Tabora.[24] From these accounts, one gains an impression that this reflects an administrative awareness that fishing and beekeeping were entrenched in Ugalla and that they could not be prevented, and were important productive activities. Furthermore, recognition was given to the value of local knowledge as fishermen were seen as a potential source of information concerning what was taking place in the area (e.g. poaching). Indeed against the context of the enclosure of vast areas of land from which people had been forcibly resettled, as forest reserves, native rights to forest produce became a political issue in the 1950s (Rodger 1954: 46–47).

In the event, once its boundaries had been agreed, the Ugalla River Game Controlled Area was gazetted in 1954.[25] Three classes of resource user were permitted to use the area on licence: 'beekeepers', 'fishermen', and 'sports-hunters' (i.e. international trophy hunters).[26] Sports-hunting in particular was felt to provide 'justification for the formation of these areas'.[27] Nevertheless, the way the interests of African people were taken into account by civil servants responsible for gazetting Ugalla River Game Controlled Area reveals that the conservation process was nothing as simple as a clear-cut exclusion of local interests. Within the colonial administration there were different views and the goal of conservation had to be negotiated vis-à-vis recognition of the importance of the beekeeping and fishing 'industries' for the region and for the livelihoods of the people concerned. Thus, we can see how contradictions between visions of the future of the locality held by rural people on the one hand, and an external conservation vision on the other, were becoming embedded in the way Ugalla was transformed into a protected area.

People's Return to Ugalla: A Bid to Resettle the Land

Once Tanganyika received its Independence from Britain in 1961, new social actors emerged. They included African bureaucrats in local government, refugee fishermen and women from Burundi, Zaire and Malawi, sports-hunting Tanzanian citizens and Burundian ivory traders. In the case of Ugalla, local leaders articulated their claims through a language of African rights and 'tradition'. They argued that Ugalla had been forcibly taken by the colonial government, and that it was part of their traditional homeland, being the place they had originated from, closely associated with their tribal identity (*'Ugalla'* meaning place/territory of the Galla). Their claims went beyond the demand to make a livelihood; they argued that they wished to return to live there, in effect challenging Ugalla's status as a conservation area.

We can trace claims to resettle Ugalla back to the 1940s. In 1994, when I interviewed a man, *Mtemi* (Chief) Nsokoro Mvula, who had been a chief until the 1960s, he said that he had first sought to mobilize people to return when his father, a chief from Ugalla, had died. He said that it was necessary to bury his father in Ugalla, and they would have to offer libations (*tambiko*) to him each year. Also, he himself no longer wanted to be subordinate to the host chief in the resettlement area. *Mtemi* Mvula's account is confirmed by archival documentation. In 1948, the District Commissioner agreed that *Mtemi* Mvula and his supporters could move to settle in the far western portion of Ugalla; subsequently he drove the Chief there in his motor vehicle. *Mtemi* Mvula represented this as a victory, but the agreement was in keeping with a perceived need for new resettlement concentrations due to labour demands in the Lupa Gold Mines and along the Mpanda Line Railway.[28]

This background to issues of identity and claims to the territory is significant for an understanding of how, during the 1960s, a group of people managed to position themselves as those with 'true' rights over the territory of Ugalla. The manner in which they located Ugalla at the centre of local culture and politics was a strong counter-narrative to expert representations of Ugalla as a 'natural' environment for wildlife conservation. Here we see that the resettlement history was in no way forgotten and became an important means for people to legitimate claims to the area. For example, a communiqué from the Executive Officer for Tabora District Council – for the first time in Kiswahili not English – to his counterpart in Mpanda, reveals local people's use of the idiom of 'tradition' and resettlement history in supporting their claim. The administrative document is headed 'the Wagalla to return to their traditional homeland'.[29] It goes on to state that the leaders, David Yongolo and Abdurahaman Kaponta, claimed rights as 'chiefs' to move to Ugalla with their people, arguing that it used to be their country, that if they moved they would be near traditional fishing and honey-hunting grounds, and also that their present agricultural land was infertile.

At this time, African officials in local government voiced support for people trying to return to Ugalla, causing immediate concern within the Game Department (still dominated by Europeans). As Game Department officials realized, local political alliances were emerging and it appeared that the Tabora District Council had the power to let these people in. 'There seems to be a certain faction ... who are keen to re-settle the area comprising the Ugalla River Controlled Area ... As we are all aware, this is one of the show pieces of this part of the world, and any settlement would in fact be disastrous. I wonder if we could possibly make this into a Native Authority National Park?'[30] The language of these administrative documents underlines a tension between the British civil servants still in senior positions in the Game Department and African members of the District Council who represented the case of people from the area.

Two groups attempted to return to Ugalla in the early 1960s: one of 547 people from Mpanda (Uruwira resettlement concentration) under the leadership of Kaponta, and the other of 500 people from Tabora under the leadership of Yongolo. They went to Ugalla accompanied by a game officer, tsetse officer, agricultural officer, and local government officer. Kaponta chose the site of Kasekela outside the southern boundary of the GCA, and Yongolo a site called Igombe outside the northern boundary (the only portions of Ugalla GCA not also gazetted as forest reserve where settlement was prohibited). In both areas, the administrative officers demarcated a line beyond which settlement and agriculture could not take place.

Local versions of events recorded during my fieldwork gave a different representation of what took place. Apparently, *Mtemi* Ngugula Yongolo tried to force everyone to settle at Igombe, his 'hereditary chiefdom', by tying a rope across the road to prevent people from going further. Both archival administrative documents and oral historical accounts refer to a line of demarcation representing a boundary. For administrators accompanying people to the area, this line was to stop them degrading resources in the game reserve. For the Chief, the line was to consolidate a settlement and therefore to provide political meaning to his office. Here we have two different meanings given to the action that took place; further meanings arose from the people's own understanding of the attempt to resettle the area that generated conflict with both main parties.

When *Mtemi* Yongolo and a group of people finally settled at Igombe they found they had become accustomed to welfare provisions in the resettlement area and that wild animals ate their crops and were dangerous. Thus, they did not have the ability to cope with the new situation that confronted them. Present popular narratives convey the perception that once settled they realized Ugalla was no longer good for people, emphasis being given to the power of animals. Apparently, the wife of Yongolo died when in the new settlement. Because she was a Muslim, they went to seek immediate assistance of a ritual leader for the funeral. When they returned they found that hyenas (this may allude to a transmogrified form taken by witches) had taken her body and pulled it to pieces. The interpretation of this event is used to convey an image of the reclaimed land having been 'only good for animals'.

The bid to resettle the land provoked an immediate response from the Game Department. In 1965, the administration elevated the status of Ugalla River from a 'game controlled area' to a 'game reserve'. This category of protected area prevented people from living within the boundaries (Government Notice 281 & 282, June 1965).

The change in status of Ugalla to a game reserve had the consequence of once again making beekeeping and fishing a highly visible 'problem' for the administration. Because productive activities were not permitted in a game reserve, people's right to keep bees and to fish was called into question. It was eventually decided that fishermen could continue to use the reserve because a special exception had been made for them to fish in the Ugalla River when the area was gazetted in 1954. However, because beekeeping and honey-hunting activities had been on a much smaller scale and carried out by fishermen and hunters in the 1940s and early 1950s, no precedent had been made for beekeepers in the prior legislation. A legal misrepresentation of local history was used against the beekeepers: 'their claim that they have got permanent camps along the river is not justified. The area, to a distance of 30–40 miles on each side of the river was cleared of permanent habitation in 1950 (sic.) as an anti-sleeping sickness measure ... anyone found in the reserve will be prosecuted'.[31] Eventually, in 1967, officials at the Beekeeping Section in Tabora gained special permission for '*bona fide*' beekeepers to keep their beehives in Ugalla.[32]

In the actions of people seeking to return to Ugalla, we see that the resettlement history comes full circle. Faced with direct experience of living in the area, people were confronted with the hardship this could entail and with the daily reality of land given over to wildlife. In this respect, the area's status as a game reserve devoid of settlement became consolidated through these events, as did acceptance that beekeeping and fishing were key activities that could occur in the area. This situation has continued up until the present day, when beekeeping, fishing and tourist (i.e. non-resident) sports-hunting are the main legitimate activities that take place in Ugalla Game Reserve, with settlement and other forms of productive resource use being banned, although they may take place illegally.

Beekeeping and Fishing in Present Day Ugalla Game Reserve

Along a track that runs the length of the Ugalla River from east to west, and in the forest away from the river are beekeeping and fishing camps. These camps consist of a shelter, which is typically made of poles thatched with grass or palm fronds, and places to dry fish or to process raw honeycomb. To the outside eye, the camps appear impermanent, for the shelters become dilapidated during the rains when the men have returned to their villages. However, each camp bears the name of a village area that existed prior to resettlement in the 1920s, and many of the people who work in these camps are first, second, third, or even fourth generation descendants from sleeping sickness evacuees. These lines of descent can be traced through the many relationships of kinship and affinity that exist between men who work in the reserve (see Fisher 1997b). (Some women do go to the river but very few; numbers are difficult to estimate, but maybe 20 women as compared to 300–500 men in a given season.)

Each seasonal camp has a boundary, which is not marked but is part of the local knowledge people hold about the area. Government officials do not know these boundaries, but the camps have recognized leaders who will mediate if and when disputes arise. Such knowledge of the environs of camps in Ugalla is extensive, albeit localized. Although some people may work in the area for a short period, others build up this knowledge over many years, being part of their family history, their memories, their skills, and their working relationships. People also have an intimate knowledge of the movements and activities of others – game officials, tourist hunters, poachers, and so on – in the Reserve, which emerges in claims and counter-claims concerning who is carrying out illegal activities and who has a right to work in the area.

As occupations, beekeeping and fishing are very different in character and organization; nonetheless, their development has been closely linked. As one man described:

> once a fisherman meant all, like one person, because a fisherman he was a Galla, and a honey-hunter he was a Galla, and a hunter likewise. Now a honey hunter fished and for bees he looked for honey in the forest, in the trees ... and he hunted as well. These three things they went together (Fisher, field diary 1993).

This is an example where resettlement has spawned the re-working of collective identity, which is given fresh meaning in the present day.

Another man described how:

> Beekeeping is of recent times ... in the past there was a sea of fish, but we went on as fishermen and in the end people became beekeepers ... this is why we say that fish gave birth to bees .... The people who were moved ... [due to sleeping sickness] ... they went to Kasontwa to find trees and honey hunt and to get two or three buckets of honey and that was all. This honey was eaten and the price was very low, three shillings for a whole bucket ... [The beekeeping co-operative] ... it was like an injection, you brought twenty buckets with you and one of your companions would see and say I will bring forty buckets, and someone else would take sixty ... The hives became many because of the money, it was food (income), it became a proper market (Fisher, field diary 1993).

Today, some men manage to combine beekeeping and fishing, but many channel their resources into one activity. Beekeeping is primarily carried out by people from the region; very few immigrants keep bees unless they marry into a beekeeping family or are close to a successful beekeeper. There are many reasons for this, including a fear of bees and the fact that it takes many years to gain skills, labour and beehives to be successful on an annual basis.

Beehives are a substantial form of private property, from which, in a good year, a significant income can be gained. For beekeepers working in Ugalla, individual rights are held over the trees the beehives are placed in. These rights can exist for a man's working life and pass between generations. Given that a successful beekeeper can own several hundred beehives, and that there will be a group of beekeepers working in each camp, many of whom will be related, intricate relationships grow up between people, trees and the land contained within Ugalla Game Reserve. These relationships are played out through men's experience of working and living together in close proximity through a season and often over many years.

Fishing is somewhat different in that there are many in-comers and it is easier for officials to intervene at camps in accessible riverside locations. Also, people can fish for a season, working as labourers to generate an income, without necessarily returning the following year. Nonetheless, many men are from families who have fished on the Ugalla River for decades, being descended from sleeping sickness evacuees, who may retain lively memories of life in a resettlement concentration. Other fishermen are refugees (some are fisherwomen), who live in nearby refugee camps, and who originated from neighbouring African countries; still others are immigrants from elsewhere in Tanzania.

In the interactions and disputes that arise between long-established fishermen and beekeepers and the more recent in-comers, tribal identity and the history of resettlement emerge in claims over the right to work in the area. Those who are long established allege that they have a 'true' right to fish or keep bees in Ugalla based on ties to the locality and from being of Galla origin. People also use the resettlement history to locate themselves within the genesis of Ugalla as a game reserve for conservation purposes. Indeed, past relations with representatives of the colonial administration are evoked in claims based on the argument that permission to fish soon after resettlement was an endorsement of special status and rights.

Claims over rights emerge when people feel that use of the reserve or the specific camps and sections of the river is threatened. For example, when refugees first started to fish at the river, it provoked 'Wagalla' fishermen to try to exclude them from access to the river. These refugees had very different and highly productive fishing skills and they alleged that existing fishermen were backward and used magic for fishing. This magic was seen as unacceptable to refugees who described themselves as modern, and likened existing fishing practices to hoe cultivation while they, the refugees, were used to using ploughs (Fisher, field diary 1994). In turn, existing fishermen alleged that they used the river in a traditional way that did not lead to over fishing and where traditional fishing leaders could control the activities that took place.

Despite periodic disputes between different groups of resource users, the manner in which beekeeping and fishing are institutionalized in Ugalla Game Reserve demonstrates the success with which people have managed to build up seasonal productive activities. This should not imply an absence of conflict between local people and the wildlife authorities, far from it. Periodic allegations are made by representatives of the state or international agencies that beekeepers or fishermen are being environmentally destructive, degrading the woodland, threatening wildlife, or over-fishing. Typically, short-term attempts to control or exclude people are made through burning camps, new permit systems, and restrictions on resource use. In actual fact, the legal status of fishing and beekeeping remains ambiguous.

The capacity of rural inhabitants – both long-standing inhabitants and newcomers – to carry out seasonal forms of productive activity in Ugalla is largely due to the success of local forms of organization in encompassing change. These forms of organization have roots in the way seasonal productive activities have historically developed as skills exclusive to certain groups of people (c.f. Abrahams 1967a; Roberts 1970; Unomah 1973; Cory n.d. [Mss.EA]). In more recent times (post-1950s), this organization has enabled people to establish formal co-operatives, which help them to market produce and to represent themselves as different collectivities who have livelihood interests in the Reserve. Local development associated with beekeeping and fishing from Ugalla is no mean feat, particularly if one considers that dried fish is marketed throughout the region and that honey and beeswax are sold to Europe through the fair trade market (Fisher 1997a, 2000).

In the way that people make their livelihoods in Ugalla, it can be said that the ex-resettlement area has been reconstructed in social terms. This is manifest through local knowledge, the existence of private property, recognized use rights to land and natural resources, and people's daily experience of living and working in Ugalla Game Reserve.

The fact that there is no permanent settlement or agriculture, and that the forest has regenerated over past settlement sites, has favoured a view of the area as 'natural' with conservation value. This provides an image, values and scientific justification for conservation policies. Ironically, lack of settlement, together with what is considered to be good quality woodland and water, also make the area particularly attractive to beekeepers and fishermen. In the case of both conservation and local livelihood activities, resettlement policies have generated unintended consequences in the present day. In the process of using the area for livelihood purposes, people's experiences of resettlement, their tribal associations to the territory, and long-term interaction with representatives of the state, have become reworked and been given different meanings over time. This has enabled them to maintain a claim to the area, despite restrictions on resource use.

Conclusion

This chapter has examined how forced resettlement from Ugalla in western Tanzania in the 1920s acted as a critical event, which generated a very different historical path for the area and people than might have been the case had the human population remained undisturbed. People were originally resettled as a public health measure; nonetheless, lack of human habitation coupled with regeneration of the forest and an increase in wildlife, opened up the possibility for wildlife conservation as a new form of land utilization. After Ugalla was gazetted as a protected area, conservation and restricted forms of local resource use, beekeeping and fishing, became closely linked. Thus, there was never a simple exclusion of all local livelihood activities, and certain groups of people have maintained access to natural resources, even though the nature of the human activities and of people's presence in the area has radically changed over time.

In viewing forced resettlement as a critical event, we have seen how tradition has reflexively been used to generate counter-tendencies to dominant forms of land use and administration in order for people to make claims, to maintain associations to their purported homeland, and to gain access to valued natural resources. This particular resettlement was, and continues to be, a social process that did not simply finish when people were moved away from their land. People developed different connections to the area. These led to the establishment of beekeeping and fishing as the two legitimate forms of local resource use in the area.[33]

Analysis of resettlement over a long time perspective has provided a window which has enabled us to see how different actors situate themselves in relation to one another and how they respond to the critical event and its outcomes. It may be the case that no single group controls these outcomes; and, as a consequence, people continue to negotiate access and rights to the place and its natural resources. These processes of negotiation may or may not generate conflict at a given time. The consequences of resettlement in Ugalla have cut across history in complex and discontinuous relationships, linking environment, people and politics in ways that are not typically captured by technical experts seeking to resettle people or to conserve the environment. It is to this social character of resettlement schemes, and the linkages between conservation and resettlement in a part of East Africa, which this chapter has tried to call attention.

Notes

1 I would like to extend my thanks to Alberto Arce and John Fisher for reading and commenting on earlier drafts of the text.

2 The term 'displacement' is used to refer to the removal of people from an area of land, while 'resettlement' refers to the planned relocation of people to another area.

3 A more detailed account appears in Fisher, E. (1997) 'What Future for the *Shamba la Bibi*? Livelihoods and Local Resource Use in a Tanzanian Game Reserve'. Unpublished doctoral thesis, University of Hull, UK. Archival sources are the Tanzania National Archives (TNA), colonial administrative files held by the Wildlife Division of the Ministry of Tourism, Natural Resources and Environment in Dar es Salaam (WD), and colonial records kept in Rhodes House at Oxford University (RH), East Africana Section, Dar es Salaam University Library (EA).

4 For details of the Tabora-Ufipa epidemic see Maclean 1927a/b, 1929a/b, 1930 a/b/c; Swynnerton 1923–4; Fairbairn 1948.

5 'Resettlement concentration' is the term used in archival literature from this period. The idea was that habitations should be concentrated together within a single area in order to 'push back' the bush and therefore eliminate the presence of carriers of the sleeping sickness trypanosome, game and tsetse fly.

6 A comment in the Annual Report for Tabora sub-District (1926) suggests that the authorities did not themselves know how many people were moved: 'decided to evacuate the whole of affected area ... [Ugalla River] ... 3,000 – 5,000 removed – estimate at probably half' (TNA.mss 1733/9 969).

7 TNA.mss.21711, 21709, 11307, 11515, 21710, 21712, 31731.

8 Bagshawe, Provincial Commissioner, Western Province Annual Report, 1932 RH.mss.Afr.s.3059.

9 RH.mss.Afr.s.3059.

10 Bagshawe, Report on Native Affairs for the Western Province, 1935 (page not given). RH.Mss.Afr.279–306.

11 Bagshawe Personal Diary 9.10.34, RH.mss.Afr.s.279–306, Vol XVII; for a romanticized account see Carnochan and Adamson (1935, 1937).

12 Tabora sub-District Annual Report, 1926, TNA.mss.1733/20 (105); Bagshawe Papers, diary 1932–3/1934–5, RH.mss.Afr.s.279–360.

13 Bagshawe Diary 9.1.36; see also 4.9.33; 20.9.33; 16.11.35; 9.1.36; 5.11.36; 27.6.36, RH.mss.Afr.s.279–306. Annual Report for Western Province, 1934, RH.Mss.Afr.s.3059.

14 Maclean, Abstracts from Tanganyika Diary, RH.Mss.Afr.s.622; TNA.19931; Appendix D, TNA.1733/9 (69); TNA.21712.

15 Cameron, 1877; Bohm 1888; Reichard 1890; Blohm 1931, 1933 a/b; Tabora Provincial and District Books. See also Shorter 1968; Abrahams 1967b.

16 Western Province Annual Report, 1936, RH.mss.Afr.s.3059; Gillman 1933 (64), RH.mss.Afr.s.1175; Burtt 1936, RH.Mss.Afr.s.1263 (1/2); Potts 1925–1952, RH.Mss.1259.

17 Tabora sub-District Annual Report, 1926, TNA.mss.1733/20 (105); WD.mss.ugr letters 21.9.54/7.10.54/25.7.55/11.10.55.

18 TNA.mss.23892 (Volumes 1 and 2).

19 Tabora District Book (microfiche: R.H. n.d.). Also, TNA.mss.19931 vol. 1 and Government Notice 213, updated in the Wildlife Ordinance of 1940.

20 WD.mss.ugr. Letter from G.H. Swynnerton, Game Warden, Arusha, to the Honourable Member for Agriculture and Natural Resources, DSM, 2.4.53.

21 WD.mss.ugr. Letter from D.K. Thomas to the Game Warden in Arusha, 16.8.52 (also replies 9.11.53/ 16.9.52/ 17.9.52/ 26.9.52).

22 Letter from the Provincial Commissioner, Western Province, to the Game Warden, Arusha, 19.3.53.

23 WD.mss.ugr. Letter from D.K. Thomas, the Game Ranger of Tabora Range, to the Game Warden, Arusha, 2.10.52.

24 WD.mss.ugr. Letter from the District Commissioner, Mpanda to the Provincial Commissioner, Western Province, 3.12.53. WD.mss.ugr. Letter with minutes of a District Commissioner's Conference, Tabora, 28.1.54.

25 Government Notice No. 83, formalized in the 1958 Fauna Conservation Ordinance, 7th Schedule, No. 48. WD.mss.ugr. Letters from the Provincial Commissioner of Western Province to the District Commissioner, Mpanda, 14.7.53 and 22.7.53.

26 WD.mss.ugr: minutes of a District Commissioner's Conference, Provincial Commissioner of Western Province to the Game Warden, Arusha, 28.1.54.

27 WD.mss.ugr Letter from D.K. Thomas, Tabora Game Ranger to the Game Warden, Arusha, 31.6.54.

28 TNA.Mss.10599.

29 WD.mss.ugr. Letter from the Executive Officer, Tabora District Council to the Executive Officer, Mpanda District Council, 20.11.62. Also, letter from the Executive Officer, Mpanda District Council to the Executive Officer, Tabora District Council, 28.2.63.

30 WD.mss.ugr. Acting Chief Game Warden to the Permanent Secretary, Ministry of Agriculture, Forests and Wildlife, 1.6.64.

31 WD.mss.ugr. A.G. Juyawatu, the Game Warden to Mr Ntenga, the Senior Field Officer of the Beekeeping Section, Tabora, 29.12.64; Report from Mr Ntenga, the Senior Field Officer, Beekeeping Section, Tabora to the Principle Secretary for the Ministry of Agriculture, Forests and Wildlife; Ntenga, personal communications during 1993 and 1994; WD.mss.ugr. Letter from J.N. Kundaeli, the Principle Game Warden to the Regional Game Warden, 19.4.66.

32 WD.mss.ugr. Statement of the 18.8.66 by the Game Warden, Tabora.

33 Nonetheless, resettlement events which took place more than seventy-five years ago have not been forgotten. Even today, certain groups of people contest control over resources and the meanings associated with the place, using past linkages as the basis of their claims.

References

Abrahams, R. G. 1967a. 'The Peoples of Greater Unyamwezi, Tanzania'. In D. Forde (gen.ed.) *Ethnographic Survey of Africa*. London: International African Institute.

—— 1967b. *The Political Organisation of Unyamwezi*. Cambridge: Cambridge University Press.

AWSID (Admiralty War Staff Intelligence Division). 1916. *A Handbook of German East Africa*. London: Her Majesty's Government.

Bennett, N.R. (editor). 1970 [1895]. *Stanley's Dispatches to the New York Herald 1871–1872, 1874–1877*. Boston, USA: Boston University Press.

Blohm, W. 1931. *Die Nyamwezi, Land und Wirtschaft*. Hamburg: Friederichsen, De Gruyter & Co. M.B.H.

—— 1933a. *Die Nyamwezi: Gesellschaft und Weltbild*. Hamburg: Friederichsen, De Gruyter and Co.

—— 1933b. *Die Nyamwezi: Texte (Berichte und Mitteilungen in der Sprache der Eingeborenen)*. Hamburg: Friederichsen, De Gruyter & Co.

Bohm, R. 1881–1883 *Bericht Aus Kakoma. Mitt Eilungen der Afrikanischen Gessellschaft en Deutchland*. 3: 2–16.

Burton, R. 1860. *The Lake Regions of Central Africa*. 1st edition, Vols. 1 & 2. London: n.d.

Cameron, V.L. 1877. *Across Africa*. Volumes 1 & 2. Virtue & Co. Ltd (n.d.)

Carnochan, F.G., and Adamson, H.C. 1935. *The Empire of the Snakes*. London: Hutchinson & Co. Ltd.

—— 1937. *Out of Africa*. London: Cassell & Company Ltd.

Clyde, D.F. 1962. *History of the Medical Services in Tanganyika*. Dar es Salaam: Government Press.

Comaroff, J., and Comaroff, J.L. 1992. *Ethnography and the Historical Imagination*. Boulder, Colorado: Westview Press.

Cunnison, I. 1951. *History of the Luapula: An Essay on the Historical Notions of a Central African Tribe*. Cape Town, London, New York: Oxford University Press, the Rhodes Livingstone Institute.

Fairbairn, H. 1942. 'The Agricultural Problems Posed by Sleeping Sickness Settlements'. *The East African Agricultural Journal*, 9: 20–32.

—— 1944. *Sleeping Sickness in Tanganyika*. Dar es Salaam: Government Printer.

—— 1948 'Sleeping Sickness in Tanganyika Territory'. *Tropical Diseases Bulletin*, 45: 1–17.

Fisher, E. 1997a. 'Beekeepers in the Global "Fair Trade" Market'. *The International Journal of Sociology of Agriculture and Food*, 6: 109–60.

—— 1997b. 'What Future for the Shamba la Bibi? Livelihoods and Local Resource Use in a Tanzanian Game Reserve'. Unpublished PhD thesis: University of Hull, UK.

—— 2000. 'Forest Livelihoods: Beekeeping as Men's Work in Western Tanzania'. In C. Creighton and C.K. Omari (eds). *Gender, Family and Work in Tanzania*, pp. 138–76. Aldershot: Ashgate.

Fisher, E., and Arce, A. 2000. 'The Spectacle of Modernity: Blood, Microscopes and Mirrors in Colonial Tanganyika'. In A. Arce and N. Long (eds). *Anthropology, Development and Modernities: Exploring Discourses, Counter-tendencies and Violence*, pp. 74–99. London & New York: Routledge.

Glover, P. E. 1939. 'A Preliminary Report on the Comparative Ages of some Important East African Trees in Relation to their Habitats'. *South African Journal of Science*, 36: 316–27.

Hatchell, G.W. 1949. 'An Early Sleeping Sickness Settlement'. *Tanzania Notes and Records*, 27: 60–64.

—— 1956. 'Resettlement in Areas Reclaimed from Tsetse Fly'. *Tanzania Notes and Records*, 53: 243–446.

Heelas, P., Lash, S. and Morris, P. 1996. *Detraditionalization: Critical Reflections on Authority and Identity*. Oxford: Blackwell.

Iliffe, J. 1979. *A Modern History of Tanganyika*. Cambridge: Cambridge University Press.

Jackson, C. H. N. 1936. 'Some New Methods in the Study of Glossina Morsitans'. *Proceedings of the Zoological Society, London* 1936: 15–20.

Long, N. 1997. 'Agency and Constraint, Perceptions and Practices. A Theoretical Position'. In H. de Haan and N. Long (eds). *Images and Realities of Rural Life: Wageningen Perspectives on Rural Transformations*, pp. 1–20. Assen, The Netherlands: Van Gorcum.

Maclean, G. 1927a. 'Tanganyika Territory Annual Medical and Sanitary Report for 1925'. *Tropical Diseases Bulletin*, 24: 563.

—— 1927b. 'Tanganyika Territory Medical and Sanitary Report for 1926'. *Tropical Diseases Bulletin*, 24: 954–6.

—— 1929a. 'The Relationship between Economic Development and Rhodesian Sleeping Sickness in Tanganyika Territory'. *Annals of Tropical Medicine and Sleeping Sickness*, 23: 37–46.

—— 1929b. 'Tanganyika Territory Annual Medical and Sanitary Report for 1927'. *Tropical Diseases Bulletin*, 29: 695–6.

—— 1930a. 'Sleeping Sickness Measures in Tanganyika Territory'. *Kenya and East African Medical Journal*, 3: 120–26.

—— 1930b. 'Tanganyika Territory Annual Medical and Sanitary Report for 1928'. *Tropical Diseases Bulletin*, 27: 809.

—— 1930c. 'Tanganyika Territory Annual Medical and Sanitary Report for 1929'. *Tropical Diseases Bulletin*, 26: 696.

—— 1931. 'Tanganyika Territory Annual Medical and Sanitary Report for 1930'. *Tropical Diseases Bulletin*, 27: 809.

—— 1932. 'Tanganyika Territory Annual Medical and Sanitary Report for 1931'. *Tropical Diseases Bulletin*, 28: 342.

—— 1933. *Memorandum on Sleeping Sickness Measures*. Pamphlet no. 8. Dar-es-Salaam, Tanzania: Tanganyika Territory Medical Department, Government Printer.

Mblinyi, M. 1989. 'Women's Resistance in "Customary" Marriage: Tanzania's Runaway Wives'. In A. Zegeye and S. Ishemo (eds). *Forced Labour and Migration: Patterns of Movement within Africa*, pp. 61–72. London: Hans Zell Publishers.

McCracken, J. 1987. 'Colonialism, Capitalism and the Ecological Crisis'. In D. Anderson and R. Grove (eds). *Conservation in Africa: People, Policy and Practices*, pp. 63–78. Cambridge: Cambridge University Press.

Milne, G. 1936. 'Report on a Soil Reconnaissance Journey in Parts of Tanga, Central, Eastern, Lake and Western Provinces of Tanganyika Territory'. Dar es Salaam: Unpublished, N.

Moffett, J.P. 1939–1940. 'A Strategic Retreat from Tsetse Fly: Uyowa and Bugomba Concentrations 1937'. *Tanzania Notes and Records*, 7–10: 2–5.

—— 1958. *Handbook of Tanganyika*. Dar-es-Salaam, Tanzania: Government Printer.

PAWM. 1992. Planning for Priority Game Reserves: Status Reports for Mkomazi, Maswa, Ugalla, Rungwa-Kizigo and Uwanda, Planning and Assessment for Wildlife Management, Department of Wildlife, Ministry of Tourism, Natural Resources and Environment. Unpublished, Dar-es-Salaam, Tanzania.

—— 1994. Ugalla Game Reserve Management Plan, Planning and Assessment for Wildlife Management, Wildlife Division, Ministry of Tourism, Natural Resources and Environment. Unpublished, Dar-es-Salaam, Tanzania.

Phillips, J. 1931. 'A Sketch of the Floral Regions of Tanganyika Territory'. *Transactions of the Royal Society of South Africa*, XIX: 59–70.

Potts, W. H. 1937. 'The Distribution of Tsetse Flies in Tanganyika Territory'. *Bulletin of Entomological Research*, 43: 365–74.

Ranger, T. O. 1969. 'The Movement of Ideas 1850–1939'. In I.N. Kimambo and A.J. Temu (eds). *A History of Tanzania*, pp. 161–88. Nairobi, Kenya: Heinemann, Educational Books.

Ranger, T. 1983. 'The Invention of Tradition in Colonial Africa'. In E. Hobsbawm and T. Ranger (eds). *The Invention of Tradition*, pp. 211–62. Cambridge: Cambridge University Press.

Reichard, P. 1890. 'Meine Erwerbung des Landes Uganda'. *Deutsche Kolonialzeitung*, 7: 77–9.

—— 1892. *Deutsch Ostafrika: das Land und seine Bewohner*. Leipzig: publisher not documented.

Roberts, A. D. 1968. 'The Nyamwezi'. In A. Roberts (ed.). *Tanzania before 1900*, pp. 117–53. Dar-es-Salaam, Tanzania: East Africa Publishing House for the Historical Association of Tanzania.

—— 1970 'Nyamwezi Trade'. In R. Gray and D. Birmingham (eds). *Pre-Colonial African Trade: Essays on Trade in Central and East Africa before 1888*, pp. 39–74. Oxford: Oxford University Press.

Rodger, B. 1954. Unyamwezi Development Report. Dar Es Salaam: Unpublished.

Rodgers, W. A. 1980. Wildlife and Natural Resource Conservation Planning. Department of Zoology, University of Dar-es-Salaam, Tanzania (unpublished): Unpublished Report for Tabora Rural Integrated Development Programme (UK Overseas Development Administration).

—— 1982 'An Ecological Survey of Ugalla Game Reserve'. *Tanzania Notes and Records*: 36: 1.

Rodgers, W. A., and Nicholson, B.D. 1973. 'National Projects – Wildlife Development and Managment Priorities in Tanzania'. Dar-es-Salaam, Tanzania: Mimeo Report, Game Division.

Sayers, G. F. 1930. *The Handbook of Tanganyika*. London, UK: MacMillan and Co. Ltd.

Shorter, A. 1972. *The Kimbu*. Oxford: Oxford University Press.

Smith, F. G. 1994. *Three Cells of Honey Comb*. Western Australia (Nedlands): Private Publication.

Swynnerton, C.F.M. 1923–4. The Relationship of some East Africa Tsetse Flies to the Flora and Fauna'. *Transactions of the Royal Tropical Medicine and Hygiene*, 17, 128–41.

Thomas, D. K. 1961. 'The Ugalla Game Controlled Area'. *Tanganyika Notes and Records*, 57: 227–30.

Unomah, A. C. 1973. 'Economic Expansion and Political Change in Unyanyembe, c. 1840–1900'. Ph.D Thesis, University of Ibadan, Nigeria.

Archival References

Tanzania National Archives, Dar Es Salaam (TNA)

2551. Report of the District Political Officer, Tabora to the Secretary of the Administration, Dar Es Salaam, 1919.

42398/2. Annual Veterinary Report for Tanganyika Territory, 1920.

42398/2. Unyamwezi Development Report, by Rodger, J.T.R.C., 1954.

31351. Compulsory Resettlement of Africans: General, n.d.

11825. vol.1. Swynnerton, C.F.M. n.d.

7538, vol.2. The Game Experiment, Shinyanga, Jackson, 11.5.46.

21261. Tanganyika Territory Legislative Council, the Sleeping Sickness Problem in the Western and Lake Provinces, and in Relation to Uganda, 1933.

11307. Sleeping Sickness: General Principles of Treatment, Quarantines, 1927–1933.

11515. Sleeping Sickness Outbreak Western Province, Volumes 1, 2, 3, 1927–1932.

21709. Sleeping Sickness Concentration Committee: Minutes of Meetings.

21710. Sleeping Sickness Concentrations: Medical Arrangements Concerning, n.d.

21711. Sleeping Sickness Concentrations: Educational Arrangements Concerning, n.d.

21712. Sleeping Sickness Concentrations: Economic and Development Concerning, n.d.

31731. Resettlement as a Preventative Measure Against Sleeping Sickness.

10599. General Correspondence on Sleeping Sickness, n.d. Volumes 1 & 2.

40876. Tsetse Operations in Western Province, n.d.

11551 Game Reserves Volumes 1 and 2.

19931 Volume One, Game Reserves in Connection with Tsetse Research and Anti-Sleeping Sickness Measures.

10599 Labour Arrangements.

1733/4 (52), Annual Report for Tabora District, 1925.

1733/9 (69), Annual Report for Tabora District, 1926.

1733/20 (105), Tabora sub-District Annual Report, 1926.

1733/-(5). Tabora District Annual Report 1921.

21475. Annual Report for Western Province, 1932.

Maclean, G., 1929: 122; Appendix D, Report on Sleeping Sickness, Tabora District Annual Report, 1926, TNA.Mss.1733/9 (69).

Letter from Hatchell, District Officer Ufipa to the Provincial Commissioner, Mwanza 27.8.29. in 11771 Volume 2.

TNA Mss.4283982 Administrative Arrangements.

Tanzania Wildlife Department Archives for Ugalla Game Reserve (WD)

Letter from the Director of Veterinary Services to the Honourable Member for Agriculture and Natural Resources, 19.7.1949.

Letter from the Director of Veterinary Services to the Honourable Member for Agriculture and Natural Resources, 19.7.49 (TNA.Mss.11234).

Letter from the Provincial Commissioner, Western Province to the Game Warden, Arusha, 16.9.52.

Letter from D.K. Thomas, the Game Ranger of Tabora Range to the Game Warden, Arusha, 2.10.52.

Letter from the Director of Veterinary Services to the Game Warden, Tenguru, 31.10.52.

Letter from D.K. Thomas to the Game Warden in Arusha, 16.8.52 (also 16.8.52/ 16.9.52/ 17.9.52/ 26.9.52/9.11.53).

Letter from the Acting Director of Veterinary Services to the Game Warden, Tenguru, 5.1.53.

Letter from D.K. Thomas, the Game Ranger for Tabora Range to the Game Warden, Arusha, 30.1.53.

Letter from the Provincial Commissioner, Western Province to the Game Warden, Arusha, 19.3.53.

Letter from G.H. Swynnerton, Game Warden, Arusha, to the Provincial Commissioner, Tabora, 2.4.53.

Letter from G.H. Swynnerton, Game Warden, Arusha, to the Honourable Member for Agriculture and Natural Resources, DSM, 2.4.53.

Letter from the Honourable Member for Agriculture and Natural Resources to the Game Warden, Arusha, 4.5.53.

Letter from the District Commissioner, Mpanda to the Provincial Commissioner, Western Province, 3.12.53.

Letter with minutes of a District Commissioners Conference at which the Game Ranger, Tabora was invited to attend, Provincial Commissioner of Western Province to the Game Warden, Arusha, 28.1.54.

Letter from D.K. Thomas, Tabora Game Ranger to the Game Warden, 31.6.54.

Letter from the Provincial Commissioner of Western Province to the District Commissioner of Mpanda, 14.7.53.

Letter from the Provincial Officer to Member for Local Government (n.d.(1960s) TNA Mss.4283982).

Ker & Downey Safaris Ltd. Letters 21.9.54; 7.10.54; 25.7.55; 11.10.55.

Letter from the Executive Officer, Tabora District Council to the Executive Officer, Mpanda District Council, 20.11.62.

Letter from the Executive Officer, Mpanda District Council to the Executive Officer, Tabora District Council, 28.2.63.

Letter from the Executive Officer, Tabora District Council to the Executive Officer, Mpanda District Council, 20.11.62.

Letter from the Executive Officer, Mpanda District Council to the Executive Officer, Tabora District Council, 28.2.63.

Acting Chief Game Warden to the Permanent Secretary, Ministry of Agriculture, Forests and Wildlife, 1.6.64.

Letter from A.G. Juyawatu, the Game Warden to Mr Ntenga, the Senior Field Officer of the Beekeeping Section, Tabora, 29.12.64

Report from Mr Ntenga, the Senior Field Officer, Beekeeping Section, Tabora to the Principle Secretary for the Ministry of Agriculture, Forests and Wildlife, n.d. (also personal communications to Fisher during 1993 and 1994).

Letter from J.N. Kundaeli, the Principle Game Warden to the Regional Game Warden, 19.4.66.

Statement of the 18.8.66 by the Game Warden, Tabora.

East Africana Section, University Library, Dar Es Salaam (EA)

Cory, H. n.d. The Initiation and Organisation of some of the Sukuma Nyamwezi Secret Societies, Mss.105.

Rhodes House, Queen Elizabeth House, Oxford (RH)

Bagshawe, F. J. E., 1932–1933 Private Diary, RH.Mss.Afr.s.279–360. Volume XIX, November 10th 1932 to December 6th 1933.

—— 1934–1935 Private Diary, RH.Mss.Afr.s.279–306. Volume XVI October 9th 1934 to December 31st 1935.

—— 1936a Private Diary, RH.Mss.Afr.s.279–306. Volume XVII January 1st 1936 to July 22nd 1936.

—— 1936b Private Diary, RH.Mss.Afr.s.279–306. Volume XVIII July 23rd 1936 to December 31st 1936.

—— 1937a Private Diary, RH.Mss.Afr.s.279–306. Volume XIX January 1st 1937 to August 21st 1937.

—— 1937b Private Diary, RH.Mss.Afr.s.279–306. Volume XX August 22nd 1937 to December 31st 1937.

Annual Provincial Reports for the Western Province, 1932, 1934, 1936, Mss.Afr.s.3059.

Report on Native Affairs for the Western Province, 1935. Mss.Afr.s.3059.

Photographs, Urambo Sleeping Sickness Concentration 1936 and Uyowa and Bugoma Concentrations 1937. Mss.Afr.s.3059.

Burtt, B. 1936 An Enumeration of the Principal Vegetation Types, Unpublished 1932–1936, Mss.Afr.s.1263(1/2).

Gillman, C., 1923–1925 Private Diary (64), RH.Mss.Afr.s.1175. Volume VII February 15th 1923 to June 28th. 20.2.34, 1.3..34, 1.4.34. Also 24–25.5.34, and Gillman's notes and views on a Tabora Reclamation (Tsetse) Conference, 1937, RH.Mss.Afr.s.1175, Vols. VII, X, XI, XII, XIV.

—— 1933a Geographical Notes on Tanganyika Territory, RH.Mss.Afr.s.1175. Volume XI, Notes on a Rapid Transverse, August 7th and 8th 1933.

—— 1933–1936b Private Diary, RH.Mss.Afr.s.1175. Volume XII, April 30th to June 10th 1936.

—— 1937–1940 Private Diary, RH.Mss.Afr.s.1175. Volume XIV, February 1st 1937 to November 30th 1940.

—— 1933 1175 Volume XX

Hutchins, E. E. Morogoro District, Vol.1, Part A, Sheets 25–26, August 1931, film no.MF15.

Johnstone, J. R. (administrator in western Tanzania) 1933–1940 Bits and Pieces, or, Seven Years in the Western Province of Tanganyika Territory.

Longland, F. 1927 Tabora Provincial Book, Reel No. 19 & Tabora District Book, Reel No. 20.

Maclean, G. 1926–61a Abstracts from Tanganyika Diary, RH.Mss.Afr.s.622.

Potts, W. H. 1925–1952 Tsetse Research in Tanganyika Territory, RH.Mss.1259.

Swynnerton, C.F.M. RH.Mss.Afr.s.1897. Tsetse Research in Tanganyika 1919–1938.

8

The Influence of Forced Removals and Land Restitution on Conservation in South Africa

Christo Fabricius and Chris de Wet

Introduction

South Africa is one of the few countries where the forced removal of people from protected areas has been followed by land restitution. This chapter addresses the impact of past forced removals and subsequent land restitution on biodiversity conservation and the sustainable use of natural resources in South Africa. It focuses on protected areas from which people were removed and on unconserved land, on which people were settled. It questions some of the assumptions about the positive and negative impacts of land restitution on conservation, and highlights emerging challenges to conservationists and land beneficiaries.

It is widely accepted that the policies of pre-democratic South Africa resulted in the extensive removal of over 3.5 million people to divide the country geographically along racial and ethnic lines (Surplus People Project 1985). As part of this ideology of centralized control, people were also removed from land containing rich biodiversity resources in order to incorporate such land into provincial nature reserves or national protected areas.

More recently, a process has begun to return land to its rightful occupants since the democratically elected government came into power in 1994 and embarked on a land reform programme. This set the scene for a slow process of land restitution, whereby displaced people could reclaim land from which they had been forcibly removed after 1913, the year in which the Native Land Act was passed which divided South Africa into separate, racially based territorial areas. A Land Claims Court was established to deal with claims, with non-government organizations such as the National Land Committee and its affiliates (with strong links to the Department of Land Affairs) playing a prominent role in helping communities advance their claims. Amongst the first claims to be settled were those to state-owned land, including protected areas from which people had been removed.

The displacement and restitution processes have had far-reaching socio-economic effects and have also affected conservation in all its facets. The conservation community in South Africa feared that restitution in National Parks and nature reserves would signal the demise of protected areas, as they knew them. One of the first claims to the conservation estate was to a portion of the Kruger National Park (KNP) at Pafuri, from which the Makuleke community had been forcibly removed in 1969. Most of their land was incorporated into the KNP and the Madimbo Corridor, a military cordon along the Zimbabwean border. A senior conservation official made a telling media statement after the lodging of the Makuleke land claim:

> 'The claim ... could set a precedent that will tear Kruger (National Park) apart ... If the Makuleke claim is upheld in respect of land inside the Park, all conservation areas will be under threat; Conservation status will not be worth the paper it is written on' (quoted in Steenkamp 1999).

Rumours spread that the country's conservation heritage would be sacrificed to satisfy the land hunger of previously disadvantaged people.

The most prominent land claims linked to protected areas have been: a) the Makuleke land claim (Northern Province); b) the San claim, to a part of the Kalahari Gemsbok National Park[1] (Northern Cape Province); c) the Riemvasmaak claim, to a section of Aughrabies National Park (Northern Cape Province); d) a claim to part of St. Lucia National Park[2] (KwaZulu Natal); e) Ngome Forest (KwaZulu-Natal); and claims to f) Dwesa, g) Cwebe and h) Mkambati Nature Reserves[3] in Eastern Cape Province. Additional protected areas which may be affected by similar claims include the Great Fish River Reserve Complex (Eastern Cape Province), Madikwe Game Reserve (North-West Province), and Blyde Canyon (Mpumalanga Province). Information about these latter cases is largely undocumented and the likely impacts on sustainable resource use and people's quality of life vary markedly and are difficult to predict.

People forcibly relocated to make way for conservation in South Africa shared a number of characteristics. Poor, with limited formal education, they lived in underdeveloped parts of the country with weak formal infrastructure. These small communities therefore lacked political leverage. The areas they occupied were attractive to conservationists, being rich in biodiversity. The inhabitants made extensive use of natural capital (wild plants and animals) in their livelihood strategies, possessing a well-developed ecological knowledge because of their dependence on natural resources. The removals had a profound influence on their lifestyles and well-being. The lost value of grazing land and crops alone at Lake St. Lucia was estimated at R9600 (approximately £960 in December 1999) per household per year (Forse 1998).

Displacement generally left the relocated people worse off in every respect (Surplus People Project 1985; Ramphele 1991; de Wet 1995). Often compensated inadequately or not at all, they were further impoverished due to diminished access to natural resources in the areas to which they had been moved. Cernea (1997) mentions loss of access to common property resources as one of the principal risk-factors contributing to socio-economic impoverishment in resettlement projects worldwide. Such impoverishment is even more pronounced when people have to move from resource rich areas such as those targeted for conservation. In South Africa, very little infrastructural development took place in the resettlement areas. Many communities became geographically fragmented during the process as they were moved to several different areas. They became more centrally and politically controlled by the apartheid government, since their resettlement was accompanied by an erosion of local institutions as the government used the Bantu Authorities system to tighten its control over rural politics. Settlement areas were more densely populated, less productive and poorer in biodiversity than the land from which people were removed (Fabricius and Burger 1997). Against this backdrop, the first part of this chapter explores the impact of people's removal from protected areas on the biological, social and institutional facets of conservation.

Did Forced Removals have a Positive Impact on Conservation?

The official rationale for these forced removals was that they were essential for the establishment of a representative and ecologically viable network of protected areas (Wynberg and Kepe 1999). Possible positive impacts of forced removals on conservation are a) the expansion of the country's conservation estate; and b) reduced consumptive use and decreased land deterioration, especially in high-biodiversity areas from which people were relocated.

No evidence of negative impacts on conservation prior to people's relocation could however be found in any of the cases studied, and no lasting impacts are evident. Although this does not constitute irrefutable proof that people had no negative impacts prior to relocation, many of the areas from where people were removed are currently acclaimed as biodiversity-rich localities or 'biodiversity hot-spots'. For example, the Pafuri area in the northern parts of Kruger National Park, from where 2000 people were removed in 1969, contains 75 per cent of the Park's biodiversity (Steenkamp 1999), and the grasslands and forests at Dwesa and Cwebe on the Wild Coast are regarded as areas of endemism and high species richness (Lieberman 1997).

The evidence suggests that removals from conservation areas took place in terms of wider governmental considerations relating to the administration and control of rural areas, rather than principally for conservation-specific reasons. The wave of forced removals from protected areas between 1966 and 1975 coincided with a series of 'black spot' removals during the heyday of apartheid. Forced removals also coincided with a strengthening of the popular belief amongst conservationists that people and conservation are incompatible (cf Spinage 1998). This belief was particularly strong in relation to communal land and was fuelled by Hardin's (1968) 'Tragedy of the Commons' essay. This paradigm is now widely believed to be flawed, as Hardin was referring to 'open access' resources, i.e. a specific type of common property resource that is exploited in the absence of local institutions, sanctions and monitoring (Oström 1990).

The Negative Conservation Impacts of Forced Removals

The main negative conservation impacts of forced removals from protected areas are that they contributed to unsustainable resource use outside protected areas, because of increased pressure on natural resources in areas already degraded due to over-population. People's expulsion from biodiversity-rich areas led to their attitudes to conservation and conservationists becoming increasingly negative, with a measurable increase in poaching and unprecedented incidents of natural resources being vandalized, often accompanied by land invasions. There was very little if any support for conservation and even less respect for conservationists in rural areas occupied by black South Africans. Forced removals also contributed to a loss of traditional ecological knowledge (see below).

Unsustainable Resource Use outside Protected Areas

People inhabiting protected areas were invariably removed onto communal land that was already densely populated, placing additional pressure on the natural resource base outside protected areas. At Dwesa, the removals went hand in hand with a gradual clamp-down on the harvesting of forest and marine resources in the Reserve, with negative rather than positive results for conservation (see next section). This put intense pressure on the small pockets of forests outside the Reserve and on the limited mussel beds adjacent to it. At Cwebe, people had historically preferred to use the productive grasslands inside the protected area closer to the coast for grazing. After displacement, larger areas of poorer quality grasslands were required to sustain livestock (Fay 1999). At St. Lucia, displaced people started moving into the Dukuduku State Forest that was more fragile than the grasslands of the Eastern Shores area in Lake St. Lucia National Park from where they had previously been removed (Barker 1997). At the Kalahari Gemsbok National Park, the removed San people joined a 'coloured' community in an informal settlement at Mier and started farming with livestock (Chennels 1998). Overgrazing resulted in shifting sand dunes and loss of species richness on the unstable soils (Van Rooyen 1999).

Negative Attitudes towards Conservation and Conservationists

The historical and contemporary association of conservation with injustice and suffering because of relocation has had lasting consequences. At Dwesa and Cwebe Nature Reserves, people occupying the Reserves were removed in the 1920s and 1930s but had controlled access to forest resources after removal. In 1981, Transkei Nature Conservation took over administration of the Reserves and a total embargo was placed on all forms of natural resource use (including seawater and sand) inside the protected areas (Timmermans 1999). Neighbouring communities became involved in protracted conflicts with conservationists. Forest products were illegally harvested at greater intensities than before, illegal grazing became a regular occurrence, and communication between conservationists and all neighbouring people (not only those who were displaced) broke down. Law enforcement and extension initiatives on communal land outside the protected areas had little effect, and the unprotected forests outside the Reserves soon became degraded.

People's frustration boiled over in 1994. In an act of protest, they invaded the protected areas in large numbers and wantonly cut down trees, set the grasslands alight and extensively vandalized the inter-tidal shellfish beds while conservationists helplessly looked on. Poaching of wildlife increased dramatically and red hartebeest became locally extinct while the blesbok herd was reduced from 300 to four animals (Timmermans 1999). Spotted genet, African wild cat and blue duiker skins are readily available in *muti* (traditional medicine) shops in the neighbouring town of Willowvale and it is widely accepted that they originate from Dwesa (C. Fabricius, pers. obs.).

At Mkambati, protest has taken a less visible form. Local youths would hunt wildlife with dogs and fire-arms at night while others would set wire snares, knowing that being apprehended would result in arrest or physical abuse. This practice, known as *ukujola*, is locally interpreted to mean 'taking by stealth or cunning that which is rightfully yours'. Commercial poachers from other villages further afield are discouraged by the villagers because of their impact on local wildlife resources, but are not reported to the authorities. Burning is used to attract wildlife closer to the perimeter fence, which is countered by burning by conservationists to draw wildlife towards the centre of the Reserve. This burning and counter-burning has caused a reduction in species diversity in the grasslands, with certain grass species disappearing temporarily because of continuous firing (Kepe et al. 1999). Reports about the impact of hunting at Mkambati vary: there are unverified reports of an entire herd of blesbok being driven over a cliff in an act of vandalism (senior nature conservation official, pers. comm.), while other reports claim that wildlife has shown a steady increase between 1982 and 1987, despite local hunting (Kepe et al. 1999).

At St. Lucia, relocated people settled with others in the Dukuduku forest and refused to move. This community has subsequently grown and is a major threat to the conservation of the forest, and has so far aggressively resisted efforts to relocate them. When the government refused to grant a mining permit for a large titanium mine inside St. Lucia National Park, the resettled people generally expressed themselves in favour of the mine rather than of conservation (Barker 1997). Elsewhere, at Ngome forest in KwaZulu-Natal, people who were relocated from the Ntendeka Wilderness Area in 1966 were compensated in 1997 by the purchasing of four agricultural farms with taxpayers' money for their use. Nine hundred people nevertheless invaded the wilderness area in November 1997, damaging the ecosystem and causing a major problem for the new government, as the Minister of Water Affairs and Forestry had to apply for a court order to evict them. These people are now part of the community neighbouring Ntendeka (Barker 1998) and are bound to impact negatively on the conservation of the wilderness area in future.

Such negative attitudes of resettled people towards conservation, and stereotyping of conservationists by local communities and *vice versa*, will not be reversed overnight. The development of relations of trust is an important challenge to beneficiary communities and conservationists.

Loss of Knowledge and Social Capital

Much traditional ecological knowledge has been lost because of forced removals, as people (especially the younger generation) became detached from natural resources and spent even more time away from rural areas as migrant labourers. Local institutions became eroded because of increased central political control, and many communities became fragmented in the process of being relocated to different corners of Southern Africa. The Riemvasmaak community, which was forcibly removed from an area that was part of Aughrabies National Park, was split into three groups. One group was relocated to Welcomewood in the former Ciskei homeland and another to Khorixas in Namibia, while the third group remained within the Northern Cape Province (Lund 1998).

The Effects of Land Restitution on Conservation

Despite widespread fears that land restitution would signal the end of many of South Africa's parks and nature reserves, land restitution has had a decidedly positive impact on conservation. Such positive impacts relate to the expansion of conservation areas, the emergence of new conservation models, the unfolding of new tourism opportunities, improved relations between conservationists and local stakeholders, changes in attitudes, policies and approaches by conservation professionals and institutions, the emergence of new institutions and strengthening of existing ones, and the revival of traditional ecological knowledge and the development of complementary skills in communities.

However, as is discussed later, there is a real possibility that these gains may not be sustainable in future.

Expansion of Conservation Areas

One of the surprising features of land restitution in protected areas is that, in almost all instances, relocated people chose to remain in the resettlement areas. This made it possible for the land that had historically belonged to the relocated people to remain formally conserved. The reasons for this include mediation by NGOs, compromises by government departments seeking for politically expedient solutions, and communities being offered attractive incentives not to reoccupy their land. Such incentives included compensation, attractive lease agreements which gave them immediate and on-going access to capital, infrastructure development and training. Facilitators and conservation bodies also made (as yet unfulfilled) promises of new wealth from tourism, but economic opportunities might yet emerge as developers begin to approach resettled communities with tourism development proposals.

There were also important disincentives to move. People have developed new roots and social linkages in the resettled areas and appear to be unwilling to be resettled once more, while many of the communities consist of predominantly old-age pensioners and children because of the migration of able-bodied men and women in search of jobs. Hardships were also anticipated in protected areas, for example lack of infrastructure, and the problem of dangerous animals (Fabricius 1999; Fabricius, Koch and Magome in press).

One of the important catalysts for constructive change was the transformation of conservation management structures. This resulted in new policies that forced conservation field staff and middle management to negotiate and acknowledge the rights of communities. For example, the South African National Parks (SANP)[4] 'transformation mission' reads as follows: 'The transformation mission of the South African National Parks is to transform an established system for managing the natural environment to one which encompasses cultural resources, and which engages all sections of the community' (http://www.parks-sa.co.za/). The media 'hype' around restitution in protected areas has also inspired all parties to behave in a reconciliatory manner.

Many protected areas were in fact expanded during the land restitution process, by incorporating 'unused' state land such as military areas. The Kruger National Park was increased by 3800 hectares when the Madimbo Corridor, a former military area of high conservation value, was included in the new Makuleke area of the Park (Steenkamp 1999). More significantly, the agreement enhanced the prospects of establishing a transfrontier Park between South Africa, Mozambique and Zimbabwe. The boundary of Cwebe Nature Reserve was redefined during the negotiating phase to include a strip of coastal land occupied by holiday cottages (Fay 1999). At the Kalahari Gemsbok National Park, an additional 25,000 hectares of undeveloped and partly-degraded commercial farmland adjacent to the Park is being claimed by the !Khomani San, who intend establishing a wildlife ranch on the land.

Although the Mier area will not be contractually incorporated into the Park, the proposed wildlife ranching enterprise will be more compatible with conservation than current livestock ranching. Gemsbok and springbok (arid-adapted antelopes) are less prone to predation by wild carnivores, thereby reducing the likelihood of conflict, and move farther away from drinking points than sheep, thus having a more benign impact on the vegetation. The land being claimed inside the Park will be used for traditional and cultural purposes only, and will remain an integral part of it (Chennels 1998; SANP press release, 9 April 1999).

The only known case where there is a significant probability of a net loss to conservation because of land restitution is at Riemvasmaak, from which communities were removed in the mid-1970s. After the removals, part of Riemvasmaak was used as a military training area, while part of it was added to the Aughrabies National Park. A Trust was elected in 1994 and negotiated on behalf of the community. The land, with the exception of the approximately 4000 ha incorporated in the Park, was returned to its original occupants in 1995 (Lund 1998). The SANP's vision was that the restituted land should be jointly managed as a conservation area. Most of the claimants are however against conservation as a land use option, prefer to farm with livestock, and are not interested in tourism and conservation. Alternative land is currently being sought for the community to farm on (joint press release issued by the SANP and Riemvasmaak Trust, 4 February 1998). Until such time, the future of Riemvasmaak as a conservation area is uncertain and negotiations around joint management or alternative land use have recently broken down.

The new agreements resulted in apparent win–win solutions for communities and conservationists but whereas the 'win' for conservationists is clear and measurable, the 'win' for communities is less clear. In particular, it is unclear how on-going arbitration and monitoring will take place to ensure that all parties are complying with formal agreements. It is also uncertain whether the emerging community institutions will be capable of managing earned revenue and how the inevitable intra-community conflict, stemming from newly-found 'wealth', will be managed. Another unknown is whether conservation field staff will continue to comply with new conservation policies once the land restitution process attracts less media and other attention.

Emergence of New Conservation and Land Tenure Models

New options and models have emerged as part of the land restitution process. Lease agreements have been entered into, whereby land which has been handed back to communities on condition that its conservation status does not change, will be rented from them by the State. This is the current solution at Dwesa (Timmermans 1999), while the Makuleke portion of Kruger National Park has been reclassified as a Contractual Park, co-managed by SANP and the Makuleke (Steenkamp 1999).

Many of the claims have culminated in collaborative management agreements, whereby communities (in theory) accept shared responsibility for the management of protected areas (e.g. at Kruger, St. Lucia, and Dwesa). In some cases it has been agreed that limited portions of conservation land will be used for commercial tourism by the beneficiaries, and this is in most instances combined with joint (i.e. equal partner) management (Chennels 1998; Forse 1998; Steenkamp 1999; Timmermans 1999).

It has to be noted that these models are a result of compromise on the side of both parties, but more so on the side of communities. In all instances one of the conditions of the agreement is that the conservation status of the land may not change, and communities are in many cases not allowed to use the resources consumptively, e.g. by hunting them. Proposals by the Makuleke community to hunt elephant in the contractual Makuleke region of the Kruger National Park are being resisted by some conservationists (Macleod 2000). Although the SANP initially opposed the proposal, it seems to have shifted its position, arguing that if a management plan for the section of the Kruger Park that has been restituted to the Makuleke includes hunting as a form of resource utilization, then hunting in the contractual park will be allowed (SANP press release, 25 January 2000).

Some of the private wildlife ranchers adjacent to Kruger are however being permitted to profit from elephant hunting, albeit on a limited basis. Formal agreements are typically reviewed after 25 years, and the Makuleke might very well negotiate for a hunting permit when the current lease expires. Whether or not communities will be agreeable to renewing the agreements at the end of this period will depend on whether a) their expectations have been met, and b) relationships of trust have developed between them, conservation professionals and private sector partners.

New Tourism Opportunities Unfolding

In most of the cases studied, tourism has been held up as an important source of jobs and revenue. The unfolding tourism ventures capitalize on opportunities to combine both cultural and nature tourism in the same Park: e.g. a San cultural village as well as guided trails with Bushmen/San as interpreters have been proposed for the Kalahari Gemsbok National Park (Chennels 1998). It has also become clear that international tourists are willing to pay for conservation by paying above-average prices for experiences that will contribute to the preservation of particularly large mammals such as the 'big five' (elephant, rhinoceros, buffalo, lion and leopard). The Makuleke community plans to establish an up-market tourism lodge in the Pafuri section of Kruger National Park (Steenkamp 1999). It is predicted that the benefits to conservation of such a venture as well as the romanticism of the Makuleke story, will act as draw-cards for tourists. Land beneficiaries are being trained in hospitality skills, business management and tour guide skills: e.g. a training centre has been set up for members of the Makuleke community in their village where they study for National Diplomas (Steenkamp 1999). New tourism alliances are emerging between Parks authorities, the private sector and the beneficiary communities: e.g.

the Makuleke have publicly advertised for proposals for joint tourism ventures in the Kruger National Park, and the beneficiaries at Dwesa and Mkambati are being linked to Spatial Development Initiatives spearheaded by the Department of Trade and Industry (Kepe et al. 1999; Timmermans 1999).

It must however be realized that while tourism is an attractive option, it will not address the acute short-term financial needs of communities. Investments in tourism typically take 25 to 30 years to realize returns, and poor rural people are seldom prepared to wait that long for development to become a reality (Magome et al. 1999). Beneficiary communities need to be trained in entrepreneurial skills to be less reliant on tourism alone and to make the most of the diverse income-generating opportunities associated with owning land in protected areas. These include lease fees for land, contracting for certain of the Park's services, employment in the Park, live game sales, and toll fees for entering certain sections of the Park.

Improved Relations

Relations at two levels of the conservation hierarchy have improved as a result of land restitution. At the senior management level, officials have started taking local communities seriously in response to increased media attention and political pressure. The appearance of senior conservation officials at community meetings has become commonplace, and national, provincial and local politicians have in many instances become involved in negotiations and interactions. At the field level, protected area managers have begun to recognize their neighbours as essential role players and have started treating them with more dignity than before. Joint management structures have been formed for Dwesa Nature Reserve, and Kruger, Kalahari Gemsbok, Aughrabies and Lake St. Lucia National Parks (Chennels 1998; Steenkamp 1999; Timmermans 1999).

There are however early warning signs that relationships might not remain friendly and sincere. Conservationists had been hopeful that poaching would decrease with the improvement in relationships, but this does not seem to be taking place. Communities at Dwesa continue to harvest mussels and other shellfish in the nature reserve illegally and at unsustainable rates (Timmermans 1999). Another source of contention is the domination in some instances by conservationists and NGOs of joint management committees (Reid 1999), causing communities to feel vulnerable and powerless (Chennels 1998). In addition, community institutions are frequently embroiled in internal conflicts over leadership, with Communal Property Associations and Trusts struggling to get off the ground (Timmermans 1999).

Conservationists and communities have come to realize that participatory management is a long and continuously evolving process, with a real danger that the weaker partner will be passively carried along. Mechanisms need to be sought to facilitate and monitor the active participation of beneficiary communities in conservation management.

Changes in Conservation Attitudes, Policies and Approaches

Provincial and national conservation agencies have had to adjust their policies and practices in response to the new powers and legal status gained by communities. The SANP, which manages all national parks, has established a Social Ecology Department to improve relations with communities and has incorporated participatory management into its policies. SANP has formulated a new policy not to oppose land claims, but rather to assist with the process. Provincial conservation agencies such as the KwaZulu-Natal Conservation Service and the Eastern Cape Ministry of Economic Affairs, Environment and Tourism have stated their intention to encourage mutually beneficial partnerships with communities, to assist them in running their own parks and tourism ventures.

Conservationists with different backgrounds and value systems tend to disagree on the appropriateness of strategies and actions, and factions are emerging in many agencies. In all conservation institutions there are proponents and opponents of collaborative management, at every level in the hierarchy. Because collaborative management is risky and experimental, this approach is prone to being discredited; its opponents can easily gain the upper hand. The erosion of communities' powers and the reversal to more autocratic approaches is thus a real possibility.

Formation of New Institutions

In the process of formulating and implementing land claims, communities have begun forming new institutions to deal with the complexities of the land claims process. Examples of such institutions are: trusts to manage funding, e.g. at Makuleke, Dwesa, Kalahari Gemsbok National Park, and Aughrabies National Park; Common Property Associations that have been established at Dwesa, Makuleke and elsewhere; conservation/monitoring/development committees, e.g. at Dwesa, Mkambati and Makuleke; joint management committees such as at Makuleke, Dwesa, St. Lucia, Aughrabies and others (Forse 1998; SANP press release, 9 April 1999; Steenkamp 1999; Timmermans 1999).

Non-government organizations and charities have played an important role as technical advisors and facilitators, e.g. the Transkei Land and Soil Organization (TRALSO) at Dwesa and Mkambati, Friends of Makuleke and Mafisa at Kruger, and the Southern African San Institute at Kalahari Gemsbok National Park, to mention a few. These NGOs were strengthened and grew with the assistance of donors such as DFID (UK) and GTZ (Germany).

These new institutions are necessary for common property management to be effective, but in the early stages they invariably lack the financial and technical capacity to play their rightful roles. They are even at this early stage embroiled in internal conflict and their members are experiencing communication problems because of a lack of infrastructure (Fabricius et al. in press). Some of them (e.g. Common Property Associations) have had to be abandoned because of skills shortages and their inappropriateness to rural lifestyles (Timmermans 1999). Two of the essential characteristics of lasting common property institutions, i.e. *sanctions*, and *monitoring* (Oström 1990), are in many instances absent, and the establishment of lasting institutions will require greater financial and other commitment from government. There is also a need for improved integration across different institutions and sectors to contribute collectively to the strengthening of local institutions.

Revival of Traditional Knowledge

The extent to which relocated people have retained their traditional knowledge about plants, animals and ecosystems is remarkable. Members of the Makuleke community are able to identify and rate medicinal plants inside the Kruger Park (Barbara Wilson, Geography Department, Rhodes University, pers. comm.), while older people at Dwesa have a sophisticated knowledge of useful plants and grasses (Fay 1999). Such older people also show an advanced understanding of the succession of grassland to forest and the role of nurse plants in establishing new forests. At Mkambati, local hunters' knowledge and ability to use fire to manipulate the movements of wildlife is highly developed (Kepe et al. 1999). In the Kalahari Gemsbok National Park, San trackers are able to monitor the movements and behaviour of wildlife using a specially-developed hand-held computer with icons depicting species and their activities. S.A. National Parks issues certificates of competence to trackers who have passed a stringent test, and older trackers have started training younger apprentices (Liebenberg 1997).

It is however unclear how this knowledge will be put to practical use in conservation management. Conservationists are sceptical about uneducated people's ability to manage natural resources. As one senior conservation researcher put it: 'I, with a PhD, don't even know for certain how to manage the Park. How will these people [with reference to the Makuleke community] be able to do it?' (Anonymous SANP official, pers. comm.).

Conclusions and Future Challenges

Forced relocation has had few if any benefits for conservation in South Africa and has caused great hardships for relocated communities. Land restitution, on the other hand, has not had the predicted negative effect on conservation in the short term. In fact land restitution has been beneficial in many respects and there has been a net short-term gain for conservation in almost every case study investigated.

We predict that forced relocation to make way for protected areas is not something of the past. Farm labourers are currently being threatened by the expansion of the conservation estate in the Northern and Eastern Cape, and the implications of land purchases for this group are often glossed over by conservation planners (cf. Kerley and Boshoff 1997). It is also quite likely that transfrontier conservation areas ('Peace Parks') will result in people being resettled (Koch, 1999).

The lessons are clear: forced relocation is almost invariably detrimental to conservation. Conservation models which do not involve relocation, such as Schedule 5 Parks (parks with people living in them) and Biosphere Reserves (core conservation areas surrounded by privately-managed land subject to land use restrictions), are difficult to implement. It takes a long time to formalize agreements and the institutions required to manage them are often unstable in the early stages of their establishment. Such innovative models nevertheless appear to be the most sustainable in the long term and are characterized by fewer of the conservation problems associated with models involving relocation (Archer 1999).

We recommend the following course of action to address the concerns raised in this chapter. Researchers, conservationists and communities should monitor the ecological, social, economic and institutional impacts of restitution. The process is poorly documented and no monitoring programmes appear to be in place. The income-generating opportunities for local communities arising out of Parks should be diversified beyond eco-tourism, so that communities can receive short-term benefits in addition to the long-term benefits from tourism. Government departments, assisted by NGOs, need to find innovative land and resource tenure solutions, in order to deal with problematic cases such as Riemvasmaak, where the community is divided and where current land use options appear to be unsustainable (M.T. Hoffman, pers. comm.). It is also important that the capacity of local institutions should be developed by donors, government and NGOs to enable them to monitor the adherence of all parties to the land claim agreements, create local rules to govern natural resource use, enforce local rules and impose fair sanctions, administer funds and resolve and manage internal conflicts. Legal mechanisms need to be developed to accommodate the needs of farm-workers when protected areas are expanded or when new Parks are proclaimed.

These recommendations are not a panacea, and the suggested courses of action face problems such as shortages of resources, personnel and training, resistance by groups whose interests are being threatened, and slow reform of administrative systems. International aid agencies need to anticipate these problems and fund programmes aimed at alleviating them.

Land reform in South Africa (of which restitution is a key component) will of necessity involve resettlement. For black South Africans to return to the land from which they have been expelled, or for them to gain access to what has since 1913 been 'white-owned' land, they will have to move from where they currently are living (de Wet 1997). Resettlement, even when voluntary, and even when involving the gain of land, is inherently problematic. The risk of various types of socio-economic disruption and impoverishment is high (Cernea 1997). In the majority of cases across the world, new land settlement has had mixed fortunes at best and, even in successful cases, the attainment of self-sufficiency can take a number of years (Scudder 1991). Resettlement is a long-term process which can take up to a generation for settlers to re-establish themselves in a socio-economically sustainable manner and to develop the institutions necessary to do so in a new setting (Scudder 1993).

In most instances of land restitution in protected areas in South Africa the relocated people have chosen to remain in the new settlement areas. Regaining access to their former land therefore does not involve them being resettled a second time. However, many of the causes of social and economic disruption inherent in resettlement are still present. People are faced by a series of simultaneously occurring changes in power relations, territorial access and occupation, social relationships, administrative structures and access to resources.

This is particularly problematic in the context of conservation areas. People are in most instances not simply returning to their former land to do with it as they wish. They have to enter into joint management ventures with conservationists who are often seen as being implicated in their having been driven off the land in the first place. These officials may in a sense also feel dispossessed of what they have come to see as their land and may feel concerned about having to share it with people they believe are unable to manage biodiversity resources of global significance.

New local-level institutions and procedures have to be developed, without there necessarily being consensus as to the goals or ground rules. There is thus a high risk of the current experimental institutional arrangements becoming dysfunctional or even disintegrating, and of relations around resource usage being characterized by on-going conflict. This will have disastrous results for both conservation and income generation.

If the gains that appear to have been made for conservation from the process of land restitution in South Africa are to last, then the trends identified in this paper need to be monitored. Valuable lessons can be learnt from resettlement elsewhere in the world, and from situations where political and administrative change has taken place rapidly. A detailed and sensitive monitoring of resettlement and land restitution in relation to conservation in South Africa may in turn be valuable in dealing with and understanding similar situations elsewhere.

Notes

1 Since the time of going to press, the Kalahari Gemsbok National Park has been renamed Kgalagadi Transfrontier Park.

2 Since the time of going to press, St. Lucia National Park has been renamed Greater Lake St. Lucia Wetland Park.

3 In this chapter, the term 'reserve' refers to a protected area and not to Bantustan/homeland which was part of the apartheid political machinery.

4 The acronym for South African National Parks has recently been changed from SANP to SANParks.

References

Archer, F. 1999. *Participatory Decision-making at Richtersveld Contractual Park: myth or reality?* Draft report to the Evaluating Eden Project. London: International Institute for Environment and Development.

Barker, N. 1997. 'Return to the Battle-front at St. Lucia'. *Mail & Guardian*, 20 November.

―― 1998. 'Land Row Threatens Rare Forest'. *Mail & Guardian*, 8 January.

Cernea, M. 1997. 'The Risks and Reconstruction Model for Resettling Displaced Populations'. *World Development*, 25: 1569–87.

Chennels, R. 1998. 'Case Study: The San-Kalahari land restitution claim'. In *Workshop on Land Claims on Conservation Land: Record of the Proceedings*, pp. 44–7. Pretoria: IUCN South Africa Country Office.

de Wet, C. 1995. *Moving Together, Drifting Apart. The Dynamics of Villagisation in a South African Homeland*. Johannesburg: Witwatersrand University Press.

―― 1997. 'Land Reform in South Africa: A vehicle for justice and reconciliation, or a source of further inequality and conflict?' *Development Southern Africa*, 14: 355–62.

Fabricius, C. 1999. 'Evaluating Eden: Who are the winners and losers in Community Wildlife Management?' In D. Eldridge and D. Freudenberger (eds). *People and Rangelands: Building the Future*. Proceedings of the VIth International Rangeland Congress, pp. 615–23. CSIRO, Australia.

Fabricius, C. and Burger, M. 1997. 'Comparison between a Nature Reserve and Adjacent Communal Land in Xeric Succulent Thicket: An indigenous plant user's perspective.' *South African Journal of Science*, 93: 259–62.

Fabricius, C., Koch, E. and Magome, H. in press. *Community Wildlife Management in Southern Africa: Challenging the assumptions of Eden*. Evaluating Eden Series. London: International Institute for Environment and Development.

Fay, D. 1999. 'Local Knowledge of Grass and Grassland Management in Dwesa-Cwebe'. Unpublished manuscript.

Forse, W. 1998. 'St. Lucia Eastern Shores: Determining the land value in conservation areas'. In *Workshop on Land Claims on Conservation Land: Record of the Proceedings*, pp. 51–6. Pretoria: IUCN South Africa Country Office.

Hardin, G. 1968. 'The Tragedy of the Commons.' *Science*, 162: 1243–8.

Kepe, T., Cousins, B. and Turner, S. 1999. *Resource Tenure and Power Relations in Community Wildlife Contexts: The case of the Mkambati area on the Wild Coast of South Africa*. Draft report to the Evaluating Eden Project. London: International Institute for Environment and Development.

Kerley, G. and Boshoff, A. 1997. *A Proposal for a Greater Addo National Park*. Report no. 17, Terrestrial Ecology Research Unit, Dept. of Zoology, University of Port Elizabeth.

Koch, E. 1999. *Fences of Fire, Man-eating Lions and a long Bicycle Ride: Do Transfrontier Conservation Areas open the Way for improved rural Livelihoods as Part of Southern Africa's Peace Dividend?* Draft report, Evaluating Eden project. London: International Institute for Environment and Development.

Liebenberg, L. 1997. *Integrating Traditional Knowledge with Computer Science for the Conservation of Biodiversity*. Cybertracker Software, Constantia.

Lieberman, D. 1997. 'The Ethnobotanical Baseline.' In *Indigenous Knowledge, Conservation Reform, Natural Resource Management and Rural Development in the Dwesa and Cwebe Nature Reserves and neighbouring Village Settlements*, pp. 40–88. Interim report for the Human Sciences Research Council, Institute of Social and Economic Research, Rhodes University, Grahamstown.

Lund, F. 1998. *Lessons from Riemvasmaak for Land Reform Policies and Programmes in South Africa. Volume 2: Background Study*. Programme for Land and Agrarian Studies, Research Report no. 2. University of the Western Cape, Belville, South Africa.

Macleod, F. 2000. 'Elephants to be hunted in Kruger?' *Mail & Guardian*, 21–7 January.

Magome, H., Grossman, D., Fakir, S. and Stowell, Y. 1999. *Partnerships in Conservation: The State, Private Sector and the Community at Madikwe Game Reserve, North-West Province, South Africa.* Draft report to the Evaluating Eden Project. London: International Institute for Environment & Development.

Ostrom, E. 1990. *Governing the Commons. The Evolution of Institutions for Collective Action.* Cambridge: Cambridge University Press.

Ramphele, M. 1991. *Restoring the Land: Environment and Change in Post-Apartheid South Africa.* London: Panos Publications.

Reid, H. 1999. 'Contractual National Parks and the Makuleke Community'. Unpublished manuscript, Durrell Institute of Conservation Ecology, Canterbury.

Scudder, T. 1991. 'A Sociological Framework for the Analysis of new Land Settlements'. In M. Cernea (ed.). *Putting People First: Sociological Variables in Rural Development,* pp. 148–87. New York: Oxford University Press, for the World Bank.

—— 1993. 'Development-induced Relocation and Refugee Studies: 37 years of change and continuity among Zambia's Gwembe Tonga'. *Journal of Refugee Studies,* 6: 123–52.

South African National Parks (SANP). 1998. Joint press release with Riemvasmaak Trust, 4 February.

—— 1999. Press release, 9 April.

—— 2000. Press release, 25 January.

Spinage, C. 1998. 'Social Change and Conservation Misrepresentation in Africa'. *Oryx,* 32: 265–76.

Steenkamp, C. 1999. *The Makuleke Land Claim: Power Relations and CBNRM.* Draft report to the Evaluating Eden Project. London: International Institute for Environment and Development.

Surplus People Project. 1985. *The Surplus People.* Johannesburg: Ravan Press.

Timmermans, H. 1999. *Power Relations and Perceptions at Dwesa and Cwebe Nature Reserve on South Africa's Wild Coast.* Draft report to the Evaluating Eden Project. London: International Institute for Environment & Development.

Van Rooyen, A.F. 1999. 'Rangeland Management at the Mier Communal Area, Northern Cape'. PhD dissertation, University of Natal, Pietermaritzburg.

Wynberg, R. and Kepe, T. 1999. *Land Reform and Conservation Areas in South Africa: Towards a Mutually Beneficial Approach.* Johannesburg: IUCN.

9

How Sustainable is the Communalizing Discourse of 'New' Conservation?

THE MASKING OF DIFFERENCE, INEQUALITY AND ASPIRATION IN THE FLEDGLING 'CONSERVANCIES' OF NAMIBIA[1]

Sian Sullivan

Introduction

> We have also come to understand and realize that many of the ... people who came to introduce the [1996 Nature Conservation Amendment] Act to us, are the former all-white employees of your Ministry who as individuals resigned from Government to venture into private sector businesses.

The above quote is from a June 1999 letter to the Minister of Environment and Tourism, Namibia. It was written by two residents of southern Kunene Region, who recently each applied for formal Permission to Occupy Land (PTO) leases to establish campsites and thereby capitalize on a post-independence increased flow of tourists to this wildlife-rich area. Their immediate complaint is that the granting of these applications has been put on hold following a request to this effect by the local 'conservancy committee'. More revealing, however, is the rationale behind their complaint: that how can this hold on local entrepreneurial activity be justified when national policy vis à vis conservation in communal areas has been driven largely by expatriates, many of whom are themselves currently employed in the private sector. This is coupled with serious, albeit contested,[2] allegations levelled at the 'legality and authority' of the conservancy committee.

Namibia's conservancy policy for communal areas was developed as the basis for community-based natural resource management (CBNRM) through devolved management of wildlife without moving people from the land (Nujoma 1998). Communal area residents, as conservancy members, can benefit from, and have management responsibilities over, animal-wildlife. To be registered as a wildlife management institution, a conservancy requires a defined boundary and membership, a representative management committee, a legal constitution and a plan for the equitable distribution of benefits (MET 1995a and b). Like the much publicized CAMPFIRE programme of Zimbabwe – blueprint for USAID-funded CBNRM programmes throughout southern Africa and elsewhere[3] – the assumption informing conservancy policy is that 'conservation and development goals can be achieved by creating strong collective tenure over wildlife resources in communal lands' (Murombedzi 1999: 288). This 'new' conservation thus is driven by: acknowledgement of the costs experienced by farmers living alongside wildlife in these areas; a need to counter the alienating effects of past exclusionary conservation policies; realization of the lack of economic incentives for local people to maintain a benign relationship to animal-wildlife; and recognition of the economic development needs of rural populations. The primary 'facilitators' of CBNRM tend to be NGOs. In the Namibian case, a key player has been the NGO IRDNC (Integrated Rural Development and Nature Conservation) which is considered to have 'a particular onus ... to facilitate conservancy registration and development' (Durbin et al. 1997: 5).

Namibia's conservancy policy has been heralded as the most progressive initiative of its kind in southern Africa (Mafune 1998). In September 1998 Namibia became the first country worldwide to be honoured for a *people*-centred environmental initiative with a World Wide Fund for Nature (WWF) Gift to the Earth Award (Sutherland 1998). It is claimed that conservancies will improve livelihood sustainability through diversification of incomes (Ashley 1997; Hulme and Murphree 1999); that they are based on a participatory decision-making process that is empowering to women (Jones 1999a: 302); and that they will 'empower poor, disadvantaged rural people' (Jones 1995; Ashley 1998 in Callihan 1999).

As identified by recipients, however, this 'new' conservation can be also viewed as a continuation of past conservation policies: in terms of who is driving and implementing policy and in the ways in which local difference and aspirations are masked by the associated 'communalizing' rhetoric (see Table 9.1 for recent critique). Displacement in these contexts becomes something more subtle than the physical eviction of peoples from their land in the name of conservation. It is about the manner in which local, multi-layered narratives of, and rights to, land and resources are displaced in global discourses that survive only by excluding such complexity; and about how local

Table 9.1 Recent Critiques of USAID-funded Southern African CBNRM Programmes

Reference	Location and programme	Comment
Marindo-Ranganai and Zaba 1994: 8	Zimbabwe, CAMPFIRE	Re: views of Tembo-Mvura gatherer-hunters of north Zimbabwe: 'CAMPFIRE is a programme for the Chikunda and the Safari people. They are the ones who gain from it. What CAMPFIRE does is to stop us from hunting so that white people can come from far away to kill animals for fun. We have heard that these people pay money but we have never seen any of it... All the village wildlife committee is made up of the Chikunda'.
Patel 1998: 22, 41	Zimbabwe, CAMPFIRE	Villagers from five districts considered the wildlife sector to remain in the control of 'a distant "white" force, in which the safari operator and his clients wield the ultimate power', thus bolstering 'the economic and political power of minority whites in Zimbabwe' rather than constituting meaningful local empowerment.
Wels, 1999: 20–21; Dzingirai, 1995: 4 in Wels, 1999: 21	Zimbabwe, CAMPFIRE	There is a noticeable trend towards the construction of fences by safari and hunting operators around hunting blocks so that 'clients can hunt freely and safely without having to worry about human habitation' and to 'prevent animals from damaging property and crops and humans themselves outside the hunting area'. Justifiable associations between fencing and alienation and exclusion have led to perceptions on the part of villagers 'that the safari operator wanted to create a private farm out of their land, ... to prevent people from accessing ... resources ... [and] to reintroduce white colonialism'.

Table 9.1 continued

Reference	Location and programme	Comment
Matenga 1999: i	Zambia, ADMADE	'[I]s just a modernization project in the wildlife sector designed not to improve economic livelihoods of local communities but to defuse local opposition towards national wildlife conservation'. Matenga concludes that 'while these projects were in theory supposed to empower the local communities through their participation in the management and sharing in the benefits of wildlife related activities, their participation has proven to be elusive … leading to their disempowerment economically, socially, psychologically and politically'.
Taylor 1999: 10	Botswana, NRMP	'One of the expatriate NRMP team members in Botswana admitted informally that their real aim is conservation, and community development is included as a means to achieve this'.

differences can constitute distinct relations of disadvantage, enhanced in ways that are masked by such normalizing discourses.

In this chapter, and drawing on fieldwork in Namibia since 1992, I use the particular context of the establishment of Namibian communal-area conservancies to draw attention to several issues underlying 'community-based conservation'. My discussion begins with an alternative framing of the conservancy model as representing a continued concern for preserving threatened large mammal species and 'wilderness', where the blatant exclusion of people from resources is no longer acceptable. Divergences between conservationist and local priorities are apparent in the different ways that debate regarding conservancy establishment has been articulated: namely, that instead of being pursued as a policy enabling greater community rights to animal-wildlife it has been appropriated locally as a forum for expressing and contesting claims to land. I move on to explore assertions of the success of CBNRM initiatives in Namibia under the rubric of conservancy formation: specifically, that the anticipated diversification of incomes will improve livelihood sustainability; that decision-making processes are representative and participatory; and that conservancies per se provide an enabling environment for empowering structurally disadvantaged people.

Conservancies and Continuities: Moulding Wildlife Conservation to a Post-apartheid Context

The term 'conservancy' emerged in the 1970s in an apartheid-structured South Africa to describe the consolidation of exclusive rights over animal-wildlife among co-operating white settler farmers, largely through the employment of game guards to militate against 'poaching' on freehold land by black African 'neighbours' (Wels 1999). Furthering the 'ecological apartheid' of the protected area system, conservancies were seen in this context as the only 'viable alternative for the *salvation* of wildlife on private land' in a context where it was considered that '[f]ailure to provide security and management for wildlife on private land must, inevitably, lead to its demise' (Collinson 1983: 167, in Wels 1999: 12).

In Namibia, the conservancy concept similarly emerged in the context of freehold farmland. Here, since 1968 and subject to conditions set by the MET (Ministry of Environment and Tourism) (particularly with regard to fencing) European settler farmers have had legal rights to consumptively and otherwise utilize animal-wildlife on their farms (Jones 1995: 4). Under these circumstances landowners 'realised that it is advantageous to pool their land and financial resources to make available a larger unit on which integrated management practices can be carried out' (Jones 1995: 4; Barnes and de Jager 1995). Some twelve conservancies existed on freehold land in 1999 which, while acknowledged and supported by the MET, were without legal status (Jones 1999c: 11).

Alongside this co-ordinating of wildlife access and management by settler farmers on freehold land, conservationists were voicing increasing concern regarding the future of animal-wildlife in Namibia's communally-managed indigenous 'homelands'. A particular focus of this anxiety was the Kaokoveld of north-west Namibia; the imagined 'last wilderness' of South African environmentalists (Reardon 1986; Hall-Martin et al. 1988; but see Bollig 1998), and the world-famous birthplace of Namibian community-based conservation. Here, large-scale losses in the 1970s and 1980s of internationally-valued large mammal species, particularly desert-dwelling elephant (*Loxodonta africana*) and black rhino (*Diceros bicornis bicornis*), provided an impetus to enlist local support for conservation (Owen-Smith 1995). Initially, this was led by individuals spearheading a privately-funded conservation charity, the Namibian Wildlife Trust (NWT). These included Mr G. Owen-Smith, now IRDNC's co-Director and Project Executant.

The reasons for the 1970s and 1980s wildlife losses in Kaokoland are many and complex. In the 1960s the area was exploited as something of a private hunting reserve by top government officials, including Cabinet Ministers in the South African government (Reardon 1986: 13). In the 1970s, it appears that 'the majority of men appointed to safeguard the Kaokoveld embarked on a hunting frenzy' (Reardon 1986: 13).[4] In the late 1970s and early 1980s devastating drought caused wildlife losses, both directly and through stimulating local 'poaching' in attempts to counter erosion of pastoralist livelihoods. Organized illegal trafficking in ivory and horn during the 1980s, known to have been pursued as a 'deliberate policy of the various organs of the South African state' (Ellis 1994: 3), also may have reduced Kaokoland's elephant and rhino populations. The situation in north-west Namibia was exacerbated by regional warfare between South Africa, Namibia and Angola. This made firearms available, often via distribution by the South African Defence Force (SADF) to local people as a means of fostering tensions between different groups, thereby compromising regional and national opposition (Fuller 1993: 81).

In other words, the ultimate causes of wildlife losses appear beyond the control of local people. Nevertheless, it is they who were constructed as a locale of responsibility for protecting regional wildlife populations. A network of paid male 'community game guards' (CGGs, formerly 'auxiliary' game guards) was created, appointed with the help of local headmen and oriented towards protecting the region's threatened large mammal species. This initiative generally is credited with creating empowerment and a sense of 'ownership' over wildlife (Durbin et al. 1997: 13) and it is this 'participation' of local people which is considered to have enabled recovery of wildlife populations during the late 1980s. Undoubtedly the CGGs contributed to wildlife population increases (otherwise related to improved rainfall and a relaxing of combat activities in the area), but as much by extending the *policing* and *anti-poaching* role of MWCT (the then Ministry of Wildlife, Conservation and Tourism, now the MET) as by enhancing local participation. Similarly, the employment offered by the CGG system was perhaps as important as any attributed 'empowerment' over wildlife: unsurprisingly, CGGs became less effective after the mid-1980s in areas where salaries and rations, as well as supervision by the MWCT and NWT/IRDNC, were reduced (Durbin et al. 1997: 20). Assertions of 'success', in terms of both wildlife increases and local empowerment, thus depend on what are malleable interpretations of context.

Following independence in 1990, the apparent success of the north-west Namibian CGG system was invoked by the MWCT and IRDNC in applying the conservancy concept to communal areas (MWCT 1992; Jones 1999a). The Nature Conservation Amendment Act of 1996 thus significantly alters the 1975 Nature Conservation Ordinance by devolving *proprietorship* over wildlife, and concessionary rights over commercial tourism, to people on communal land (MET 1995a and b). I emphasize the term proprietorship because, as elsewhere (Neumann 1997; Madzudzo 1999; Matenga 1999), the ultimate ownership of wildlife remains with the state (MET n.d.: 9; *The Namibian* 1999a). As detailed above, proprietorship is conditional on registration as a conservancy with a defined boundary and membership, a representative management committee, a legal constitution and a plan for the equitable distribution of benefits.

The employment of male CGGs remains a defining component of the wildlife-rich emerging conservancies in Namibia's communal areas.[5] Although they are viewed by NGOs and donors as a 'primary link' between 'communities' and the formal conservation authority (Durbin et al. 1997: 15), their major functions, like game guards on both protected areas and private conservancies, are wildlife monitoring, policing and anti-poaching (confirmed by local views in Mosimane 1996: 15–16, 29–30; Powell 1998; *The Namibian* 1999a).

Recently, consultants for WWF have recommended that CGGs be equipped with firearms, suggesting that wildlife protection activities in Namibia's communal areas might become increasingly militarized (Durbin et al. 1997: 18). Ironically, given the language of devolving rights to resources to local 'communities', it seems that CBNRM also intensifies policing of animal-wildlife in communal areas. More serious are the implications of what amounts to arming civil society in the name of wildlife conservation (Leach 1999).[6] That this is occurring in Namibia is evidenced by the locating of armed guards to protect the IRDNC-supported conservancy office in Sesfontein/!Nani|aus, southern Kunene Region, following recent local protest to circumstances surrounding conservancy establishment in the area (Sullivan in press a).

It has been observed that 'similarities in institutional arrangement between conservancies that have developed on freehold land and those on communal land are striking', with both measuring 'up well against the principles for designing long lasting common property resource management institutions' (Jones 1999c: 13). Given the historical evolution of the conservancy concept, the legacy of exploitative policies supporting state and settler interests, and extremely restricted access to alternative models for 'self-determination' among communal area inhabitants, however, it is hardly surprising that 'joint solutions' for the conservation of wildlife in communal areas have emerged which are in line with existing ideas of conservancies promoted by the MET. While the legislative situation may be more progressive than elsewhere, continuities with past priorities are clear. Conservancy establishment in communal areas remains '*land acquisition for conservation* in the non-formal sense' (Jones 1999b: 47 emphasis added), with a focus on effective protection and policing of an internationally-valued wildlife of large and dangerous mammals. 'Rural development' and 'empowerment' in these contexts appear circumscribed: constrained to providing effective protection for a handful of species which are potentially harmful to local residents and their other economic activities; and dependent on deals struck up with outside tourism and hunting operators, often outfits whose claims to capitalize on wildlife and wilderness are those considered legitimate by agencies and individuals advising 'communities' (also Mosimane 1996: 37). In this sense, CBNRM *in practice* maintains the interests of conservationists, tour-operators, hunters and tourists; i.e. those conventionally associated with 'touristic' enjoyment of, and financial benefits from, wildlife and 'wilderness'.

Claims to Land, Claims to Wildlife: Objectives and Interests framing Policy Appropriation

Conservancy legislation is asserted as devolving 'a large measure of authority, and responsibility over wildlife and the right to benefit from wildlife use to *landholders* themselves, both freehold and communal' (Jones 1999c: 13 emphasis added). Observations of parallels in the development of conservancies on different categories of land (see above), and references to communal area residents as 'landholders', however, obscure substantial structural differences regarding land distribution and rights. Specifically, that a minority of settler freehold farmers have inalienable rights to a major proportion of the most productive land in southern and central Namibia.[7] Moreover, their title to land means that they effectively and legally own the capital constituted by their land and the resources on it, including 'huntable game'. With the human population density of commercial farmland being under a third that of communal areas (Moorsom 1982; Adams and Werner 1990; Central Statistics Office 1994), and with the former hosting some 70 per cent of the nation's 'game' (Jones 1995: 4), these relationships clearly are grossly unequal. Moreover, for freeholders, ensuring returns on their wildlife capital is by no means dependent on their membership of a conservancy.

Conservancy policy since its inception thus has been understood and appropriated by local people in communal areas primarily as a land issue, and secondarily as a wildlife management issue, with local meetings dominated by debate regarding claims to land rather than to wildlife (also see Taylor 1999: 10). Three further reasons have fuelled this situation. First, discussions over establishing conservancies have provided a much-needed outlet for debate regarding land redistribution in the context of speculation and optimism ushered in by an independent Namibia. Second, because two criteria for gazetting a conservancy are that its physical boundaries and community membership be defined, the situation is treated as one of establishing rights to land areas even though legally a 'community' is only establishing rights to returns on animal-wildlife in those areas. Third, and related to this, because there has been a lack of an overriding legal procedural basis for establishing tenure rights to land in communal areas, the conservancy option has become the only means by which people can gain any apparent security to land. This, together with a constitutional context in which Namibian citizens can move to wherever they wish on communal land (with the unmonitored *proviso* that they observe the customary rights of existing inhabitants) (GRN 1991: 28–9), enhances anxiety over claims to community 'membership'.[8] The exponential rate at which conservancies are now being formed thus might be an attempt on the part of communal area inhabitants to establish rights to land and resources in the absence of any other legitimate way of doing so (cf. Shivute 1998; Inambao 1998a); as well as reflecting the 'marketing' of the concept and a capitalizing on opportunities presented by donors and NGOs. Elsewhere, and reflecting ambiguities in how the conservancy policy is understood, it appears that people have been unable to use conservancy policy to ensure that they retain access to natural resources other than animal-wildlife (Powell 1998: 120).

In recognition of the importance of secure land tenure to support rights to wildlife resources, policy-makers in the MET, as well as implementing conservation NGOs, anticipated and hoped that 'the conservancy approach, even if embedded only in wildlife legislation, could help shape appropriate [land] tenure reform' (Jones 1998: 5; also Durbin et al. 1997: 10). Indeed, the National Land Policy tabled in 1997 included an option for 'legally constituted bodies and institutions to exercise joint ownership rights over land', implying that a community which defined itself as a conservancy could register tenure rights to the land defining the conservancy's territory (GRN 1997: 9). The recently tabled Communal Land Reform Bill, however, appears not to support this option (GRN 1999). While stating that regional Land Boards 'must have due regard to any management and utilization plan framed by [a] conservancy committee' (GRN 1999: Section 31(4): 20) the Bill does not explicitly vest conservancies with tenure rights other than those set out in the Nature Conservation Ordinance, i.e. to wildlife and wildlife-related revenues.[9] Elsewhere, the Bill appears to focus on the individualization of land-holdings: in providing for the registration of farming and residential *units* 'in the name of the person to whom it was allocated' (GRN 1999: Section 25: 14); and in the granting, by a Land Board, of leasehold tenure to individual applicants (GRN 1999: Section 30: 19). It remains to be seen how an essentially individualizing land policy trajectory (Shigwedha 2000) will affect the establishment and maintenance of 'community-held' communal area conservancies.

Diversification of Incomes will improve Livelihood Sustainability

Community-based conservation and community-based tourism generally are considered able to improve 'livelihood sustainability'. It is thought that revenue from consumptive and non-consumptive uses of wildlife will enhance livelihoods by diversifying sources of income. And that this will be sustainable because tourism, worldwide and in Namibia, is a growth industry;[10] and because '[o]nce income is derived by local communities from the use of wildlife, they develop a vested interest in conserving game animals' (Jones 1995: 9), whereby environmental degradation, namely erosion of biodiversity and habitat integrity, is reduced. CBNRM thus relies on an economizing framework to justify projects and policy aimed at the 'sustainable use of natural resources' as a means of rural development (Ashley and Garland 1994; Ashley et al. 1994; Ashley 1995, 1997; Callihan 1999; Jones 1999c following Murphree 1993).

But it is unlikely that revenue from wildlife and/or tourism can constitute a particularly large source of income for all members of a 'community' at household and individual levels (Hackel 1999). This is without projected increases in rural (human) populations.[11] Again, this reflects a structural situation whereby population densities throughout the communal areas generally are higher than in the commercial farming areas. Thus, average benefits per capita are likely to be always much lower for people in communal areas. Table 9.2 indicates that per capita income from the consumptive and non-consumptive uses of wildlife in Namibia's communally-managed areas generally is low. In Table 9.2, the highest recent annual per capita income, by an order of magnitude, is that recorded for Torra, Namibia's 'flagship' conservancy. Here, income per inhabitant, in 1999, worked out at approximately N$1041.39 or US$132.32 (£87.33).[12] An additional N$363.32 or US$46.09 (£30.42) per inhabitant was received in wages to community members employed by Damaraland Camp. Callihan (1999: 10) points out that wages are likely to constitute the main source of income extended via the establishment of conservancies: this of course will depend on a conservancy's ability to secure enterprise investment and is relevant primarily for conservancy members who are offered employment ('trickle-down effects' notwithstanding). The next highest conservancy per capita annual income is substantially lower at N$150 or US$19.06 (£12.77). A comparison with the government old-age pension of N$160/*month* indicates that the *relative* annual per capita contribution provided by conservancies also is low.[13] The use of surrogate monetary values for resources consumed directly is misleading. For example, the figure of US$25,000 calculated for the value of meat consumed in Kunene Region in 1993 (Jones 1999c: 2) is spurious considering the manner by which local people have been alienated from the consumption of 'bushmeat' throughout this century and criminalized should they hunt for their own use.

Despite low per capita returns, CBNRM discourse often goes further than arguing that incomes from wildlife and tourism can diversify livelihoods. For example, it is suggested that returns on wildlife will encourage people to

Table 9.2 Recent Figures for Income Received in Communal Areas from both Consumptive and Non-consumptive Uses of Wildlife

Year	N$$^{a-1}$	n	N$/cap.$^{a-1}$	Details
Purros community:				
via tourism, partnerships and joint venture agreements				
Since 1993?	6,000	75 adults	80	Bednight-levy from Skeleton Coast Fly-in Safaris. This company has a PTO for their campsite which is located within the proposed Purros conservancy.
Since 1993?	60,000	75 adults	800	Estimated tips to Himba settlement from visiting tourists.
Epupa Falls, Kunene Region:				
via tourism, partnerships and joint venture agreements				
1998–1999	12,000	?	?	Payment from Epupa and Hotspring Campsites, run by Kaokohimba Safaris.
1998–1999	10,000	?	?	Payment from Epupa Camp to community. Amounts to <3% per tourist bed-night.
1998–1999	8000	?	?	Payment from Omarungu Camp.
East Caprivi, Mudumu National Park: Lianshulu Lodge and neighbouring villages:				
via tourism, partnerships and joint venture agreements				
1994	26,000	3581?	7.26 (35/household)	Distributed by Lianshulu Lodge to 5 neighbouring villages (comprising 746 households with an average of 4.8 people per household in Caprivi Region).
?	62,000?	3581?	17.31	40 local staff employed at Lianshulu and paid between N$600–2500/month. Based on 40 x a mid-figure of 1550 between these two salaries: probably an over-estimate.

Table 9.2 continued

Year	N$^{a-1}	n	N$/cap.^{a-1}	Details
Etendeka and neighbouring 'communities', Kunene Region: via tourism, partnerships and joint venture agreements				
1996	40,000	4,500	8.89	Bed night levies collected by Etendeka Mountain Camp and distributed to 5 neighbouring communities (Sesfontein, Warmquelle, Kowareb, Omuramba, Otjikowares). Nb. The payment of levies to neighbouring 'communities' has not been paid in recent years due to ongoing dispute regarding conservancy development among recipient communities.
Torra conservancy, Kunene Region: via tourism, partnerships and joint venture agreements				
July 1998–June 1999	400,000	552	724.64	Income received by 120 households (with an average of 4.6 per household for Kunene Region) from joint business venture with Damaraland Camp (run by Wilderness Safaris).
July 1998–June 1999	174,846	552	316.75	Bed-night levies collected from Damaraland Camp. Nb. in Roe et al. (2000: 51–52) this is calculated as a per capita income of N$582 based on distribution between 300 conservancy members rather than a total population which they record as around 500 people.
July 1998–June 1999	200,000	552	362.32	Local salaries from Damaraland Camp (14 employed plus casual labour).
July 1998–June 1999	23,812	552	43.14	Costs of training offered to employees by Damaraland Camp.
via trophy-hunting agreements				
July 1998–June 1999	17,000	552	30.80	Hunting concession negotiated with Savannah Safaris.
July 1998–June 1999	120,000	552	217.40	Fees for animals shot paid by Savannah Safaris.

Table **9.2** continued

Year	N$ᵃ⁻¹	n	N$/cap.ᵃ⁻¹	Details
Kunene campsite in proposed Marienfluss conservancy: via tourism, partnerships and joint venture agreements				
?	6000	?	?	Bed-night levy from Skeleton Coast Fly-in Safaris.
?	50,000	?	?	Recommended tips from guests to campsite employees.
Salambala conservancy, East Caprivi: via tourism, partnerships and joint venture agreements				
1999	1,000	7,000	0.14	Earnings by LIFE-funded campsite; income after salary costs of N$1,000 removed.
via trophy-hunting agreements				
2000 & 2001	135,000	7,000	19.29	Two-year hunting concession negotiated with Wésé Adventures (+ staff employed and meat from elephant hunts distributed).
Nyae Nyae conservancy, eastern Otjozondjupa Region: via trophy-hunting agreements				
2000	260,000	2,000	110	Hunting contract with La Rochelle, including N$40,000 for game translocation, at the discretion of the conservancy.
2001	270,000	2,000	115	As above
2002	280,000	2,000	120	As above
Khoadi‖hôas conservancy, Kunene Region: via tourism, partnerships and joint venture agreements				
2000	294,000	1200	245	Grant from the newly initiated Game Products Trust Fund (GPTF) (primarily built on income received from the auction of stock-piled ivory) to be used for the construction of alternative water-points for use by elephants, away from farms.

Notes: N$ = Namibian dollars (US$1 = N$7.87 in January 2001); n = numbers of individuals in area; N$/cap.ᵃ⁻¹ = Namibian dollars per capita per year.

Sources: Central Statistics Office (1994: 11, 16); Jones (1999b: 4, 1999c: 78–9, 1999d: 8); Inambao (1998b); *The Namibian* (1999b); Roe at al. (in prep. 28, 52, 57, 60–61, 102, 116–17, 122, 124).

dis-invest in other means of livelihood, particularly livestock and cultivation, thereby reducing the 'degrading' effects of these forms of land-use while sustaining incomes (Ashley 1995, 1997; references in Powell 1998: 121; Callihan 1999). Thus for north-west Namibia Hulme and Murphree (1999 after Jones 1999a) maintain that 'the economic incentives created by devolving proprietorship over wildlife and tourism have led to people in this area re-evaluating the relative roles of wildlife and agriculture (domestic livestock and crops) in local development'. However, if per capita incomes from community-based wildlife and tourism initiatives remain low, and even without cultural influences over choice of livelihood, it is unlikely that people will view wildlife as an alternative to their usual means of livelihood. Instead, it might be anticipated that people will direct income and/or increased decision-making power deriving from CBNRM towards livelihoods over which they have direct control and ownership, and via which they are more likely to raise their individual material standards of living (as observed in Nabane 1995; Jones 1999c: 31; Murombedzi 1999).

Again, while some communal areas of Namibia appear ideal for enhancing livelihood opportunities through capitalizing on animal-wildlife this is by no means evenly distributed. Kunene and northern Erongo Regions in north-west Namibia are characterized by diverse landscapes, a spectacular wildlife of large mammals, and relatively low human population densities. Under donor-led framings of community-based conservation, these constitute perfect conditions for the evolution of so-called '5-star conservancies' (Durbin et al. 1997; Jones 1999c). Not surprisingly, therefore, this area has been a focus of NGO and donor support for the establishment of conservancies: seven out of fourteen registered conservancies are found in this area (see Figure 9.1). Critique perhaps is particularly unwelcome in this context because these circumstances appear so ripe for 'success'. At the same time, widely publicized elaborations of success based on these situations present an unrealistic picture of the possibilities for the national conservancy policy to improve livelihoods in the country's communal areas as a whole.

Also obscured are concerns at national level to increase *user-accountability* for the costs of maintaining public-sector services and national assets in remote and difficult environments. This is clear in the context of water provision, for which a community-based system of water-point committees is being advocated – ostensibly as a means of empowering communal area farmers, but basically by encouraging their participation in funding and maintaining boreholes (Africare 1993; Tarr 1998). CBNRM similarly involves a shifting of costs and responsibilities to local levels: in the policing of people's activities in relation to wildlife; in the funding of community institutions designed to manage wildlife and related revenues; and in the day-to-day experience of living with large and sometimes dangerous mammals (see Table 9.3). MET and IRDNC employees also have argued that revenue accruing to conservancies from wildlife could be mobilized to fund other sectoral developments such as school-building (see statements in Gaisford 1997: 124). This implies a vision that conservancies could carry the costs of public-sector development beyond wildlife conservation.

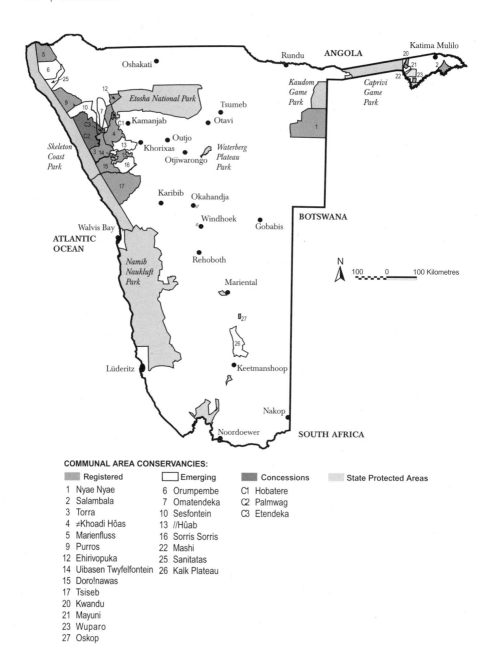

Figure 9.1 Map of Namibia Showing Locations of Registered and Emerging Communal Area Conservancies.

Source: www.dea.met.gov.na/programmes/cbnrm/cons_guide.htm downloaded and edited on 7 February 2002.

As Durbin et al. (1997: 17) state, and in accordance with the USAID's LIFE programme objective that at least five conservancies will become self-sustaining by 2002 (Callihan 1999), the 'expectation is that conservancies, once financially viable, will take on the payment of the game guards, some of the staff and equipment such as vehicles and/or radios required to support them'. To date, these have been paid for by NGOs and via the major donor-funded national CBNRM programme (LIFE), and these costs tend not to appear in calculations of income received to date by conservancies. An indication of the amounts of money involved in establishing Namibian CBNRM is indicated in the extent of its funding: some US$25 million was received from 1993 to 2000 (Callihan 1999: 6–7),[14] of which US$14 million was channelled to IRDNC between 1992 and 1999 (Durbin et al. 1997: 28). It is envisaged that the running costs of conservancies will be transferred to local conservancy institutions as communities are able to 'wean' themselves off NGO support (Jones 1999a: 300; Durbin et al. 1997). It is probable, however, that rather little income will remain after the running costs of the conservancies have been covered.[15] Logically, this amounts to a situation whereby the conservancy finances the costs of conserving an animal-wildlife accessed and valued by conservationists, ecotourists and trophy-hunters, while receiving very little additional income for the efforts of its members. The phasing out of donor-funding thus raises significant questions for the 'sustainability' and development claims of these conservation ventures. Requiring further problematization is the dependence on increases in tourism income for calculations of the sustainability of CBNRM. It is by no means certain that tourism will remain a consistent growth industry for a variety of reasons (Gaisford 1997; Infield and Adams 1999; Moyo 1999). The situation is tragically but forcefully brought home by current conflict in Caprivi Region, which fuelled cancellations by tourists and caused the temporary closure of lodges in the area and the retrenching of many of their employees (Inambao 1999, 2000b).

Participation, Representation, Empowerment – and Inequality

CBNRM is credited with providing an enabling context for the development and empowering of democratic local institutions (Jones 1999b). Donors and implementing organizations are under pressure to demonstrate the success of their activities in these terms. Recent dynamics in Namibian CBNRM, however, indicate that problems are emerging which relate to a 'massaging out' of conflict and complexity in CBNRM implementation and reporting (Sullivan in press a).

The initiation of dialogue with rural 'communities' regarding wildlife conservation in communal areas in a post-independent Namibia began with the conducting of several 'socio-ecological' surveys by the MET. These generally are credited with assessing the attitudes of communal area residents to wildlife, identifying problems *and seeking joint solutions* (Jones 1999c: 1). As Jones (1999b: 3; also 1999a) states, '[t]he conservancy approach was not imposed

Table 9.3 Recent Reported Incidents of Animal-Wildlife Impacts on Communal-area Residents, and of Local Protest against Animal-Wildlife

Reported impacts	Reference
A 1998 survey of residents in Caprivi's Kwandu area (adjacent to West Caprivi Game Reserve) indicated that 74% had experienced crop losses due to wildlife in the last five years.	Mosimane 1998 in Callihan 1999
In Kwandu, West Caprivi, four people were killed by crocodiles in the first three months of 1998. In 1999 some 450 cases of elephants destroying property were recorded by the MET whose Minister recently asserted that compensation cannot be paid for such damage because of the frequency of such occurrences.	*The Namibian* 1998 Maletsky 2000
In southern Kunene, conflicts between people and elephants are prominent at settlements along the Ugab River. Elephants moved to this river in 1994 having not been known in the area for some 50 years. Far from experiencing this new constraint on livelihoods in a passive way, and without compensation for their troubles, many people living in the area have rebuilt their homes on hillsides away from the river and avoid danger through not venturing out at night. Wind-pumps and gardens have been damaged by elephant. Early in 1999 a child was killed while crossing the river on route to the school in Ani-gab, having unwittingly disturbed a group of elephants concealed in *Tamarix usneoides* thickets along the river.	Pers. comm. with inhabitants of Ani-gab, Gudipos and ‖Gaisoas (Sullivan and Ganuses forthcoming)
'Cases of marauding desert elephants "bothering" communal farmers have in recent weeks been reported in the Kunene area, while in the Caprivi two villagers, one a Namibian and the other a Zambian, drowned after a hippo attacked their dug-out canoe along the Linyanti-Chobe River. 'Other fatal attacks on humans reported in the Caprivi have involved attacks by elephants, crocodiles and lions. 'In recent weeks there has also been an upswing in the number of livestock… that have been killed and eaten by lions and hyenas at a number of villages.'	Inambao 2000d

Table 9.3 continued

Reported impacts	Reference
In 1999 the MET recorded almost 450 cases of elephants destroying people's property. More than half of those cases (260) occurred in the Kavango region where two people were killed by migratory elephants.	Maletsky 2000
'A total of 140 cases of destruction were reported in the Kunene region, with one fatality recorded. In the Omusati, Oshana, Ohangwena and Oshikoto regions, 48 cases were reported.'	
Incidents of local protest against animal-wildlife	
In Mukwe district, western Caprivi, and following a lack of response from MET officials, farmers recently warned that 'they would take up fire arms to protect their produce from marauding elephants which have been destroying their mahangu [millet] crops'.	*The Namibian* 1998
In June 1998 it was reported that fires were started in Mamili National Park, Caprivi, by villagers living around the park as a means of injuring wildlife in the expectation that such animals would be slaughtered and the meat distributed to neighbouring villagers – some 110 buffalo were affected by the blaze.	Inambao 1998b

from outside, but developed from a joint recognition of problems and solutions between communities, government and NGOs'. Above I traced the evolution of the conservancy model in Namibia and suggested that its uptake is unsurprising because communities did not have access to alternative models (cf. Powell 1998: 117). What I would like to raise here are implications of a situation whereby initial meetings regarding communal area conservancies took place with individuals who were not necessarily representative of the wider 'community' and the diverse interests embodied by community members.

For example, in 1994 a two-week 'socio-ecological' survey of southern Kunene region was conducted to introduce the idea of establishing locally-managed conservancies to rural communities. A major meeting took place in Sesfontein/!Nani|aus, a relatively large settlement in southern Kunene Region. Shortly after this, I interviewed people from some 20 per cent of 'households' in the settlement (Sullivan 1995). Of the 28 individual and small-group discussions no adults had attended the public meeting. In fact, the majority did not even know that the meeting had taken place and certainly did not realize they had a right to attend and contribute to discussion. This survey was primarily of Damara people, the major group in a location shared with Herero and some Nama and Owambo. Otherwise it included men and women, young and old, and rich and poor. Significantly, the survey suggests that the then Development Committee of Sesfontein, in whom the MET had vested responsibility for informing the wider community of the meeting, had not fulfilled this responsibility (also see Jones 1996; Mosimane 1996: 29; Gaisford 1997; Powell 1998). Relying on local institutions thus is by no means a guarantee that 'community-based' 'joint solutions' will be reached in a consultative and representative manner. Making claims to this effect, however, sidesteps the importance of *evaluating* the process in communication with the range of individuals comprising 'communities' in the broadest sense.

Further, recent analyses are revealing a number of instances where axes of shared differences are exacerbated in CBNRM initiatives, despite their stated focus on equality, representation and empowerment (Marindo-Ranganai and Zaba 1994; Taylor 1999; Twyman in press). If some groups are marginalized despite the inclusive rhetoric of 'community-based natural resources management' then an important issue becomes how to enhance a context for dialogue and negotiation which is more empowering to those groups. A first step might be a commitment to exploring what it is about the economic and symbolic relationships people have, or are perceived to have, with the wider landscape that structures either the occlusion or the elevation of particular groups in CBNRM initiatives. Thus if 'livelihoods are not just about subsistence but also represent notions of identity and provide continuity with the past' (Twyman in press: 10), then engaging with these symbolic complexities might constitute a significant approach to addressing aims of both empowerment and livelihood sustainability (Sullivan 1999, forthcoming).

An obvious issue here is the way in which conservation projects in southern Africa revolve around a limited wildlife of large mammals, inextricable from constructions of a white South African masculine identity linked economically and psychologically to hunting (Mackenzie, 1987; Ellis, 1994; Carruthers, 1995; Skidmore-Hess, 1999; Wels, 1999). Both Leach (1999) and Ellis (1994), for example, make clear the gender implications of links between conservation, firearms, masculinity and warfare. As Ellis (1994: 55) states, there is a long-standing association 'between game parks and military men all over Africa'. Historically and today, amongst European settler and African societies, women have been the 'decorative fringe' to men as hunters and conservationists such that they are conceptually, and sometimes literally, excluded from discussion. Given that symbolically gendered associations with environment and wildlife are so strong, conferring 'distinct relations of disadvantage' for women (McNay 1992; Jackson 1997), it is perhaps surprising that until recently they have been afforded relatively little attention in wider CBNRM discourse.

A number of incidents suggest that these associations conferred a less than enabling context for the participation of women in instituting communal-area conservancies in Namibia. At the final workshop of the 1994 southern Kunene 'socio-ecological survey', for example, all Damara and Herero women who attended the meeting were physically excluded from participating by being obliged to sit outside the shelter in which the meeting was held (Sullivan 2000a). This was justified by MET convenors on the strength that they were working within the constraints of the (male) traditional leadership. Notwith-standing the extent to which current forms of this traditional leadership are a construction of Namibia's colonial history (Krieke 1991; Fuller 1993; Lau 1995), this is somewhat ironic given that the purpose of the meeting was to try to begin a process of new institution-building, enabling better representation and participation in the decentralization of decision-making power (also see Nabane 1995: 12; Gaisford 1997: 32; Matenga 1999; Wels 1999). Interesting in this regard is that Damara do not necessarily observe strict divisions of labour and decision-making along gender lines, and during my own fieldwork people were often quick to draw a distinction in this regard between themselves and Herero. A question here, therefore, is *whose* 'traditional' sensibilities the MET and conservation NGOs were trying to observe.

Namibia's LIFE project is forging attempts specifically to involve women in CBNRM initiatives, through their employment by IRDNC as 'community resource monitors' 'to better exploit natural resource management oppor-tunities and to facilitate the flow of information' regarding resource manage-ment issues (Wyckoff-Baird and Matota 1995: 1). Unqualified claims for the success of women's activators (e.g. Durbin et al. 1997: 40) leave open ques-tions as to the extent to which women are integrated into existing conser-vancy committees, and whether the separation of positions along gender lines compromises the contribution of women to conventionally male domains of decision-making (Nabane 1995).

A second question regards the influence of ethnicity in conceptions of, and claims to, land, resources and decision-making power. Ethnicity is a hoary issue in development debates and especially so in a context such as Namibia, where a unifying ideology of nation-building has been critical in structuring a 'struggle for independence' from the 'divide and rule' policies of an apartheid administration. Further, the former apartheid state tended to reify the static ethnic categories imagined by a missionary and colonial ethnography's 'excessive preoccupation with ethnicity and cultural distinctiveness' (Fosse 1992: 3; Fuller 1993), contributing further to a shying away from the implications of ethnic differences. In considerations of representation in local-level institutions, and in understanding issues infusing use of, and competing claims to, natural resources, however, ethnicity becomes a crucial axis of difference. Particularly important is a recognition that in areas of historically overlapping and contested claims to land it tends to be the same groups who are marginalized from decision-making on account of both culturally-influenced associations with resources, and perceptions of these associations by others (e.g. Marindo-Ranganai and Zaba 1994; Mosimane 1996; Sullivan 1999, 2000a, in press a; Taylor 1999; Gordon and Sholto Douglas 2000; Suzman 2000; Twyman in press).

For example, in north-west Namibia some Damara people travel substantial distances to gather specific resources and many trace ancestral associations in the wider landscape to areas far afield from current settlement locales (Sullivan 1999; Sullivan and Ganuses forthcoming). As has been pointed out to me, if these are important to people in the establishment of conservancy boundaries then they will come up in debate regarding where these boundaries are established (pers. comm. Tagg, 1999). But if the conservancy committee is not representative of these wider issues and practices of resource use and landscape history, then it is highly unlikely that they will feature in boundary debates. The probable outcome of such a situation is that individuals will procure resources much as they have always done, across boundaries not of their choosing and into areas where restrictions may be operative, because these practices remain important in affirming 'who they are'. As long as collectors avoid large mammals, it is unlikely that anyone will take much notice: but one could hardly describe this as a situation which empowers people's diverse interests in land and natural resources. Further, continuing frustration with (perceived) consistent exclusion from CBNRM debates and conservancy establishment is likely to fuel incidents of protest and conflict. This is what seems to have occurred recently at Sesfontein, north-west Namibia. Here, accusations that IRDNC worked primarily with one group over another erupted in protest, involving the enacting of a symbolic burial for the NGO marked by a grave-site. An armed guard subsequently was stationed in the settlement's IRDNC-supported conservancy office to protect it from an unsupportive faction in Sesfontein, despite the location of the office on community land and next to inherited gardens (Sullivan in press a).

Conclusion: 'Donor assistance has been significant, but donor agendas have not dominated' (Jones 1999c: 3)

A recent review of CBNRM in Namibia concludes with the exhortation to 'beware the dominance of donors and the arrogance of academia in trying to categorise and judge the lives of rural Africans and the work of the people at the coalface of conservation' (Jones 1999c: 36). Inappropriate mining metaphors aside, I would suggest that a categorizing and consequent homo-genizing of diverse groupings of people is a key element of a donor-fuelled communalizing discourse. Further, given that most evaluation of CBNRM pro-jects is donor-led and written by a relatively small group of consultants, who in many cases are intimately involved with the formulation and implementation of national CBNRM programmes, I would argue that academic research actu-ally has a crucial role to play – particularly in problematizing criteria for reck-oning the 'success' of projects, and in highlighting issues of representation and revealing alternative perspectives (Brosius et al. 1998). Interestingly, much critique of a communalizing development discourse is being led by scholars from the south (e.g. Escobar 1996; Matenga 1999; Murombedzi 1999). Aca-demic, actor-oriented research is a route whereby long-term and detailed work, exploring local diversity and multiple voices, can make explicit contradictions and tensions between an essentializing ideology of 'community' and local aspir-ations and differences. Admittedly, however, a major challenge facing aca-demic researches which reveal alternative and occluded narratives is an embracing of the responsibility to make these researches available and access-ible to national and international policy discourses (with the attendant criticism this may entail) (Sullivan 2000b, in press b).

As Matenga (1999: 15) points out, a gloss of success in the marketing of southern African CBNRM programmes makes it rather hard to criticize the famed and 'outstanding' (Durbin et al. 1997: 5) CBNRM projects of the region. Clearly, it is preferable that local people benefit from the animal-wildlife with which they live instead of remaining alienated from these resources in a 'fortress conservation' of the colonial past. But underneath the rhetoric, CBNRM is not the radically and qualitatively different approach to conservation that it claims to be. Escobar (1996) argues that a language of emancipation and democratiza-tion is inseparable from a northern modernizing development discourse which asserts conformity and control through donor-funding to the countries of 'the south'. In the case of conservation, a cavalier coinage of the term 'community' is a means of extending the modernizing agenda of the so-called 'Washington consensus' of the World Bank and International Monetary Fund (Peet and Watts 1996), via the various international conventions relating to environment and development and via implementing agencies such as WWF-US and USAID. Through these processes 'communities', as depoliticized and undifferentiated entities, 'are finally recognized as the owners of their territories (or what is left of them), but only to the extent that they accept seeing and treating territory and themselves as reservoirs of capital' (Escobar 1996: 57). In the case of conser-vation in Africa, this means that support is only available to 'communities' if they agree to construct themselves as 'suitable' custodians of internationally-valued biodiversity, particularly animal-wildlife.

A middle class of 'the developed world', collectively the 'virtual con-sumers' (Kiss 1999: 8) of an exotic and spectacularly imaged fauna of 'the south', appears concerned with the pending loss of a 'global resource' of wildlife and 'wilderness'. While now stressing that local people should benefit from this wildlife, a number of perhaps unrealistic, and generally unvoiced, expectations remain: that African communal area residents should continue to live with a sometimes dangerous wildlife on 'their' land; that efforts should be made to foster the increase of populations of these same dangerous, but threatened, species; that this should occur over and above investment in alternative sources of livelihood; that, as donor-funding is phased out, revenue received from conservation efforts should be used to finance newly created communal-area wildlife management institutions; and that a primary responsibility of these institutions should be the negotiation of business agreements which allow private safari operators continued access to the wildlife resources on which their profits depend.

But is it reasonable to expect that a structurally entrenched rural poor should continue to service the fantasies of African wilderness projected by environmentalists, conservationists, tourists and trophy hunters? Or that a communalizing discourse equating rural development and 'empowerment' with wildlife preservation and foreign tourism will be 'sustainable', given both the constraints it imposes on individual aspiration and the dissatisfaction it produces in people who feel excluded? If the world's wealthy wish to retain an ideal of African wildlife and 'wild' landscapes then perhaps we should put our money where our collective mouth is: through direct payment for the service of maintaining wildlife (Simpson and Sedjo 1996; Kiss 1999). In Europe, and under certain conditions, land-use is manipulated through the payment of economically realistic subsidies to individual farmers (for example, under the European Union's arable payment scheme). In some cases this includes 'setting aside' land rather than working or converting it to alternative uses. If conservation boils down to economic incentives, I suggest that it will be 'sustainable' only if accompanied by a 'consumer pays' approach which is honest about the distribution of both interests in, and the costs of, wildlife conservation. This implies nothing short of a secure commitment to substantial and long-term (upwards of several decades) international subsidies directly to local land-users, of amounts realistic enough to compensate for the opportunity costs of not converting either land to alternative uses or large mammals to cash. Failing this, it seems logical that policing and law enforcement, whether by government officials, NGO employees or CGGs, will remain the foundation on which preservation of an internationally-valued animal-wildlife depends. So, what else is 'new'?

Notes

1 My thanks go to Kathy Homewood, James Fairhead, Eugene Marais, Guy Cowlishaw, Martin Evans, Debby Potts, Bill Adams, Mike Taylor, Keith Leggett, Rob Gordon, Heena Patel, Richard Paklepppa, Martin Evans, Brian Jones and Rick Rohde, who all commented on an initial draft of the paper; the views presented remain my own. I'm also grateful to Peter Udovch and Fiona Flintan who made several references available to me. Fieldwork in Namibia was conducted with support from the ESRC, the UCL Equipment Fund and the Nuffield Foundation and the paper is written as part of a British Academy Post-Doctoral Research Fellowship.

2 I do not intend to 'unpack' the validity or otherwise of all the claims made in this letter, but to acknowledge the significance of local observations concerning implementation of conservancy policy. The letter has been described as containing 'untrue or irrelevant allegations' (email to author from B.T.B. Jones, 26 October 1999) and as 'probably libellous' by one of the Directors of IRDNC, one of Namibia's main community-based conservation facilitating NGOs (email to author from M. Jacobsohn, 25 October, 1999).

3 In southern Africa, USAID funds CBNRM programmes in Botswana (Natural Resources Management Programme, NRMP), Zimbabwe (Communal Area Management Programme for Indigenous Resources, CAMPFIRE), Zambia (Administrative Management Design, ADMADE) and Namibia (Living In a Finite Environment, LIFE).

4 As also occurred in East Caprivi (Mosimane 1996: 7).

5 Fifty-one CGGs employed by IRDNC in 1997 (Durbin et al. 1997: 13).

6 Especially given that conflict in Caprivi and reports of arms distributions in Kunene Region suggest that Namibia is not immune to the instability and violence located in various parts of Africa in recent years (Maletsky and Amupadhi 1999; Amupadhi 2000; Amupadhi and Ngasia 2000; Inambao 2000a).

7 Nb. Sections 14(2) and 20 of the Commercial (Agricultural) Lands Act (GRN 1995) provide the government with rights to expropriate, with suitable compensation and under certain circumstances, land otherwise under freehold tenure.

8 President Nujoma's recent offer for Africans throughout the continent and overseas to settle in Namibia's 'vast landscapes' might further exacerbate insecurity regarding rights to communal land (*The Namibian* 1999b).

9 Nb. The 1997 draft *National Land Policy* apparently makes provision for a second Bill which 'will set out forms of family, group and community ownership' (Jones 1999b: 57). This has not yet appeared and it is difficult to see how these will mesh with the most recent *Communal Land Reform Bill*, the remit of which is 'to provide for the allocation of rights in respect to communal land' (GRN 1999: 2).

10 Recently Namibian tourism contributed approximately 5% of GDP and 12% of foreign exchange earnings (after mining and agriculture) and is the only sector experiencing strong growth (Gaisford 1997).

11 The average national population growth rate is calculated as 3.33% (Dewdney 1996).

12 US$1 = approximately N$7.87 in January 2001.

13 Nb. Namibia and South Africa are unique in the distribution of state pensions: income from CBNRM programmes elsewhere is likely to make a proportionately greater contribution to household livelihoods (pers. comm. Debby Potts, Dept. of Geography, SOAS, London).

14 A further US$12 million from USAID has been approved to carry the Namibian CBNRM programme from late 1999 to 2004 (Callihan 1999: 6–7). Jones (1999b: 57) states that the LIFE programme received approx. US$14 million from July 1993 until August 1999, administered primarily by WWF-US. IRDNC received Swiss Francs 2,794,550 from WWF-Intern towards its work in Kunene Region between 1996 and 2001 (Jones 1999b: 76).

15 The LIFE programme estimates that US$28,000 per year is required to run a conservancy while average income will be around US$28,600 plus wages accruing to individuals working for wildlife-related tourism ventures. This is calculated on the basis of an income of 2 x US$13,000 (from both a joint venture lodge operation and a trophy hunting contract), plus US$2,600 from a community campsite. An additional US$18,000 is the approximate figure calculated for wages to members of the community from enterprises established with foreign investments (Callihan 1999: 22).

References

Adams, F. and Werner, W. 1990. *The Land Issue in Namibia: An Enquiry*. Windhoek.

Africare. 1993. 'Rural Water Supply Maintenance in the Kunene Region, Republic of Namibia'. Report to the Africa Bureau of the United States Agency for International Development (USAID). Windhoek.

Amupadhi, T. 2000. 'Protect Your Citizens, Marchers Urge Government', *The Namibian*, 24 February, Windhoek.

Amupadhi, T. and Ngasia, H. 2000. 'Horror Weekend in the North-East'. *The Namibian*, 28 February. Windhoek.

Ashley, C. 1995. 'Tourism, Communities, and the Potential Impacts on Local Incomes and Conservation'. *DEA Research Discussion Paper* No.10. Windhoek.

—— 1997. 'Wildlife Integration for Livelihood Diversification (WILD) Project Plan', Draft document. Windhoek.

—— 1998. 'Intangible Matters: Non-Financial Dividends of Community Based Natural Resource Management in Namibia'. Report for the LIFE (Living in a Finite Environment) project, World Wildlife Fund, Windhoek.

Ashley, C. and Garland, E. 1994. 'Promoting Community-Based Tourism Development: Why, What and How?'. *DEA Research Discussion Paper* No.4. Windhoek.

Ashley, C., Barnes, J. and Healy, T. 1994. 'Profits, Equity, Growth and Sustainability: the Potential Role of Wildlife Enterprises in Caprivi and other Communal Areas of Namibia'. *DEA Research Discussion Paper* No. 2. Windhoek.

Barnes, J. 1995. 'Current and Potential Use Values for Natural Resources in Some Namibian Communal Areas: a Planning Tool'. *DEA Working Paper*. Windhoek.

Barnes, J., and de Jager, J.L.V. 1995. 'Economic and Financial Incentives for Wildlife Use on Private Land in Namibia and the Implications for Policy'. *DEA Research Discussion Paper* No. 8. Windhoek.

Bollig, M. 1998. 'Power and Trade in Precolonial and Early Colonial Northern Kaokoland.' In P. Hayes, J. Silvester, M. Wallace and W. Hartmann (eds). *Namibia Under South African Rule: Mobility and Containment 1915–46*, pp. 175–93. Oxford, Windhoek and Athens.

Brosius, P., Tsing, A.L. and Zerner, C. 1998. 'Representing Communities: Histories and politics of community-based natural resources management'. *Society and Natural Resources*, 11: 157–68.

Callihan, D. 1999. 'Using Tourism as a Means to Sustain Community-Based Conservation: Experience From Namibia.' Report for the LIFE (Living in a Finite Environment) project, Windhoek.

Carruthers, J. 1995. *The Kruger National Park: A Social and Political History*. Pietermaritzburg.

Central Statistics Office. 1994. *1991 Population and Housing Census*. Windhoek.

Collinson, R.F.H. (ed.) 1983. 'Feasibility Study on Utilisation of Wildlife Resources in Zambia'. Report for Motshwari Game (Pty) Ltd. Lusaka.

Dewdney, R. 1996. 'Policy Factors and Desertification – Analysis and Proposals'. Report prepared for the Namibian Programme to Combat Desertification (NAP-COD) Steering Committee. Windhoek.

Durbin, J., Jones, B.T.B. and Murphree, M. 1997. 'Namibian Community-Based Natural Resource Management Programme: Project Evaluation 4–19 May 1997'. Report submitted to Integrated Rural Development and Nature Conservation (IRDNC) and World Wide Fund for Nature (WWF). Windhoek.

Dzingirai, V. 1995. ' "Take Back Your CAMPFIRE". A Study of Local Level Perceptions to Electric Fencing in the Framework of Binga's CAMPFIRE Programme'. Centre for Applied Social Sciences (CASS). Harare.

Ellis, S. 1994. 'Of Elephants and Men: Politics and Nature Conservation in South Africa'. *Journal of Southern African Studies*, 20(1): 53–69.

Escobar, A. 1996. 'Constructing Nature: Elements for a Poststructural Political Ecology'. In R. Peet and M. Watts (eds). *Liberation Ecologies: Environment, Development, Social Movements*, pp. 46–68. London.

Fosse, L.J. 1992.'The Social Construction of Ethnicity and Nationalism in Independent Namibia'. *Discussion Paper* No. 14, Social Sciences Division, Multi-disciplinary Research Centre, University of Namibia, Windhoek.

Fuller Jnr, B.B. 1993. 'Institutional Appropriation and Social Change Among Agropastoralists in Central Namibia 1916–1988'. PhD dissertation, Graduate School, Boston.

Gaisford, W. 1997. 'Healing the Crippled Hand: Tourism and Community-Based Tourism as Sustainable Forms of Land Use and Development in Eastern Tsumkwe, Namibia'. MPhil Thesis, Dept. of Environmental and Geographical Science, University of Cape Town.

Gordon, R.J. and Sholto Douglas, G. 2000. *The Bushman Myth: The Making of a Namibian Underclass.* Second edn. Boulder, Colorado.

GRN. 1991. *Constitution of the Republic of Namibia.* Office of the Prime Minister. Windhoek.

—— 1995. *Agricultural (Commercial) Land Reform Act.* Windhoek.

—— 1997. *National Land Policy.* Windhoek.

—— 1999. *Communal Land Reform Bill.* Windhoek.

Hackel, J.D. 1999. 'Community Conservation and the Future of Africa's Wildlife'. *Conservation Biology,* 13(4): 726–34.

Hall-Martin, A., Walker, C. and Bothma, J. du P. 1988. *Kaokoveld: the Last Wilderness.* Johannesburg.

Hulme, D. and Murphree, M. 1999. 'Communities, Wildlife and the "New Conservation" in Africa'. *Journal of International Development,* 11: 277–185.

Inambao, C. 1998a. 'Torra on Threshold of Brighter Future'. *The Namibian,* 9 September Windhoek.

—— 1998b. 'Hope for Buffalo at Mamili'. *The Namibian,* 8 June, Windhoek.

—— 1999. 'Caprivi Attack "will have lasting Effect on Tourism"'. *The Namibian,* 12, August, Windhoek.

—— 2000a. 'Hundreds Cross Into Botswana'. *The Namibian,* 29 February, Windhoek.

—— 2000b. 'Ivory Sales Fund Elephant Project'. *The Namibian,* 15 November, Windhoek.

—— 2000c. '10 Lodges Close in North-East'. *The Namibian,* 15 February, Windhoek.

—— 2000d. 'Government Rules out Compensation for Wildlife'. *The Namibian,* 29 August, Windhoek.

Infield, M. and Adams, W.M. 1999. 'Institutional Sustainability and Community Conservation: A case study from Uganda'. *Journal of International Development,* 11: 305–15.

Jackson, C. 1997. 'Actor Orientation and Gender Relations at a Participatory Project Interface'. In A-M. Goetz (ed.). *Getting Institutions Right for Women in Development,* pp. 161–75. London.

Jones, B.T.B. 1995. 'Wildlife Management, Utilization and Tourism in Communal Areas: Benefits to Communities and Improved Resource Management'. *DEA Research Discussion Paper* No. 5. Windhoek.

—— 1996. 'Institutional Relationships, Capacity and Sustainability: Lessons Learned from a Community-Based Conservation Project, Eastern Tsumkwe District, Namibia, 1991–1996'. *DEA Research Discussion Paper* No. 11. Windhoek.

—— 1998. 'Namibia's Approach to Community-Based Natural Resources Management (CBNRM): Towards Sustainable Development in Communal Areas'. Paper for the Scandinavian Seminar College Project: Policies and Practices Supporting Sustainable Development in Sub-Saharan Africa. Windhoek.

—— 1999a. 'Policy Lessons From the Evolution of a Community-Based Approach to Wildlife Management, Kunene Region, Namibia'. *Journal of International Development*, 11: 295–304.

—— 1999b 'Rights, Revenues and Resources: the Problems and Potential of Conservancies as Community Wildlife Management Institutions in Namibia'. *Evaluating Eden Working Paper* No. 2. London.

—— 1999c. 'Community-Based Natural Resource Management in Botswana and Namibia: an Inventory and Preliminary Analysis of Progress'. *Evaluating Eden Series Discussion Paper* No. 6. London.

—— 1999d. 'Community Management of Natural Resources in Namibia', *Drylands Programme Issues Paper* No. 90. London.

Kiss, A. 1999. 'Making Community-Based Conservation Work'. Paper presented at a Society for Conservation Biology annual meeting, College Park, MD, 18 June.

Krieke, E.H.P.M. 1991. 'Historical Dynamics of Traditional Land Tenure in Ovamboland'. NEPRU Working Paper in *Proceedings of the National Conference on Land Reform and the Land Question,* pp. 545–65. Windhoek.

Lau, B. 1995. '"Thank God the Germans Came": Vedder and Namibian Historiography'. In A. Heywood (ed.). *History and Historiography: 4 essays in reprint,* pp. 1–16. Windhoek.

Leach, M. 1999. 'New Shapes to Shift: War, Parks and the Hunting Persona in Modern West Africa'. Audrey Richards Commemorative Lecture, Centre for Cross-Cultural Research on Women, University of Oxford, 28 April.

Mackenzie, J.M. 1987. 'Chivalry, Social Darwinism and Ritualised Killing: the Hunting Ethos in Central Africa up to 1914'. In D. Anderson and R. Grove (eds). *Conservation in Africa: People, Policies and Practice,* pp. 41–61. Cambridge.

Madzudzo, E. 1999. 'Community Based Natural Resource Management in Zimbabwe: Opportunities and Constraints'. Paper presented at a conference on African Environments – Past and Present, St. Antony's College, University of Oxford, 5–8 July.

Mafune, I. 1998. 'Common Property Regimes and Land Reform in Namibia: a Case Study of Skoonheid, Omaheke Region'. MPhil thesis, Dept. of Environmental and Geographical Science, University of Cape Town.

Maletsky, C. 1998a. 'Himba Seize Initiative'. *The Namibian,* 4 June, Windhoek.

—— 1998b. 'Himbas in Conservancy Move'. *The Namibian,* 29 July, Windhoek.

—— 1999c. 'NUNW Raises Fencing Fears'. *The Namibian,* 23 June, Windhoek.

—— 2000. 'No Payout for Wildlife Damage, Says Minister'. *The Namibian,* 25 February, Windhoek.

Maletsky, C. and Amupadhi, T. 1999. 'Mystery Over Pre-poll Arms Claims'. *The Namibian,* 8 October, Windhoek.

Marindo-Ranganai, R. and Zaba, B. 1994. 'Animal Conservation and Human Survival: A case study of the Tembomvura People of Chapoto Ward in the Zambezi Valley, Zimbabwe'. Harare University Research Paper, Harare.

Matenga, C.R. 1999. 'Community-Based Wildlife Management Schemes in Zambia: Empowering or Disempowering Local Communities?' Paper presented at a conference on African Environments – Past and Present, St. Antony's College, University of Oxford, 5–8 July.

McNay, L. 1992. *Foucault and Feminism.* Cambridge.

MET 1995a. *Policy Document: Promotion of Community-Based Tourism.* Windhoek.

—— 1995b. *Policy Document: Wildlife Management, Utilisation and Tourism in Communal Areas.* Windhoek.

—— 1998. *Some Facts and Figures About Communal Area Conservancies*, Windhoek.

—— n.d. *Communal Area Conservancies in Namibia: a Simple Guide*. Windhoek.

Moorsom, R. 1982. *Transforming a Wasted Land*. London: CIIR.

Mosimane, A.W. 1996. 'Community Based Natural Resource Management in East Caprivi: A case study of the Choi Community'. *SSD Discussion Paper* No. 16. Windhoek.

—— 1998. 'An Assessment on Knowledge and Attitudes About Kwandu Conservancy and the Socio-Economic Status'. Report for the Social Sciences Division, Multi-disciplinary Research Centre, University of Namibia, Windhoek.

Moyo, T. 1999. 'Namibia's Tourist Data "Not Accurate"'. *The Namibian*, 28 October Windhoek.

Murombedzi, J.C. 1999. 'Devolution and Stewardship in Zimbabwe's Campfire Programme'. *Journal of International Development*, 11: 287–93.

Murphree, M.W. 1993. 'Communities as Resource Management Institutions'. *Gatekeeper Series* No. 36. Sustainable Agriculture Programme, International Institute for Environment and Development (IIED). London.

MWCT 1992. *Policy Document: the Establishment of Conservancies in Namibia*. Windhoek.

Nabane, N. 1995. 'Gender as a Factor in Community-Based Natural Resource Management: A case study of Nongozi, Linashulu, Lizauli and Sachona Villages in East Caprivi – Namibia'. Report submitted to WWF-LIFE programme. Windhoek.

Neuman, R.P. 1997. 'Primitive Ideas: Protected area buffer zone and the politics of land in Africa'. *Development and Change*, 28(3): 559–82.

Nujoma, S. 1998. Statement by His Excellency President Sam Nujoma on the Occasion of the Launch of Communal Area Conservancies and the Gift to the Earth Award Ceremony, Okapuka. 26 September, Windhoek.

Owen-Smith, G. 1995. 'The Evolution of Community-Based Natural Resource Management in Namibia'. In N. Leader-Williams, J.A. Kayera and G.L. Overton (eds). *Community-Based Conservation in Tanzania*, pp. 135–42. Dept. of Wildlife, Dar es Salaam.

Patel, H. 1998. 'Sustainable Utilization and African Wildlife Policy: the Case of Zimbabwe's Communal Areas Management Programme for Indigenous Resources (CAMPFIRE)'. Report for Indigenous Environmental Policy Centre. Cambridge, MA.

Peet, R. and Watts, M. (eds).1996. *Liberation Ecologies: Environment, Development, Social Movements*. London.

Powell, N. 1998. *Co-Management in Non-Equilibrium Systems: Cases from Namibian Rangelands*. Agraria 138, Uppsala.

Reardon, M. 1986. *The Besieged Desert: War, Drought, Poaching in the Namib Desert*. London.

Roe, D., Grieg-Gran, M. and Schalken, W. in prep. *Getting the Lion's Share from Tourism: Private Sector–Community Partnerships in Namibia*. Report prepared for IIED and NACOBTA, London.

Roe, E., Huntsinger, L. and Labnow, K. 1998. 'High Reliability Pastoralism'. *Journal of Arid Environments*, 39: 39–55.

Shigwedha, A. 2000. 'NC Rejects Communal Land Bill, Says it Will Make the Poor Poorer'. *The Namibian*, 18 May, Windhoek.

Shivute, O. 1998. 'Conservancy Plan Sparks Tribal Row'. *The Namibian*, 16 July, Windhoek.

Simpson, R.D. and Sedjo, R.A. 1996. 'Paying for the Conservation of Endangered Ecosystems: A comparison of direct and indirect approaches'. *Environment and Development Economics*, 1: 241–57.

Skidmore-Hess, C. 1999. 'Flora and Flood-Plains: Technology, Gender and Ethnicity in Northern Botswana 1900–1990'. Paper presented at a conference on African Environments – Past and Present. St. Antony's College, University of Oxford, 5–8 July.

Sullivan, S. 1995. 'Local Participation in Community Resource Management Initiatives: Findings of Interviews with Individuals Following the MET's Socioecological Survey Meeting in Sesfontein, April 1994'. Document Prepared for B.T.B. Jones, Directorate of Environmental Affairs, Windhoek.

—— 1999. 'Folk and Formal, Local and National – Damara Knowledge and Community Conservation in Southern Kunene, Namibia'. *Cimbebasia*, 15: 1–28.

—— 2000a. 'Gender, Ethnographic Myths and Community Conservation in a former Namibian "Homeland"'. In D. Hodgson (ed.). *Rethinking Pastoralism: Gender, Culture and the Myth of the Patriarchal Pastoralist*, 142–64. London.

—— 2000b. 'Getting the Science Right, or Introducing Science in the First Place? Local "Facts", Global Discourse – "Desertification" in North-West Namibia'. In P. Stott and S. Sullivan (eds). *Political Ecology: Science, Myth and Power*, pp. 15–44. London.

—— in press a. 'Protest, Conflict and Litigation: Dissent or libel in resistance to a conservancy in North-west Namibia'. In E. Berglund and D. Anderson (eds). *Ethnographies of Environmental Under-privilege: Anthropological Encounters with Conservation*. Oxford.

—— in press b. 'Detail and Dogma, Data and Discourse: Food-Gathering by Damara herders and conservation in Arid North-West Namibia'. In K. Homewood (ed.). *Rural Resources and Local Livelihoods in Africa*. Oxford.

—— forthcoming. 'The "Wild" and the Known: Implications of identity and memory for CBNRM in a former Namibian "homeland"'. In C. Twyman and M. Taylor (eds). *Entitled to a Living: CBNRM in Southern Africa*. Oxford.

Sullivan, S. and Ganuses, S. forthcoming. *Faces of Damaraland: Life and Landscape in a Former Namibian 'Homeland'* (working title). Cimbebasia Memoir Series, Windhoek.

Sutherland, J. 1998. 'Top Award for Namibia'. *The Namibian*, 28 September. Windhoek.

Suzman, J. 2000. *'Things From the Bush': A Contemporary History of the Omaheke Bushmen*. Introduction by R.J. Gordon. Basel.

Tagg, J. 1999. Personal communication. Directorate of Environmental Affairs, Ministry of Environment and Tourism. Windhoek.

Tarr, J.G. 1998. 'Summary Report of a Retrospective Study of the Environmental Impacts of Emergency Borehole Supply in the Gam and Khorixas Areas of Namibia'. *DEA Discussion Paper* No. 25, Windhoek.

Taylor, M. 1999. '"You Cannot put a Tie on a Buffalo and Say that is Development": Differing Priorities in Community Conservation, Botswana'. Paper presented at a conference on African Environments – Past and Present, St. Antony's College, University of Oxford, 5–8 July.

The Namibian. 1998. 'Farmers Fed up with Elephants'. *The Namibian*, 13 March. Windhoek.

—— 1999a. 'Nujoma Invites Africans Across the World to Settle in Namibia'. *The Namibian*, 14 July. Windhoek.

—— 1999b. 'Residents Celebrate New Salambala Conservancy', *The Namibian*, 25 January. Windhoek.

Twyman, C. in press. 'Livelihood Opportunity and Diversity in Kalahari Wildlife Management Areas, Botswana: Rethinking Community Resource Management'. *Journal of Southern African Studies*.

Wels, H. 1999. 'The Origin and Spread of Private Wildlife Conservancies and Neigh-
bour Relations in South Africa, in a Historical Context of Wildlife Utilization in
Southern Africa'. Paper presented at a conference on African Environments –
Past and Present, St. Antony's College, University of Oxford, 5–8 July.
Wyckoff-Baird, B. and Matota, J. 1995. 'Moving Beyond Wildlife: Enlisting Women as
Community Resource Monitors in Namibia'. Paper for the Living In a Finite
Environment (LIFE) programme, Windhoek.

10

Representing the Resettled

THE ETHICAL ISSUES RAISED BY RESEARCH AND
REPRESENTATION OF THE SAN

Sue Armstrong and *Olivia Bennett*

Many people have come and gone with our words ... they never return,
and nothing ever happens (Mogwe 1992: 49).

Introduction

Botswana's aboriginal people, the San, are fighting a last-ditch battle against dispossession of their ancestral lands. This critical moment coincides with growing
resentment at the seemingly endless interest of academics and journalists in
their communities, and what they see as the persistent failure of these people to
represent the San and their concerns as they would wish them represented. Attitudes have hardened, resulting in the first attempt by San organizations throughout Southern Africa to draw up a protocol for the media and others to govern
visits to the San and coverage of their stories. This chapter examines some of the
ethical and practical dilemmas surrounding the issue of representation and *mis*-
representation by others of people whose own voices are rarely heard.

The San are, by tradition, hunter-gatherers who once roamed the African
continent from the Cape to Angola, and northeast to Zimbabwe. They are
also known as 'Bushmen', a term deriving from the first Dutch settlers in the

Cape in the 1600s. They thought the small wiry people they found roaming the *veld* were the African equivalent of the orang-utan or 'forest man' they knew from Indonesia, and christened them *bosjesmans*.

Classed as 'vermin' because they harassed the settlers, San could be shot on sight. Hunted also by black tribes who moved into their territory, the San have been progressively driven off the land everywhere by stronger peoples who have seen them as little different from the game animals who move with the seasons and leave scarcely a mark on the territory. No one has recognized their right to the land off which they live, few have shown respect for their culture, and today in Angola, Zimbabwe, Zambia and South Africa, the remaining San live in small bands on the margins of mainstream society. Only in Namibia do a few retain any scope for living according to their hunter-gatherer traditions. In Botswana, too, the people and their ancient culture are being pushed to extinction by the twin forces of prejudice and insensitive settlement programmes.

Throughout the region, the dispossession of the San's land has been a creeping phenomenon. In Botswana the process has been more recent. As the cattle industry has grown to dominate the economy, huge areas of the Kalahari have been carved up into private ranches. Many San families remained living on these cattle ranches, but as labourers with no security. When they are no longer needed, perhaps sick or too old to work, they are viewed as mere squatters on the land. Many join other destitute families who roam the Kalahari looking for work, or settle on the fringes of desert towns such as Ghanzi and Maun.

History of Dispossession

In 1961, the British colonial authorities proclaimed a huge tract of the Kalahari desert as a game reserve, specifically to protect the San from loss of land and resources. The Central Kalahari Game Reserve (CKGR) was to be a place where the San could live their traditional lives in peace, and the first President of Botswana, Sir Seretse Khama, vowed to honour this arrangement. But in modern times the recognition that the Reserve, with its abundant game and 'pristine' wilderness, represented a hugely valuable resource, especially for tourism, led the Botswana government to start removing the people to the official settlements in the mid-1990s. Several hundred bands of San inhabited the CKGR, and the government claimed that their presence was incompatible with game conservation. The removal programme contradicts official policy, which states that the local people must be actively involved in, and direct beneficiaries of, conservation activities. It is particularly inexplicable since the San's knowledge and respect for the environment are legendary. The government's justification of its removal policy is that the provision of modern services – water, education, healthcare, housing – is out of place in a Game Reserve, and that such services can only be delivered to settled groups.

The San were already on the margins of mainstream life and virtually excluded from the political process, but any possibility of their mounting united resistance to the removals was pre-empted by the way in which the government implemented the programme. Households were often approached individually, offered a deal and told their neighbours had already agreed to the terms. The implication was that they would be left alone in the wild 'un-serviced' reaches of the Kalahari, while everyone else found new opportunities in the settlements.

The CKGR resettlement is but one case, affecting two of the ten San languages groups in Botswana (some 2,000 of the more than 50,000 San in the country). The Botswana government had begun the Settlement Programme all over the country in the mid-1970s. This offered San who had been pushed off the land – primarily by cattle-farming groups – subsistence plots and houses on sites where they were promised basic facilities (permanent water, clinics, and schools) as well as help with projects to generate income. But the development scheme was based on the assumption that the San's problem was poverty, not prejudice and discrimination against their culture, and it has been 'disastrous', according to a spokesman for the First Peoples of the Kalahari, an indigenous human rights organization.[1] With the removal of the few people still living in the Central Kalahari Game Reserve the last group of San still able to hunt and gather will lose far more than prospective title to land; they will lose a way of life.

Mindful of the cruelties of apartheid in neighbouring South Africa, the Botswana government determined at independence to ignore ethnic and cultural differences and to treat all its citizens as one people. The Settlement Programme was at one point known as the Remote Area Dwellers (RADs) Programme, intended to deliver services to people of any ethnic group who lived beyond their reach. These included some of the Bakgalagadi, Herero and Hambukushu peoples. However, in the words of anthropologist Sidsel Saugestad: 'To borrow a concept from gender studies, one can say that the RAD programme, in its effort to be *culture-neutral,* had become *culture-blind.* The carefully worded neutrality of the RAD programme, trying to be all things to all people, had in effect deprived the target group of a cultural identity' (Saugestad 1994: 11). The government's reluctance – in this instance – to acknowledge ethnic identity backfired, for the San's cultural difference is at the very root of their disadvantage and poverty. The other RADs sharing the settlements, although among Botswana's poorest, still had a status higher than they, and tended to monopolize the services and any opportunities for employment. For the San in the settlements, there is practically no scope for earning a living beyond selling crafts to passing traders. Opportunities to gather *veld* foods on land around the settlements are minimal, as it is heavily over-used. Moreover, their freedom to hunt has been drastically proscribed by laws.

Nothing has been done to address the San's stigmatized status and twenty years after they were first established on marginal land off the beaten track, the settlements are places of despair and social disintegration. Alcoholism, early teenage pregnancy and family violence are rife. And people who once dealt with conflict by simply shouldering their belongings, quenching their fires and moving away from one another in the desert now resort frequently to fighting – or else apathy. Two decades of development have seen only a few San reach the level of a university education, or really prosper in mainstream society. The people are still not recognized as a separate ethnic group with specific needs, and have only marginal representation in the House of Chiefs or parliament.

Social Impoverishment

One major reason why resettlement has proved such an impoverishing experience so far is that the San are expected to join mainstream society on its terms, not their own. Integration means adopting the language and lifestyle of the Batswana, and witnessing the steady erosion of their own languages and culture. Schooling is conducted in Setswana, almost exclusively by teachers who do not speak a San language and are often unable even to pronounce the children's names correctly because of the complicated 'clicks', and so simply give them new names. San marriage rituals are not recognized, and the people are discouraged and scorned for burying their dead uncoffined, according to tradition. The loss of land and natural resources has also meant the loss of essential components of their culture, affecting their harmonious existence with nature and spiritual well-being, as well as their nutrition. The deep hurt which such routine discrimination inflicts on people is incalculable. Anthropologist Hugh Brody describes it well: 'The anger of tribal people is intense, but often directed inward. And they fall into a deep silence ... [having] absorbed the lessons of their oppressors: indigenous customs, history and ways of speech are matters of shame ... Shame and grief, accumulated from generation to generation, can tie the tongue tight' (Brody 1999: 45).

Settlement has brought radical changes in social organization and power relationships. Families and once-cohesive communities have broken up as members search for paid employment. Over the years, a 'lost generation' has developed. Many youngsters never finish school. And even those who do, find it a mixed blessing. They have little knowledge of their own culture because it was ignored by the education system, yet they often remain excluded from mainstream society because of discrimination and cripplingly low self-esteem.

Now the people are fighting what could be their final battle for the survival of their way of life as the government pushes ahead with removal of the few remaining San from the Central Kalahari Game Reserve. In fact, few but the oldest San really hanker for a return to the old ways. Many are prepared to become part of modern society, but on the basis of mutual respect, equal citizenship and opportunity. They see the issue of the CKGR as one of basic human rights. The land belongs to them, and far from giving it up for a few promises and some cash compensation, they want a stake and an active role in developments there.

History of Representation

The paradox is that the San have sunk into this state of poverty and power-lessness, yet are by no means an 'unknown' or forgotten people. They have been the subject of continual and intense scrutiny by academics and writers for centuries. But the wealth of coverage has nearly always said more about the observer than the observed. The dominant image of the San is perhaps that of the 'noble savage', exemplified in the works of Laurens van der Post. 'Perhaps', he writes, 'this life of ours, which begins as a quest of the child for the man, and ends as a journey by the man to rediscover the child, needs a clear image of some child-man, like the Bushman, wherein the two are lovingly joined in order that our confused hearts may stay at the centre of their brief round of departure and return' (van der Post 1962: 13) The extent to which this infan-talizes the San needs no spelling out. There is a reluctance in observers who see the San like this to let them 'grow up', join the modern world, and develop dreams and aspirations at variance with their mythologized image.

Van der Post singles them out from all races – of any colour – as somehow purer in their relationship to the earth. The idea that hunter-gatherers seem to represent something fundamental in our relationship to nature that has been lost by modern humans has been widely commented on. But it is also a fact that, almost universally, nomadic groups are seen as a threat to modern govern-ment, and 'encouraged' or forced to settle and adopt mainstream customs. George Monbiot describes this love/hate relationship thus:

> Civilization, from the Latin *civis*, a townsperson, means the culture of those whose homes do not move. The horde, from the Turkish *ordu*, a camp and its people, is its antithesis, which both defines civilization and threatens it. We, the stayers, detest the movers ... This is partly because we feel that they threaten us or our property. But I have come to believe that there are more substantial reasons for our disdain. We hate them because they remind us of who we are (Monbiot 1994: 1).

Another strand in representation of the San has been the view of them as curiosities: fascinating, but barely human. During the nineteenth century, they were shipped to Europe and paraded in circuses. San body parts were catalogued and displayed in many European museums. They were – as they still are – a favourite subject for anthropologists: 'In the conventional know-ledge of the eighteenth and nineteenth centuries, the "Bushmen" were among the most primitive of human "types" and would, therefore, be sure to provide a wealth of information' (Morris 1996: 68). Living people were endlessly measured, drawn and photographed, and treated as anatomy specimens without feelings or dignity.

However, not all views of them were romantic, benignly curious or, as the scientists would put it, 'objective' and 'neutral'. To those who tried to settle and farm the land, the San were a terrible nuisance, who rustled cattle, ambushed farmers, and proved 'untameable' as labour. They could be shot with impunity, even in the early decades of the twentieth century. The origin of the name Basarwa, used by the Batswana for the San, is *ba-sa-rua dikgomo*, which means 'those who do not rear cattle' – thus defining them solely in negative terms and implicitly implying poverty by Tswana standards. So even the names for the San are often 'coined by others from inside their own experience' (Mogwe 1992: 3).

Today, there is no universally acceptable collective term. Some reject the name 'Bushmen', because of its origins and the suggestion that the people are less than human. Some associate themselves positively with the name and are anxious to retain it. The organizations representing the people made a joint decision to use the collective term 'San' which, although introduced by anthropologists, does occur in several of the languages, and is more neutral than Bushmen. But people usually prefer to be called by their own names, eg Ncoakhoe for the Naro and Ju/hoan for the !Kung. A key failing with all the collective terms – RADs, San, Bushmen, Basarwa – is that they do not acknowledge that there are different groups, with their own languages, practices and history. This homogenization has been one element in the denial of a political dimension to the San, blurring any distinctive identities, different demands or varied aspirations.

The degree to which outsiders' perceptions of the San have been the dominant images was demonstrated most clearly at a 1996 exhibition in South Africa. The exhibition, entitled 'Miscast', focused on what others had done to the San and how they had represented them down the centuries. The San who saw the exhibition were deeply shocked, as it threw into stark relief this history of exploitation and misrepresentation. Yet even the exhibition, however well meant, seemed a continuation of that process. In the words of one visitor:

> We move into the next hall … The entire space is covered with enlarged newspaper and other clippings … printed onto the [floor] tiles … part of the floor shows photographs of chained, naked Bushmen with questioning, desperate eyes … Right under our feet are pictures of Saartjie Baartman from 1810, one displaying her wide, protruding buttocks, another offering a full frontal of the naked woman. We are just about standing on close-ups of her genitals. *The Bushman Venus,* the caption says, *who captivated audiences across Europe* (le Roux 2000: 189)

The intention, in spreading images across the floors of the exhibition so that visitors were forced to walk over them, was to bring home to people the degrading effect of the treatment of the San. However, for San visitors it merely reinforced their sense of shame and exclusion.

This is perhaps a classic example of the pitfalls inherent in attempts to represent the experience of others. Had the San themselves been the curators of the exhibition, would they have chosen such a device to make the point? Clearly this is a complex issue, involving questions of sophisticated museum design and visual literacy. Efforts *were* made to consult the San during the preparation of the exhibition, but these took place 'through the medium of attorneys and other agents' (Martin 1996: 9). Nevertheless, the bottom line was that they felt they were further humiliated by an exercise that was intended to support their sense of injustice. If it did anything for the San sense of identity, the exhibition served to strengthen a growing feeling that they, rather than outsiders, must be the ones to define and represent themselves.

One Voice; Many Voices

Since the mid-1980s, a number of San organizations have been founded all over Southern Africa and they are beginning to articulate for themselves what they want from development. Yet it remains extremely hard to have any influence on the decisions that affect their lives. They have so few representatives at any significant level in any of the professions or in government service that others mostly continue to speak for them. It is fair to say that most San development or networking organizations are in fact headed by non-San, although their boards are composed entirely of San.[2] Moreover, the attraction of the San for journalists, anthropologists and cultural tourists shows no sign of abating. As their living conditions continue to deteriorate, the sense of grievance is hardening. In a group discussion with Ditshwanelo (Botswana's independent human rights organization), one old man told lawyer Alice Mogwe, 'Many people have come and gone with our words ... they never return and nothing ever happens' (Mogwe 1992: 49).

The San have started to act upon their frustration at being the objects of so much attention from which they see little benefit. 'We have no voice in this country,' said a spokesperson for First People of the Kalahari in December 1996, at a seminar organized by the Media Institute for Southern Africa. Some of the barriers to communication in the modern world – with its focus on the written word – are slowly being broken down: appropriate orthographies for some San languages are being developed for the first time. However, there are other practical difficulties – the people have little chance to develop the skills of modern forms of communication. There are, for example, no San filmmakers or journalists.

But other difficulties over the issue of representation are less tangible – and the ways to deal with them less clear-cut. An obvious challenge for the San is to change the relationship between themselves and their observers, so that representations of them correspond more closely with what they feel to be true. But how? One attempt has been the drawing up of a protocol for media people and academics who visit the San. An early draft included perfectly reasonable requests for clear explanations of who the visitors are, what their purpose is, and what expectations they may have of the people they are visiting. However, some of the proposed principles – such as furnishing the community with details of any budget or funding, and the suggestion that finished articles be made available to the community before the public viewed them – were more contentious.

From a journalistic point of view, adopting what could be considered a censorial approach may be counterproductive. Trying to impose restrictive rules (that may anyway be unenforceable) will simply antagonize people, most of whom will not have set out intentionally to exploit the San. Moreover, it is impossible to suppress a story if people are determined to get it, and such rules make it more likely that reporters will write their stories from second-hand information rather than taking the trouble to visit and see the situation for themselves, hear what the people have to say, and be faced with the complexity of real life. It would be a tragedy if, in their justifiable attempt to change the *status quo*, San organizations were to overlook the power of the media, despite its many flaws, to keep the issue of oppressed people alive in the outside world.

The San Oral Testimony Project

Yet the San's demand to see any material about them before the general public does is understandable, given that so often in the past they never saw the end products of research or journalism, which were authored by outsiders primarily for non-San audiences. The Panos Institute is an information organization founded on the belief that the *way* information is generated, and by *whom*, is as important as the end product. The Institute runs an Oral Testimony Programme (OTP), which works in partnership with mainly community-based groups in parts of Africa, Asia and Latin America. It supports the gathering and communicating of oral testimonies by local people from local people on specific development issues. The primary aim is to amplify the voices of those often most directly affected by development policy, but least able to influence it – people like the San, who wish to speak for themselves rather than have their needs interpreted by outsiders.

Under Panos' OTP, we are both involved in a project in which young San men and woman from four different communities in Botswana and Namibia (with different languages) have been trained in oral testimony collection and are tape-recording interviews with other San, focusing on resettlement. Panos often trains and works with groups who have little or no experience of such activities, taking conventional oral history methodology and adapting it to the communities' circumstances and the topics in question. Since a key aim is to strengthen the capacity of partners and narrators to represent their own concerns, in their own words, the OTP emphasizes the value of the *process* of information gathering as well as the product. But this doesn't mean that we escape some of the dilemmas inherent in such work. Power relationships are still present, and more often than not tipped in favour of the interviewer and project participants and organizers.

The aim of this Panos oral testimony project is for the San themselves to gather accounts of people's first-hand experience of resettlement. The communities, working with Kuru Development Trust – a San self-help organization – will produce their own publications based on the interviews, and have plans for other information activities. Such personal accounts certainly don't replace academic or quantitative research, or indeed good journalism, but they can illuminate it, and sometimes challenge it. The project will also link the San's experience with those of other communities around the world, whose first-hand experience of rebuilding lives in the wake of resettlement has lessons for the development community. Of course these interviews will reflect subjective concerns and individual biases. But what people choose to highlight, downplay or ignore will be clues to understanding how they felt, how they have made sense of events, and what their priorities are.

This is important, for what has been missing in the debate is the San perspective on resettlement. One of the least understood yet most pernicious effects of resettlement is not economic but social and cultural impoverishment. So there has to be value in listening to people trying to articulate the invisible threads which hold societies and families together – and the way these have been altered as a result of relocation. Far from being a limitation, the subjective remembering and telling of events is a valuable resource – as the work of many oral historians has shown (Thompson 1988; Perks and Thomson 1998) However, pockets of resistance remain: some development professionals, who may have a major role in policy-making, remain sceptical about the value of subjective personal accounts.

In particular, those who elevate 'technical, expert knowledge' over 'the knowledge derived from experience [and] common sense' feel that the former, 'based on the study of a problem', has an objectivity – and therefore a validity – which knowledge derived from 'the subjective living of that problem' can never have. As John Gaventa points out, this leads naturally to the position that 'The experts may study the powerless, but must not experience the problems they face, or identify with them, for fear of losing their objectivity' (Gaventa 1993: 21–40). This is, of course, one way of justifying the phenomenon remarked upon by anthropologist Mathias Guenter in connection with a conference he attended, whereby time and again ' the task of representing the Bushmen fell, as it typically does at such gatherings of scholars, not to any of the San but to the "Sanologists", the academics who were the principal participants at the conference' (Guenther 1994: 19).

Whatever its shortcomings, the OTP is an attempt to democratize information gathering and dissemination. The resulting testimonies will be researched and authored by the San, edited by them into booklets for the community, and any Panos use of the material agreed by them. This may not always be simple. There may be extracts from the interviews that seem particularly valuable to us, producing material for an international audience alongside other accounts of resettlement, but which to the San seem of little significance or which they would rather leave unpublished. Indeed, the consultation with our San partners over the writing of this paper brought home the need to remember and respect the fact that the San's relationship with outsiders has so often been one of exploitation, persecution or prejudice. There are precious few examples of collaborative relationships run on an equal basis. Trust isn't a given: it has, quite rightly, to be earned.

Gatekeepers

There are impediments to gaining that trust. For journalists, pressures come from the demands of the industry: its prime concerns are its circulation, and its audience. Issues of cost-effectiveness, topicality, impartiality, and 'sexiness' of stories rather than 'worthiness', militate against reflecting the detail and nuances that seem important to the people themselves. And encouragement is rare from editors – the 'gatekeepers' – to put in the time and commitment to making reporting a more collaborative exercise.

Academics know well the constraints imposed by their own disciplines. And even when there is a clear commitment to tackle the issue of misrepresentation, difficulties arise, as Pippa Skotnes explains in the introduction to *Miscast*, the book of the exhibition. She writes of the numerous challenges she faced in mounting the exhibition:

> Most of these centred on my attempts to present a Bushman or San voice (or, better still, many voices). However, the processes of dispossession and marginalisation have been so successful that this was exceedingly difficult to achieve. In the end I worked through various organisations which represent San interests in South Africa, Namibia and Botswana, through legal advisers and through anthropologists (Skotnes 1996: 18–19).

'Stories and Scars'

As Skotnes and Brody have highlighted, the intense stigmatization of the San from the earliest days of their contact with others has been internalized. This in itself is another real impediment to the San representing themselves to the outside world in ways that will genuinely deepen others' understanding of their position. We have experience of this: the San participants in our OT training workshop markedly lacked self-esteem and confidence. Indeed, so powerful were the processes of dispossession that some of the younger participants initially seemed unsure of the idea that their past was worth exploring and documenting. There was ample evidence that the breakdown in continuity between generations is an important factor in the social impoverishment and confusion that the San are experiencing today. Older San who have never been to school express sadness and anxiety about the alienation of the younger generation through education. They know that far from taking pride in the fireside dancing, for example, many youngsters see it as 'primitive' and are ashamed at the near-nakedness of their mothers and sisters (particularly at performances staged for tourists). Many have no idea how to read the *veld* and to survive by hunting and gathering, and scant respect for those who still can.

Hugh Brody again writes well on this issue:

> Those who endure protracted colonial oppression internalise the attitudes of those who oppress them ... No wonder, then, that the surviving descendants of these tribes, the men and women whose oral heritages contain the stories as well as the scars of these relentless and merciless events, have an intensely difficult and complicated relationship to their own voices (Brody 1999: 45).

But Brody also points out that the silence of 'shame and grief' is being broken down, and part of that process is the assertion by such groups of the value of their own language(s) as a means of identity as well as expression. And in the course of the Panos OT workshop, as discussion deepened and confidence grew, more participants acknowledged that traditional skills and stories had a continuing value and interest, within and outside their communities, and were not merely 'primitive' customs despised in the modern world. As they began to acknowledge their own, if sometimes ambiguous, pride in the past, and the important link for the San between language and land, knowledge and rights, so their enthusiasm grew to listen to the older members of their communities, and record their stories in their own words.

Many months later, the project has taken off.[3] New interviewers have joined and when the local coordinators held sessions with interviewers and narrators to discuss and sometimes unravel the meaning of the testimonies, they were joined by San of all ages. The meetings turned into lively community discussions. While these sessions demonstrated the enthusiasm of all those involved, the limited educational background of the San and their lack of exposure to the outside world make it hard to appreciate the potential impact of their words beyond the immediate environment, or their contribution to a general development debate. Focusing their discussions with narrators and sticking with promising lines of enquiry can be hard – not least because they have a culturally different way of sharing stories or information. So one of the real difficulties of self-representation for groups like the San is that the international audiences they wish to influence are more often than not dominated by western cultural conventions – if the San do not 'speak the same language', they will have great difficulty communicating their point of view. Creating information products, even for the local market, will entail more community workshops and painstaking discussion and consultation so that everyone feels as involved in this enterprise as they have done in the narration of their experiences. And one lesson that has been painfully learnt by Kuru as well as other NGOs, is that there are huge dangers in taking things too fast – especially where the gathering of information is concerned.

For Panos, dependant on donor funding for such projects, an impediment to real empowerment of the San in ways which work for them are donors' requirements for smooth progress, the meeting of deadlines, and measurable results of their funding: so many testimonies gathered, booklets published, and new initiatives taken. Even during the OT San workshop, it was becoming apparent that in this case the *process* – so hard anyway to evaluate in conventional terms – might well be more significant than the measurable products. If this project is genuinely to contribute to the San's ability to represent themselves in ways that have meaning for them *and* for wider audiences, it must allow them the right to develop these skills, and at their own speed.

Thus the conventions of the media, of academia and of international funding – all forms of gatekeeping – constrain the San's attempts to represent themselves. And of course the San organizations themselves are gatekeepers of a sort. Here the question arises: have these organizations the right to determine policy governing information gathering and dissemination on behalf of everyone in the community? We mentioned earlier how the lumping together of the San into one group has obscured the presence of a multiplicity of views and priorities. And one of the dangers of over-reliance on representative organizations to speak for any group is that they may, wittingly or unwittingly, ignore, overlook or underplay some of the range of opinion. There can never be a line that is acceptable to all, and the question of mandate to speak always arises.

This question is made even more complex where such organizations are headed by non-San, simply because they are the ones with the skills to secure funds, write project proposals and so on. And this is absolutely not to downplay the contribution such people make, or to question their commitment. As Willemien le Roux (of Kuru Development Trust) points out, many such gatekeepers remain 'reluctantly, filling the space because the San are not yet able to verbalise [the issues] in a language that is accessible to the ones who try to enter "the gate" '. Often these representatives are caught between a rock and a hard place: trying to execute decisions made by the San, knowing some of these may prove unpopular with 'outsiders', while dealing with the suspicion of those researchers and journalists who do not respect their mandate to represent the San, even though this has often been granted on the basis of years of experience and hard-won trust. 'If people truly want to respect the San and give voice to the ones who have been silenced by the colonialism of ages, they must respect the gatekeepers the San have put up.'[4]

Of course these organizations have an inalienable right to speak up. But there is also a justifiable concern on the part of researchers about them setting the agenda or controlling the information process through, for example, the rigid application of a media protocol. Clearly there needs to be greater transparency on the part of outsiders as to their motives, but the goal should be ever greater freedom of information. The real challenge for policy-makers is to foster responsibility for, and commitment to, sensitivity and fairness in any information activity, be it academic or journalistic.

So how can the San encourage this, given that the balance of power lies largely on the side of the information-gatherers? Here the AIDS epidemic has some relevant lessons. This too is a story that requires great responsibility on the part of researchers and reporters, yet there was no way anyone could simply put the lid on coverage of the epidemic when problems of ethics and insensitivity arose. Instead there have been intense efforts by AIDS activists, including people with HIV and AIDS, to educate the media, as well as researchers and health professionals. And wherever people with HIV/AIDS have been included in fora where the epidemic is being debated, this has encouraged the development of trust between them and the media, and greater accuracy and sensitivity in reporting. However, as Willemien le Roux points out:

> One difference with the AIDS case is that everyone wishes to stop the epidemic. For the San, some of what they want to control through guidelines or policies does not have the same importance to the outside world. Yet being observed for the sake of mere observation – exposing the naked 'Bushman Venus', for example – is not the same as being observed for the sake of changing wrongs, such as reporting on the land loss of the CKGR. And it is the former which the San feel sensitive about and on which they would prefer restrictions.[5]

Of course, there *are* changes in anthropology and other forms of research: collaborative ventures are more common and participatory research more acceptable in development circles. Such disciplines have moved a long way towards closing the gap between observers and observed. So we are not claiming that the San community-based OT project is breaking new ground – nor that it is without flaws or likely to achieve all its objectives. It is merely one of many steps forward in the San's struggle to speak for themselves. As a trainee interviewer said at the end of the workshop, 'At last, we'll be writing our own history, rather than outsiders.' Indeed, it is the San themselves who are experts in their own lives and history, but who generally lack the necessary skills and resources to share it. This points to a need for others to share their expertise, whether in research, journalism, interviewing or publishing, to contribute towards the San achieving their ultimate goal of self-representation. In this instance, this would have meant having the funds, the 'language' and the skills to attend an international conference such as this one and tell their own tale of being resettled in the name of conservation – in such a way that audiences gain new perspectives and understanding of the issue.

Notes

1 John Hardbattle in conversation with Sue Armstrong, 1995.
2 At the time of writing there are also about six community-based organizations run entirely by San.
3 Oral testimonies are now being gathered throughout the region, coordinated by the Working Group of Indigenous Minorities of Southern Africa (WIMSA).
4 Willemien le Roux, personal communication to the authors, 1999.
5 Willemien le Roux, personal communication to the authors, 2000.

References

Brody, H. 1999. 'Taking the Words from their Mouths' in Tribes: Battle for Lands and Language, *Index on Censorship*, issue 4.
Gaventa, J. 1993. 'The Powerful, the Powerless, and the Experts'. In P. Park, M. Brydon-Miller, B. Hall and T. Jackson, (eds), *Voices of Change: Participatory Research in the United States and Canada*. Connecticut/London: Bergin & Garvey.
Guenther, M. 1994. 'Contested Images, Contested Texts: the Politics of Representing the Bushmen of Southern Africa'. CVA Newsletter.
le Roux, W. 2000. 'Someone's War', *Shadow Bird*. Cape Town: Kwela Books.
Martin, M. 1996. 'Foreword'. In P. Skotnes (ed.). *Miscast: Negotiating the Presence of the Bushmen*. University of Cape Town Press.
Mogwe, A. 1992. 'Who Was (T)Here First? An assessment of the human rights situation of Basarwa in selected communities in the Gantsi district, Botswana', Botswana Christian Council Occasional Paper No 10.
Monbiot, G. 1994. *No Man's Land: An Investigative Journey through Kenya and Tanzania*. London: Picador.
Morris, A.G. 1996. 'Trophy Skulls, Museums and the San'. In P. Skotnes (ed.). *Miscast: Negotiating the Presence of the Bushmen*. University of Cape Town Press.

Perks, R. and Thomson, A. (eds). 1998. *The Oral History Reader.* London: Routledge.

Saugestad, S. 1994. 'Research and its Relevance: Notes on the history of research on the Bushmen-San-Basarwa-N/Oakwe of Botswana'. Paper presented at the conference 'Khoisan studies: multi-disciplinary perspectives'. Tutzing, Germany, July.

Skotnes, P. (ed.). 1996. *Miscast: Negotiating the Presence of the Bushmen,* University of Cape Town Press.

Thompson, P. 1988. *The Voice of the Past: Oral History.* Oxford: Oxford University Press.

van der Post, L. 1962. *The Lost World of the Kalahari.* London: Penguin.

11

Negev Bedouin

DISPLACEMENT, FORCED SETTLEMENT AND CONSERVATION

Aref Abu-Rabia

Brief Historical Background

There are a number of reasons to account for the nomadism of the Bedouin. One decisive factor has been the search for grazing land and water. Another has been the need to escape retribution: when one Bedouin kills another, tribal law requires him and all his relatives to move some distance away, and seek the protection of a sufficiently strong tribe. When Islam spread in the seventh century, a wave of Bedouins came to the Negev from the Arabian Peninsula and settled. A second wave of Bedouin settlement commenced in the ninth century and lasted till the twelfth; then a third wave, again from the Arabian Peninsula, took place in the second half of the sixteenth century.

A Bedouin tribe usually comprises four extended families, each of which has a leader. In most cases the largest family in the tribe is dominant, and from its ranks the sheikh, the tribal leader, has traditionally been chosen. The tribe (*'ashira*) belongs to an association of tribes, based on territorial proximity and common lineage, or on political partnership; such an association is called a *saff (super tribe)*. In some cases a *saff* comprises several tribes that have banded together for a common goal, usually an alliance in war. In the Negev, Bedouin tribes are grouped in *saffs*.

The geographical area known as the Negev (today, the southern part of the State of Israel) comprises an area of 12,500 sq. km. In the nineteenth century, the tribal life of the Negev Bedouin was turbulent, since they lived in a state of virtual independence from the nearest Ottoman authorities, resident in Gaza (Bailey 1980: 35–80). However, with the construction of the Suez Canal (which was opened to passage in 1869), the Ottoman authorities began to concern themselves with Egypt, and to take measures to impose order in the Negev. Inter-tribal wars were ruthlessly suppressed and responsibility for the good behaviour of tribes was placed upon the sheikhs (Marx 1967: 9–10).

The Ottoman government fixed the final tribal boundaries in 1917, the last year of its rule, and these remained in effect until 1949. It is worth noting that inter-tribal borders were based on the outcomes of previous wars. Thus, large, powerful tribes were allotted large and rich territories, while weak tribes received poor lands.

Beersheba and the Negev were conquered by the British in late October 1917 and the entire country of Palestine then remained under British mandatory rule until 1948. The conquest of the Negev by the Israel Defence Forces in the 1948 war of independence created new circumstances. Most of the Bedouin tribes were expelled to Jordan, the Gaza Strip, and the Sinai. Others, fearful of the Israeli authorities, mainly because they had taken part in hostilities against the Jews, left the Negev. Some clans broke up into tribes, and the tribes into extended families; some of the larger families left the Negev, while some dispersed among different tribes. In some cases, the remnants of tribes, or even clans, joined existing tribes, completely obliterating their previous tribal frameworks. Tribes or parts of tribes either amalgamated around their former heads or around those who had connections with the authorities. Thus there came into being the nineteen tribes, whose heads were recognized as sheikhs by the authorities.

Before the outbreak of the 1948 war, between 65,000 and 100,000 Bedouin lived in the Negev, an area of approximately 12,000 sq. km, which they used for both cultivation and pasturing flocks. However, the 1954 Israeli census showed the Bedouin population there was just 11,000. All the Bedouin of the nineteen tribes were concentrated in a closed area of about 1000 sq. km in the northeastern Negev, under military rule. While some of them had actually lived in this area before, tribes that had lived elsewhere were settled on 'abandoned' lands. This measure was taken under paragraph 125 of the military emergency laws. The military government remained in force until 1966. Anyone who wanted to enter or leave the closed area had to have a special permit, and in this way the authorities established their control over the Bedouin. Even after the end of military rule, most Negev Bedouin continued to live in the closed area – and those who wanted to return to their original homes were prevented from doing so. As a result of the authorities' policy, pasturelands were overgrazed in that closed area. Of the remaining area of the Negev, part was allotted to Israeli towns and settlements, and part to a nature preserve.

In this chapter I will focus on displacement, forced settlement, the problem of land, and conservation programmes which have resulted in forced displacement of the Bedouin.

Population Data

As of today, the Negev Bedouin can be divided into two main groups, according to the manner of their settlement:

1. Residents of seven towns, established by the authorities: Rahat, Tel-Sheva [Tel al-Saba'], Kuseife [Ksifa], 'Aro'er ['Ar'ara], Segev Shalom [Shgib al-Salam], Hura, and Laqiya.
2. Residents of dispersed, unauthorized settlements: the tribes who live outside the seven towns, in concentrations of varying size. There is also a very small group of semi-nomads.

The Bedouin population of the Negev in 1999 was estimated at approximately 110,000, of whom close to 60 per cent live in the seven towns. Most of the remaining 40 per cent live in tribal settlements, in clusters of wooden, metal, or baked mud huts; in tents made of goat's hair, with additional jute bags or plastic sheeting; or in houses constructed of cement blocks or stone.

The Problem of Land

Description

The Negev landscape is partly flat, partly hilly. Most of the area is covered with a type of loamy earth known as loess, while the hilly areas are defined as basic rock. These lands are quite poor in natural vegetation (Ben-David 1982). Climatic conditions in the Negev are generally too harsh for much plant growth, and the weather varies sharply from one year to the next. Precipitation levels decline moderately from north to south and more sharply from west to east, with increasing distance from the Mediterranean barometric trough, in the direction of the arid regions of the Arava and Dead Sea Valley. Such lack of stability in precipitation, and its distribution by region and season, influence the lives of the Bedouin, and affect Negev flora and fauna. The average rainfall in the Northeastern Negev ranges from 100 to 200 mm (Even-Ari et al. 1968; Danin 1977; Abu-Rabia 1994: 8–9). Dew provides valuable extra moisture, especially on winter days without rain, and in the very hot and dry summer season. For example, the yearly quantity of dew measured in the Negev Uplands fluctuates between 26 and 36 mm. In some years, the quantity of dew is higher than the rainfall.

Ownership

One of the main problems of the Negev Bedouin has been and remains that of land. The issue of land ownership was a legal problem in the Ottoman period, when land seized by the government was considered to be its property or under its administrative control. The Ottoman regime allowed the users of land to cultivate it, in accordance with legislation providing for the payment of property tax and obtaining state permission for all transactions. In 1856, a law was enacted classifying land in five categories: private lands (*mulk*); land for agriculture or pasturage, but not for building purposes (*miri*); lands of the Muslim religious institutions (*waqf*); lands for public purposes such as crossings and public roads (*matruka*); and waste lands or lands which were not owned by anyone (*mawat*). Anyone cultivating such lands had to receive permission from the authorities. It must be remembered that it was only by the legal definition that *mawat* was not suitable for cultivation; and that because of this definition, which was finally amended by the British in 1921, it was not possible to acquire possession of these lands, even by tillage (Bahjat 1974; Aumann 1975; Stein 1984; Atran 1987).

However, the British Mandatory authorities recognized Bedouin ownership and levied taxes on cultivated land. Land was sold; surveys were carried out and the details registered in the Jewish National Fund's estate and sales books, through intermediary Arab brokers (Bresslavsky 1946: 34–91). The Bedouin did not register their land in *Tabu* (the Land Registry Office) for the following reasons: (1) fear of the burden of government taxes; (2) aversion to publicizing details about private property; and (3) failure to see the need to record land on a piece of paper as proof of ownership. In that period, proof was, literally, through the sword (*be-haq al-seif*). As a result, most land in the Negev was classified by the authorities as *mawat* (al-'Aref 1933: 235–40).

Most of the land in the Negev is currently held by the Israel Lands Authority (Marx and Sela 1980). Of the approximately 2,000,000 dunams cultivated by the Bedouin before Israel's establishment, the Authority holds some 1,820,000 dunams, the rest remaining in Bedouin hands. Practically all the land held by the Authority has been handed over to Jewish settlements in the Western Negev, with only some 200,000 dunams located in the closed area being leased to the Bedouin in small plots and by seasonal leasing agreements. In legal proceedings before the Beersheva District Court, the State won its case in a court of first instance, thereby proving its ownership of the land (Abu-Rabia 1994). Opposing this view, the Bedouin believe that they are the owners of the land they till and that they have grazing rights in additional areas. This right is based, in their opinion, on the fact that their forefathers seized the land (*hajr*) centuries ago, and that they have cultivated it without interruption for generations (al-'Aref 1933; Kressel et al. 1991).

The authorities employ a system of land expropriation for public use. The Ministry of Finance has the power to take decisions in respect of the expropriation of areas required for public usage, and the courts uphold its decisions practically automatically. This is the basis on which the Acquisition of Land in the Negev Act (Peace Treaty with Egypt) 1980 operates. Those who drew up the Act saw it as a tool for gaining control over Bedouin land, as well as establishing a *post factum* legal seal for land seizures. At one point it was proposed to apply the Act to some 300,000 dunams, the area held and cultivated by the Bedouin. The Act was intended, by combination of coercion and incentives, to move the Bedouin off the land and into towns. Early attempts to combine the two processes had been unsuccessful, and this seemed to be the way to implement government policy. According to revised plans, 65,000 dunams in the region of Tel al-Milh were expropriated to build the Nevatim airport. The evacuation of the Bedouin from this area and their hasty settlement in three towns set up for them, Kuseife, 'Aro'er and later, Segev Shalom, are an example of forced, sudden settlement. Settling the Bedouin narrowed and changed the web of their social relations, the composition of groups, and accentuated differences in wealth and styles of living.

To sum up, the Bedouin claim ownership of and tenancy on much land, and are afraid that the authorities will continue to expropriate the lands they dwell on, or over which they claim ownership. The claims against them are based on the charge that they are trespassing on state lands. On the other hand, the Bedouin are only too well aware of the government's policy of encouraging them to relocate to towns.

Agriculture and Livestock Breeding: Ecological and Political Constraints

Before the establishment of the State of Israel, the Bedouin in the Negev cultivated 2000 sq. km of agricultural land. They primarily grew wheat and barley, but also small quantities of durra, watermelon, lentils, beans, citrus trees, vines, apples, apricots, almonds, figs, and pomegranates. For water storage, they built earth mounds (*saddat*) (Abu-Khusa 1979; Orev, 1986). Today, the Bedouin cultivate approximately 400 sq. km of land where they grow wheat, barley, and fruit trees as mentioned above.

Crop yields were and still are dependent largely on annual rainfall. On average, two in every four years are considered to be drought years, one is average, and one has more plentiful rainfall. In a year of drought, the yield may be 20–30 kg of grain (barley and wheat) per dunam as against 120–150 kg per dunam in a bumper harvest, while the average year's yield lies somewhere between the two extremes (Ben-David 1988: 47). The Bedouin view agriculture as a solid economic basis that will stand them in good stead, regardless of the amount of rainfall in any specific year.

Sheep raising has been influenced by ecological and political factors alike. Ben-David (1982) notes that in 1977, the political authorities began to force sheep-breeders to adapt themselves to the new political and economic conditions. In the past, Bedouin had pastured their flocks on harvested land used for field crops and vegetables, as well as on uncultivated pasture areas on hills and rocky ground, and in *wadis* (valleys). However, when the Bedouin were transferred to a closed area following the establishment of the State of Israel, the amount of grazing areas available for their use was reduced, and this has resulted in a reduction in the size of their flocks. The upshot has been that, in the long term, political and economic conditions are leading to a lessening of the relative significance of flocks.

The ecological problem, in particular over-grazing, is a result of political considerations, which have taken the form of obstacles put in the way of sheep breeders and measures devised against them, particularly by the 'Green Patrol' – the authorities' strong-arm executive agency established in 1976, by the Nature Preservation Authority of the Ministry of Agriculture. Its main responsibilities are to watch over and supervise pasture lands, grazing, reservation of State land; to prevent black goats pasturing outside the tribal area and reduce the number of livestock; to destroy Bedouin houses outside the Bedouin towns; to impose bureaucratic restrictions on the Bedouin in order to force them to settle in the towns planned by the authorities. Most of their actions are harsh and brutal and they have received extremely wide-ranging powers to ensure that the regulations are enforced. Transferring Bedouin off disputed land and pastures and into the closed area, the stringent controls that restrict their ability and that of their flocks to roam – all this has had a strongly negative impact on sheep raising. This is true even though after 1967 restrictions on the movement of Bedouin flocks were eased and the Bedouin began to graze their sheep in the Negev, the central, and even northern regions, advancing as far as north as Hadera (near Haifa in the north of the country). Consequently there arose major problems of damage, theft, partnerships between Bedouin and Jews for illegal purposes, and the settlement by some Bedouin outside the Negev.

In 1978, the Ministry of Agriculture began to deal with the issue of pasture lands, over-grazing and conservation – imposing harsh restrictions such as reducing the number of livestock and limiting the period of the leased pasture lands outside the tribal area. Registration of livestock was organized and all the animals were tagged on the ear (*khursa*).

It was decided that the Bedouin flocks would be moved back to the Negev, and that the Ministry would be responsible for dealing with the flocks in a systematic fashion. Responsibility for these arrangements was conferred on a co-ordinating committee, which was to represent all the bodies connected with the Bedouin agricultural sector.

In summary, it may be said that migratory patterns are affected by many factors – ecological, political and economic, and therefore vary considerably from one location to another. Moreover, differences exist between the migrations of different flocks, even when the same habitat conditions apply. These result from differences in household structure, the size of flocks, their place in a family's economy, as well as personal issues. In previous times, nomadic grazing was a primary source of income and created a very considerable dependence between the Bedouin and their natural environment. The development of new occupations has reduced this dependence on the environment and on a single source of subsistence.

Today, the organization of grazing, including the surveying and location of pasturage, allotment to applicants, and inspection of grazing land are dealt with by the Bedouin Unit at the Ministry of Agriculture, in conjunction with the Green Patrol, the Israel Lands Authority, the Jewish National Fund, the Israel Defence Forces (IDF), and the Veterinary Board (Wagner 1986). Tagged sheep can roam outside the closed area for spring grazing, from the middle of February until the end of May; and for summer grazing on stubble, from the end of May until September. The authorities do not allow such free movement of goats, and additional restrictions have been imposed, such as the introduction of regulations and laws against the black goat[1] as well as the Green Patrol's crackdown on herds of goats and the owners of other flocks. The advantages[2] of the black goat have been ignored (Shkolnik et al. 1972; Even-Ari et al. 1978; Abu-Rabia 1994).

This recently imposed government system of land use ignores the Bedouin's traditional seasonal migration as a sustainable livelihood response to the environment. The Bedouin roam in certain environments in the Negev ecosystem because of the abundance of wild plants and corn which sprout from seeds shed during the previous harvest, in certain rainy years. Sometimes, in the morning, flocks graze on the green bushes that cover the hills and wadis, while in the afternoon, they return to the stubble fields. It should be noted that on the hills along the wadis, the vegetation is sparser than in the wadis, and composed mainly of shrubs. The wadis, which fill with run-off floods in winter, are rich in vegetation and provide excellent grazing areas, particularly in the spring. Feeding on a combination of green vegetation and stubble (*hasida*) provides a rich and balanced diet for the flocks. It also ensures that both types of grazing land are used, preventing over-grazing. The combination of barley stubble with green vegetation also has a stabilizing effect on milk production.

Families who harvest their lands have straw available in sufficient quantities to feed their flocks in the autumn and winter. Thus, land ownership, cultivation of the land, and the rearing of sheep complement each other. Stubble fields, straw, barley, and other grazing areas provide the flock with food, while it, in turn, fertilizes the ground. Watering sheep and goats is an essential condition for keeping livestock in the desert and the lack of water sources constitutes a limiting factor on migration and the ability to take advantage of pasturage.

Despite the difficulties, sheep rearing is still a worthwhile undertaking for anyone who has the right human resources, access to grazing land, and a good relationship with the authorities, but only on condition that it can be operated on a large scale between 150–200 heads and more. It has been demonstrated that a profitable Bedouin flock can be maintained on a permanent, non-nomadic basis, apart from a three-month stubble-grazing period, when the owners are granted exclusive rights to graze in a given area. Also, advanced sheep raising and feeding technologies can be transmitted to the Bedouin system without any particular difficulty. Thus, the Bedouin outside the seven towns continue to raise livestock, in flocks ranging in size from a few head to over two hundred. The Ministry of Agriculture estimates that the Negev Bedouin have some 200,000 head of sheep and 5000 of goats. Bedouin estimates are 230,000 sheep and 20,000 goats (Abu-Rabia 1999).

A significant improvement can be brought about in the reproductive regime of Bedouin Awassi sheep, thereby greatly increasing the quantity of meat produced by the flock. It has also been shown that under existing commercial conditions, it is possible to attain a reasonable level of profitability even with the non-improved Awassi breed, with the planned production of both meat and dairy products. Proper use of fencing and controlled grazing can increase yields, as has been demonstrated by the authorities. A demonstration 'model' farm for sheep was established (next to the new Jewish town of Lehavim) in 1982, on a 8500 dunam site intended for 1000 ewes. From the reactions of Negev Bedouin, it appears that many would be prepared to raise sheep under the conditions of a model farm (Pervolotsky and Landau 1988; Abu-Rabia 1994).

Customary Resource Management

The Bedouin greatly appreciate the plant world and their culture has developed means for preserving and protecting it. Their feeling of closeness with the plant life around them is reflected in the number of places named after plants, as well as the Bedouin custom of naming their children after plants and trees (Bailey 1984: 42–57). They have a strong awareness that trees are vital to life in the desert, for pasture, food, shade, heating, household utensils, and so on.

In their seasonal nomadic cycle through different locations, the Bedouin keep reserves of pasture for the summer. They impose heavy punishment on any shepherd who grazes that pasture before the time decided upon by the whole tribe. The Bedouin also have developed a system (*hima*) for the conservation of pasture land which can be grazed only at certain times to enable the plants to grow and disperse seed, renewing the flora. These lands are thus protected by customary tribal law. The main idea of *hima* is to prevent overgrazing and to allow regrowth of natural fodder plants by preventing destruction of young sprouts and seedlings. *Hima* also limits grazing of specific animals, so that those that close crop the plants will not ruin the pasture for larger animals such as camels, cows, horses, and donkeys. In addition to limiting grazing to specific seasons, *hima* can also permit grazing when there is an emergency, such as drought, sudden displacement, tribal war, or raid (*gazu*). *Hima* might also permit kids and lambs (*baham*) to graze in *saddat* since the small livestock do not damage the trees (Draz 1980; Shoup 1990; Abu-Rabia 1994).

Unfortunately, some of these conservation practices have ceased – a result of forced settlements, political considerations and the efforts of the Israeli authorities to compel the Bedouin to settle in towns.

Conclusion

It has been seen that at the end of the British Mandate in Palestine, most of the Bedouin population lived in the Negev desert, but after the Israelis conquered the Negev in 1948, most of the tribes were expelled to Jordan, the Gaza Strip and the Sinai Peninsula while the remaining Bedouin were confined to a closed area. Another sector of the Negev was allotted to Israeli towns and settlements, and the remainder was declared a nature preserve.

After the repeal of military rule in 1966, most Bedouin continued to live in the former closed area while others tried to use their original land for grazing, leading to conflict between the Bedouin and the authorities. Ironically, a further cause of social upheaval was the Camp David Peace Treaty between Israel and Egypt in 1980. Israeli air bases in the Sinai had to be evacuated and a site was chosen for the construction of one of the new alternative airfields on Bedouin lands in the Negev. The Israeli government enacted a special law to evacuate the Bedouin and settle them in three towns. However necessary such actions may be considered by a government faced with huge pressures on limited land resources, such a sudden upheaval in a traditional way of life cannot be viewed as other than disruptive. Although the Bedouin remain close to nature, many traditional conservation practices which prevented over-grazing and allowed renewal of the flora are now difficult to continue given the policies of the Israeli government.

Notes

1 It is claimed that black goats crop the grass very much closer than do white goats or sheep. Thus, they might destroy the pasture's ability to renew itself in the following season. This is reflected in regulations b and c:

a. Regulations re: Diseases of Livestock (Marking of Sheep), 1978.
b. Law of Plant Protection (Damage by Goats), 1950.
c. The Declaration of Plant Protection (Damage by Goats, Forbidden Areas for Feeding Goats), 1977.

2 Advantages of the black goat: see Shkolnik et al. 1972; Even-Ari et al. 1978; Abu-Rabia 1999.

References

Abu-Khusa, A. 1979. *Bi'r al-Saba' wal-hayah al-Badawiya*. Amman: Matabi' al-Mu's-sasa al-Sahafiya al-Urdunniya (in Arabic).

Abu-Rabia, A. 1994. *The Negev Bedouin and Livestock Rearing: Social, Economic and Political Aspects*. Oxford/Providence: Berg.

—— 1999. 'Some Notes on Livestock Production among Negev Bedouin Tribes'. *Nomadic Peoples*, 3(1): 22–30.

al-'Aref, A. 1933. *al-Qada bayna al-Badw*. Jerusalem (in Arabic).

Atran, S. 1987. 'Hamula Organization and Masha' Tenure in Palestine'. *Man*, 21: 271–95.

Aumann, M., 1975. 'Land Ownership in Palestine 1890–1949'. In M. Curtis (ed.). *The Palestinians*. New Brunswick: Transition Books.

Bahjat, H. 1974. 'The Social and Economic Development in the Jerusalem District. 1840–1873'. *Society and Heritage*, 1(3): 92–100. al-Bireh: The Palestinian Folklore and Social Research Society, *In'ash al-Usra*.

Bailey, C. 1980. 'The Negev in the Nineteenth Century: Reconstructing history from oral traditions'. *Asian and African Studies*, 14: 35–80.

—— 1984. 'Bedouin Place-names in Sinai: Towards understanding a desert map'. *Palestine Exploration Fund*, 1–2: 42–57.

Ben-David, J. 1982. 'Stages in the Sedentarization of the Negev Bedouin, a Transition from Semi-nomadic to Settled Population'. Ph.D. dissertation, The Hebrew University of Jerusalem.

—— 1988. *Agricultural Settlements for the Bedouin Population, Policy Proposal.* Jerusalem: The Jerusalem Institute for Israel Studies.

Bresslavsky, J. 1946. *Do You Know the Country? The Negev.* Tel-Aviv: HakibbutzHameuchad (in Hebrew).

Danin, A. 1977. *The Vegetation of the Negev,* Tel-Aviv: Sifriat Poalim.

Draz, U. 1980. *Improvement of Range Lands and Fodder Crop Production in the Syrian Arab Republic.* Rome: Food and Agricultural Organization.

Even-Ari, M., Noy-Meir, E., and Naveh, Z. 1978. 'A Letter to the *Jerusalem Post* in Favour of the Black Goat'. *Jerusalem Post,* 15 June.

Even-Ari, M., Shanan, L., and Tadmor, N. 1968. 'Runoff Farming in the Desert'. *Agronomy Journal,* 60: 29–32.

Kressel, G., Ben-David, J., and Abu-Rabia, A. 1991. 'Changes in the Land Usages by the Negev Bedouin since the Mid-nineteenth Century'. *Nomadic Peoples,* 28: 28–55.

Marx, E. 1967. *Bedouin of the Negev.* Manchester: University of Manchester Press.

Marx, E. and Sela, M. 1980. 'The Situation of the Negev's Bedouin, Appendix no. 1', in Ben-Mayer's *Team for Evacuation and Resettlement of the Bedouin.* Tel Aviv: TAHAL.

Orev, J. 1986. 'The Use of Earth Dykes by the Negev Bedouin for Runoff Utilization'. *Notes on the Bedouin,* no. 17, edited by J. Ben-David. Sede Boker: The Jacob Blaustein Institute for Desert Research, Ben-Gurion University of the Negev.

Pervolotsky, A. and Landau, J. 1988. *Improvement and Development of Sheep among Northern Negev's Bedouin.* Beit-Dagan: Administration for Agricultural Research.

Shkolnik, A., Borut, A., and Choshniak, I. 1972. 'Water Economy of the Black Bedouin Goat'. *Zool. Soc. of London Symposium,* no. 31.

Shoup, J. 1990. 'Middle Eastern Sheep Pastoralism and the Hima System'. In J. Galaty and D. Johnson (eds). *The World of Pastoralism, Herding Systems in Comparative Perspective.* New York and London: The Guilford Press.

Stein, K. 1984. *The Land Question in Palestine, 1917–1939.* Chapell Hill: University of North Carolina Press.

Wagner, A., 1986. 'A Letter to Mr. A. Katz-Oz, Minister of Agriculture, Regarding the Bedouin Livestock'. Beersheba: Ministry of Agriculture.

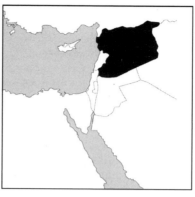

12

Customs Excised

ARID LAND CONSERVATION IN SYRIA

Jonathan Rae, George Arab and *Tom Nordblom*

Introduction

The steppe area in Syria, as in most other countries in West Asia and North Africa (WANA), is an administrative region based on an agro-ecological division of the country. Running a crude rainfall scale from high to low, zones 1–4 are designated agricultural regions where settlement and cultivation is permitted and predominant. Zone 5, which covers all land below the 200mm isohyte (the so-called 'steppe line'), is the designated steppe area or *badiyah* (Figure 12.1). Covering more than half the country, or 10.2 million hectares, the steppe has been set aside as rangelands where dryland cultivation is illegal. Most of those herding in the steppe are nomadic or agro-pastoral Arab tribes. Despite their long history in the steppe and a sophisticated resource access system, these tribes are currently excluded from government rangeland management efforts. With the recent adoptions by the authorities of centrally-controlled enclosures – otherwise termed the 'plantation concept' – in steppe conservation this exclusion has resulted in dispossession and hardship for many pastoral groups.

A strictly agro-ecological division of the country is a relatively new concept in Syria, as elsewhere in WANA. Up till 1958, the *badiyah* was a judicial division, defined largely by the expansion of cultivation, where tribal customary law held sway under the authority of local leaders. For over a century successive governments have been active in an often violent policy of expanding and centralizing power with the ultimate aim of removing the tribes completely as

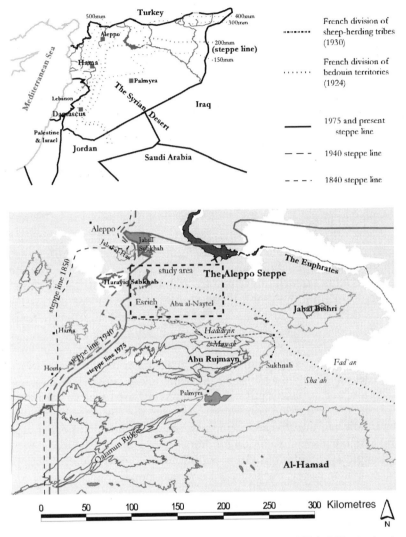

Figure 12.1 Syrian Location Map (incl. French Divisions of Tribal Territories in the Steppe)

a threat to settled authority. The Ottomans in the nineteenth century subdued the tribes and laid the groundwork for tribal administration in Syria and neighbouring regions. Between 1920 and 1944 the French briefly stalled the process, but the nationalists reinvigorated it following independence in 1944. Within fourteen years, the Syrian authorities had finally abolished all tribal rights including the formal powers of local sheikhs. Any rights or claims to steppe resources that tribal families and groups might have had under customary law prior to 1958 were lost, and all non-private steppe (98 per cent) was officially considered state land, open to all Syrian citizens without hindrance.

Settled society has long-considered the moving tribes and the nomadic economy as 'primitive'[1] and 'archaic'. And so in the post-1958 era settlement schemes expanded and for the first time the state entered into large-scale, technically led grazing management programmes. Intervention here was imperative: for the authorities saw an abused landscape still under threat, and contributing a fraction to the national economy of what it could under scientific management. As in so many other dryland regions, the authorities blamed the moving tribes and sought the introduction of new institutions and structures to supplant them and take responsibility for steppe regulation. Since the early 1980s rangeland policy has centred on the establishment of fodder shrub plantations, which in 2000 cover an estimated 200,000 ha. Apart from conserving animal feed for times of stress, the plantations are increasingly viewed as wildlife sanctuaries with some now a focus for fauna reintroduction programmes.

In theory, steppe areas deemed degraded by the Steppe Directorate of the Ministry of Agriculture and Agrarian Reform (MAAR) can be appropriated by the state for rehabilitation purposes. It is effective privatization. The site is demarcated with a trench or earth fence and sown or planted with a mix of exotic and endemic drought-tolerant shrubs. Access to the site for local herders is denied in the first three to five years of establishment, whereupon limited and prescribed entry is permitted upon payment of a fee.

As most of the steppe is technically state land, the only criterion for the appropriation of land for plantations is a technical one of what sites are thought degraded or vulnerable. By the same token, any Syrian national is formally allowed to take short-term contracts on a part of any plantation once open. The plantation concept assumes that the customary land tenure system, abolished thirty years previously, is now redundant, neither representing a viable alternative, nor a hindrance to top-down establishment and management of plantations. In a country where open access to steppe resources is state policy, dispossession is a mute point.

Full appropriation of land for plantations, and a system of fines and security measures have given government far greater central control over access than in previous schemes, even if only to relatively small areas of the steppe. But these areas, sometimes as much as 20,000 ha for a single plantation, are nevertheless large and of low productivity, so centralized monitoring and enforcement of rules is costly if not prohibitively expensive in the long term. Indeed, the problem of free riders (rule breakers typically trespassers) on Syrian plantations is clearly evident from the fines issued. In Aleppo province alone, which has 30,000 ha of plantation at five sites in the northern steppe, over 1000 fines (generally all 5,000 Syrian lira or $100 for trespass) were levied between 1994 and 1999. Many more were caught though not fined. Other more serious incidents, some arising from inter-tribal conflict over access, have been resolved at higher political levels of the province or the state. In some instances, plantations have been burnt, pulled up or overrun by livestock. The relevant authorities remain undaunted.

This chapter will first sketch an alternative reality to the assumption that all steppe land is state land. Specifically, developments in the customary land tenure system prior to 1958, and during the wilderness years that followed, will be examined. Neglect by a 'strong' government, or the socio-economic changes of the twentieth century are often cited as key factors behind the breakdown of customary legal systems around the world. Syrian government rhetoric would have you believe it as well. Behind the rhetoric, however, is not a disorganized mass of herders free of the shackles of tribalism and in need of new structures and institutions to regulate grazing. Instead, the tribal structure and legal system are shown to be adaptive and well-entrenched in local society, demanding recognition from a reluctant bureaucracy and government. In this instance, the clash between formal and customary land tenure systems is accentuated when plans include land appropriation, such as the plantation concept. This will be dramatically illustrated by a conflict between the sheep-breeding Abraz clan of the Hadidiyin tribe and the local authorities of Aleppo Province which arose over the establishment of a plantation. To situate this example, the discussion first turns to customary land tenure, its current status and recent evolution in Syria.

The Customary Land Tenure System

Key to understanding property rights in the Syrian steppe is the concept of legal pluralism. For a long time the formal legal system of the state, the *qanun*, has co-existed with tribal customary law, *'urf*. Whereas the *qanun* is by definition written, the *'urf* is largely unwritten. To a degree the *qanun* often confirms existing local custom; while it has also recognized that custom is one of the sources of Islamic law, *shari'a* (Heyd 1973: 168–9), itself an enduring pillar of the *qanun*. The moving tribes were not fully subjected to *qanun* until 1958. Till then, customary law under the authority of the sheikh and state, prevailed in all matters including marriage, divorce, homicide and property rights. In Syria and neighbouring areas these tribal legal institutions have been described as 'remarkable for their sophistication, and a central feature of the culture' (Stewart 1987: 489).

The twentieth century has seen dramatic changes in the operating environment of all Bedouin tribes, strongly influencing shifts in their economic activities as well as in their social structures and institutions. Though not altogether new to the tribes, principal among these factors are:

1. a marked rise in the power of central authority in the form of the nation state, both actively and passively undermining and or altering the role of customary structures and institutions, and;
2. a rise in technological and market forces driving agricultural expansion at the expense of better pastures, boosting sheep flocks but sending camel numbers into near-terminal decline, and spurring rapid human population growth.

In terms of land tenure these forces have opposed each other. Active government measures include an aborted attempt to settle herders (1950–74), the formal abolition in law of customary systems for resource control and access (1958), and the establishment of an open access regime with the nationalization of the steppe (1953 and 1974). Working in the opposite direction and driving changes in the customary land tenure system has been the intensification of production. The shift from camels to sheep, the growth in the sheep population, and the expansion in cultivation have raised land scarcity and shaped the evolution of the customary land tenure system towards discrete territories as a basis for access rights to natural resources, and increasing exclusivity.

Similar in sophistication to the state's own property rights system, customary regulation, '*urf*, recognizes varying levels of exclusivity for different natural resources such as agricultural fields, water, pasture, and firewood (Wilkinson 1983: 303–04). Property here is conceived not as an asset but as a socially recognized right to possess the flow of benefits that arise from the control of things and objects (Behnke 1991: 4). Initial possession is primarily through the investment of labour (such as digging a well) or occupancy (Wilkinson 1983: 303–04; Rose 1994: 12). Gaining social recognition of these rights comes through exchange, and that exchange involves the mutual ceding of rights. Rights, however, are not absolute but a function of an individual's or group's direct efforts at protection, of other groups' capture attempts, and of informal and/or government legal protection (Wilkinson 1983: 304; Rose 1994: 15). In other words, the amount of effort or resources devoted to a level of protection is related to the balance of socio-political and economic costs and benefits that the holders can expect to incur. This concept of property rights is closely related to that of transaction costs, or those costs associated with the establishment, transfer, capture, and protection of rights. New rights, or changing levels of exclusivity and individualization, are created in response to new economic conditions (Barzel 1989). Individuals or groups will choose to acquire, maintain and relinquish rights as a matter of choice, and exercise them when they believe gains from such actions exceed the transaction cost. The Bedouin customary tenure regime is neither static nor a replica of other similar regimes but evolves through local political and cultural institutions under pressure of growing resource scarcity (Platteau 1995: 36).

A starting point for customary rights in the steppe is the supposed Hadith, or saying of the Prophet Muhammad, that states 'humanity holds three things in common, water, vegetation and fire'. The ultimate ownership of these resources remains with God but their use is common to all (Wilkinson 1983: 306). Since pastures, water and firewood, are limited within and between years, the question then arises as to how competition can be regulated between individuals or groups to stem conflict and depletion.

The basis of property rights and territoriality among the moving tribes has traditionally been water. Establishing a right in water brought with it a parallel usufruct right in the surrounding pastures. Without water, grazing was usually impossible, and in the vast majority of cases this water was only available with the investment of labour in sinking or maintaining a well or cistern.

This meant that water could be owned. Water, however, could not be denied in cases of thirst, nor could surplus water be refused to prevent a pasture from being exploited. However, as Wilkinson argues 'private ownership [be it by a group rather than an individual] is an essential starting point of nomadic territorialism' (1983: 306). Rights in long-existing water sources and pastures could be captured by a group because of their 'political and military strength' (Beck 1981: 257) vis-à-vis another group, or through the fact that the resources had been abandoned.

In practice, then, possession or occupation of a site or resource is nine-tenths of customary law; investing in it and/or securing the consent of neighbouring groups gives the claim legitimacy. The term most often used by the Bedouin tribes for occupying land is 'laying hands on the land'. Investment of labour or wealth can take many forms such as digging, renovating or maintaining a well or cistern, cultivating (not necessarily only crops and trees but in theory shrubs as well), building a house, and so on. Investment is not always needed though in most instances it takes place. There are cases among the Hadidiyin tribe of the Syrian steppe where regular grazing over years has won the herders recognition from neighbours and wider tribal society, of their rights to the land. Once secured, in whichever way, the group (termed here 'the core group') then has the sole right to investment, and will usually take active measures against those violating these rights. However, the classic model remains possession followed by investment in water, given how crucial water is to pastoral activities.

An important question then arises: what defines the extent of a territory for a given water source? Here the basic rule of thumb is that the territory of a water source extends half the distance to the next working well or cistern. The starting point is usually an imaginary line and its actual location remains fuzzy though members of each group well know the general extent of their lands vis-à-vis their neighbours. A demarcated border[2] (full or partial) will only be settled if land scarcity becomes an issue in terms of increased competition for pastures or an extension of cultivation to within the border proximity, or if neighbouring groups have some other problem or feud. The experience in Syria between 1944 and 1990 is that when tribal borders need to be settled it is usually done according to custom (equal division between wells may not always be the rule) using customary judges with state authorities participating as guarantors.

In the past these water-centred territories formed the basis of what was termed a tribal *dirah* or zone of movement. Rather than an overt political region, which the territories were/are, the *dirah* is better conceived as a functional region encompassing a herder's/clan's range of movement across agro-ecological zones throughout the year. The size of the *dirah* once depended on the herding animal, with camel-herders having substantially larger *dirah* and venturing deeper into the desert than their brethren the sheep and goat herders. These latter herders were restricted by the water dependency of their animals to the desert fringes where water sources were more frequent and plentiful. With a shift from camel to truck the term *dirah* has fallen widely out of use, reflecting the ease and speed that many households now have in transporting their livestock and importing water and feed to grazing sites. Furthermore, whereas water formed the axis of control (albeit by groups rather than individuals), today where informal regulation exists, the emphasis has generally swung round to pastures per se.

This basic model is somewhat complicated by the fact that migratory production systems are organized around a principle of seasonal occupancy rather than continuous occupancy and use. This highlights another basic feature of territoriality which used to have greater relevance when there were significant numbers of camels – that of non-exclusiveness and complementarity. In short, pastures and water of the near steppe occupied by the sheepherders in winter and spring were the summer pastures for the camel-herding tribes. In other words, their *dirah* overlapped. In some instances this relationship between the sheep-herding and camel-herding tribes was generally without conflict, such as between the Hadidiyin (sheep) and the Sba'ah (camel) who had a written agreement detailing obligations. Between other groups, however, most notably the Bani Khalid (sheep) and the Rwalah (camel), there was less symmetry in movement and their relationship was more adversarial.

The replacement of camels by trucks in the mid and later twentieth century and the adoption of sheep among the old camel-rearing tribes fundamentally altered this non-exclusive and complementary relationship. The Sba'ah took to sheep early and in large numbers and increasingly came into conflict with the Hadidiyin. The problems were resolved through customary channels with two government sponsored treaties (1944 and 1956), which together divided more than 500,000 ha of the northern Syrian steppe into exclusive territories for each tribe. In the drier southern Syrian steppe around Palmyra, the Rwalah more than other tribes kept hold of their camels, though their sheep flocks also grew in number. The implications for customary land tenure in the area was, however, prematurely terminated with the departure of all but a few of the Rwalah from Syria to Saudi Arabia in the early 1960s. With land now vacant, the Bani Khalid fought off challenges from neighbouring sheep-rearing tribes and laid their hands on the area and continue to hold it to this day.

The shift from camels to sheep, the growth in the human and sheep population, and the expansion and interest in cultivation has raised land scarcity and shaped the evolution of the customary land tenure system towards discrete territories and increasing exclusivity to pastures. The notion of clearly defined boundaries is not new in customary law. Among the Ahaywat tribe in hyperarid southern Sinai, Stewart in an exhaustive study found long existing and 'sharply defined boundaries' (1986: 3) where 'other tribes are not allowed to cultivate or dig cisterns within the territory' (Stewart 1978: 2). In another study by Rae (1999) of the Hadidiyin and Haib tribes, who occupy the rich and relatively well-watered pastures of the north Syrian Steppe, a similar pattern of clearly defined territories was found. Demarcation of some of these latter territories came as part of the just mentioned treaties between the Hadidiyin and the Sba'ah. Other territories in the region followed as disputes arose sparked by increasing sheep numbers and the spread of fragmented cultivation. In most instances the disputes were settled and tribal boundaries defined with the assistance of local state authorities.[3] On other occasions, state authorities have not been involved as guarantors and tribal institutions have fulfilled their role.

Those with such rights of investment are referred to here as the *core group* of a territory. The core group obviously has rights to graze the pastures in their area: however, they also need to maintain mobility across wide areas to counteract the effects of the dynamic and risky natural environment of the steppe. But here as well, in a wider political environment conducive to informal regulation, scarcity will increase the measures of exclusivity by groups over grazing.

Customary Regulation of Grazing

Informal management of rangelands is not a case of manipulating environmental factors for these are generally outside a herder's control; what is instead important is demographic manipulation of the herding population. It is worth detailing Wilkinson's (1983: 309) 'disposable population model' for tribal grazing regulation. He suggested the idea of a core population, identified for instance with the sheikhly clan, in control of all the main points of natural resources within a given territory, and demographically related to the carrying capacity of an anticipated bad year. In good years, client, neighbouring, or other so-called 'peripheral' families and groups (the terms are not synonymous) could be accommodated while in bad years they could be prevented from entering, or be forced out if need be. In severe years the core group would have to move as well.

There are two populations that need to be manipulated, that of the periphery and that of the core. Rise in the core population and the need to redress it to suit resource availability is a continual process. Some families may settle or seek alternative opportunities, perhaps in the cities (Barth 1961: 117). However, there is also the potential of more overt action by one faction of the core group to expel another. The alternatives to these are for the core group to restrict entry by peripheral groups wanting access to pastures, for the core group to introduce new rules regulating their own sheep numbers in the territory, or for the core group to expand territorially. Examples of all these methods arose in a study of the Hadidiyin and neighbouring tribes (see Rae 1999). The control of the peripheral groups depends on the identity of the individual or group, on the availability of pasture, and the ability of the core group to adapt and enforce its rule. This last variable ranges widely between groups and territories, reflecting the historical circumstances of the territory and core group, their strength and cohesiveness, as well as their influence in the local administration.

Dispossession and Conflict: Overlap of Land Tenure Systems

The Abraz are a large clan of the Khaumah section of the Hadidiyin tribe. They dominate the steppe areas of Aleppo province and have consequently suffered from the rapid establishment of four plantations over 22,000 ha in the Aleppo steppe during the decade 1983–92. This narrative illustrating the untold stresses that plantations have on local communities concerns proposals for a fifth and largest plantation in Aleppo Province.

No rules on the location of a plantation actually exist. Their purpose as agents of rehabilitation does of course suggest that they are located in areas that the Steppe Directorate officials deem degraded. Similar to other countries in the region, Syria has never conducted long-term monitoring studies of its steppe rangelands, so measuring degradation, and knowing what counts for degradation, is difficult. The Steppe Directorate assesses degradation on criteria of shrub cover and a measure of floral composition, though no formal guidelines exist. Each provincial Steppe Directorate office is responsible for identifying sites in its region and in collaboration with the Head Steppe Directorate in Palmyra,[4] the size of the plantation. Only private steppe land cannot be included within a plantation; otherwise, all other land is technically state land, including cultivated fields and fictional co-operative pastures, and can be and is appropriated. With the site and size determined, a committee[5] is established under the authority of the Minister of Agricultural and Agrarian Reform, to produce a technical and economic feasibility study for the proposed plantation. No environmental or socio-economic impact assessment of the likely effects of plantations is required or ever carried out (which becomes clear in an example given later in this chapter). If the committee, which usually has around one month to report, gives the go-ahead, MAAR provides the Steppe Directorate with financial support for plantation establishment.

Such a top-down and technically led project has ridden roughshod over customary structures and institutions. With co-operative and licensed agricultural rights[6] largely ignored in plantation location, customary rights are never considered. Here lies the paradox of state authority on the steppe. On the one hand they covertly co-operate with tribal systems in local dispute resolution and guarantee the agreement in writing, especially when it comes to tribal rights in steppe resources. On the other hand, the authorities refuse to formally recognize customary land tenure or any other claims short of private rights, when it comes to state interventions such as plantations.

Figure 12.2 overlays the formal and customary land tenure maps for the western portion of the Aleppo steppe. Plantations in the region are highlighted and the four under discussion here can be found in the north-west corner. These latter plantations lie within the territorial boundaries of the Abraz clan and their neighbours the independent tribal group, the Haib. Two plantations, Adami and Obisan are completely within Abraz land, while Maraghah and Ein al-Zarqah also take in portions of the Haib's territory. The Abraz clan, perhaps numbering around 1500 families, can be divided in to ten groups, of which seven share in the Abraz's steppe land. Since the early 1960s and later sealed in writing with the state as guarantors in 1990, the Abraz steppe area has been divided between the seven Abraz groups (see Figure 12.2 and Table 12.1). Typically, these sub-divisions have been further divided (through investment or distribution) between each household even if no cultivation or settlement was then envisaged. The size of these family plots averages 500 ha, and gives the family an area in which to invest without risk of dispute. Almost invariably, unless the plot is cultivated, these sites are the regular wintering grounds of their tribal owners. In spring, if reasonable grazing is in reach and accessible, they stay, otherwise they seek pastures elsewhere. Some of these households are more settled than others, particularly when cultivation was permitted, and have built adobe, typically single-roomed homes. Overall, Abraz's steppe territory covers 28,500 ha of which approximately 2500 ha were cultivated prior to the steppe-wide ban in 1995 (Rae 1999).

An in-depth survey conducted among 46 herding households occupying lands around the four plantations in Aleppo province in 1996, established that upwards of 100 nomadic and agro-pastoral households in mud-brick structures and tents were forcibly evicted. Of these households, 50 from the Salabkha division of Abraz were moved for the Obisan plantation. Twenty-five households of Awwadeen lost their settlements (though none cultivated) with the establishment of the Maraghah plantation, while the same number of Awwadeen and Bu Shedid were dispossessed at the 'Ein al-Zarqah site. As for the Adami plantation, fifteen Samamra households, twelve cultivating, lost their mud-brick structures. Others who had 'owned' land at the plantation sites but were not wintering there, went uncounted in the study as did the peripheral population who frequented the areas for grazing.

With loss of land disproportionately distributed among the Abraz, some groups found themselves unable to re-accommodate those that had been dispossessed among them. The proportion of the 100 households forced to leave the region altogether is not known, though it is thought to be a small minority. Most are understood to have secured a regular winter camp location on land of their Abraz sub-section.[7] The reduction in overall land available and the crowding of households on what remained, prompted actions symptomatic of growing land scarcity. Many of the peripheral groups, which had in the past regularly grazed in Abraz territory, found their access blocked. One of these uncounted peripheral households was from the Nu'im tribe, a part of

Table 12.1 Sections of Abraz with Land in North-western Aleppo Steppe

Abraz divisions	Abraz land in north-western Aleppo Steppe		Est. sheep number
	Est. Original size (ha)	After plantations (ha)	
Kanaknah	15,860	10,025	52,000
Samamra	14,880	9,970	42,000
Salabkha	2,110	610	
Awwadeen	7,675	1,830	26,000
Mehailat	3,750	3,750	18,000
Bu Shedid	5,600	2,025	15,000
Bu Salah	1,050	615	26,000
Hessayen	480	480	

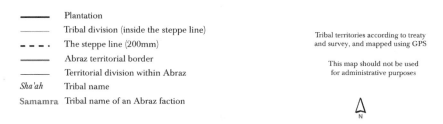

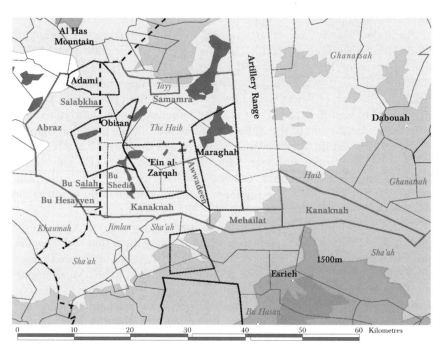

Figure 12.2 Plantation Location Overlaid on Customary Land Tenure in Western Aleppo Steppe, Syria

which is semi or fully settled on the slopes of Al Has mountain immediately north of the Abraz territory (see Figure 12.2). Having frequented the pastures of Salabkha and Samamra for more years than they can remember, three Nu'im households were told by local Abraz men to leave in the winter of 1991 and never come back. Initially, they resolved not to go but after a tussle, which resulted in the burning of a Salabkha tent, the Nu'im households promptly fled back to Al-Has. Within the year, the Nu'im paid compensation to the aggrieved Salabkha household, and since have not returned.

The authorities were blind to the enormous pressures these plantations were imposing on the customary land tenure system or the economy of the local herders. At some point the pressures would inevitably surface and the authorities would be forced to take note. The land situation for Abraz was near breaking point. In the space of ten years Abraz had lost 40 per cent of its tribal territory in the north-west Aleppo Steppe. The authorities were unaware. In 1995, the Aleppo and Hama Steppe Directorates drew up plans for the largest plantation yet, one of 50,000 ha. The site, immediately south of the other plantations in Aleppo province, was planned to take in much of what remained of Abraz territory, and much more besides. The first thing the Abraz and other locals knew of the proposed project was when tractors appeared on the steppe to demarcate the site. During previous plantation establishments, family heads and Abraz elders had voiced their complaints with the Steppe Directorate but to no avail. On this occasion the sheikh of Abraz, Faysal al Nuri of Kanaknah, took his protest to the Governor of Aleppo Province, one of the most powerful people in the country. On 13 February he wrote to the governor to clarify the land tenure issue, underline the land scarcity pressures, and describe the inevitable outcome should the new plantation go ahead:

> Once these lands are annexed and the said reserve is established ... we would no longer have lands for our sheep to graze. We were moved from 'Ein al-Zarqah and Maraghah where two reserves were established. To the north of us is the al-Haib tribe ... with whom we have a bloody dispute ... [and consequently] we are not welcome on their pasture. Moreover, the establishment of the reserve would cause hundreds of herders to move away, many of whom have houses in the area.[8]

The Governor asked Ghassan Eimesh, the director of the Aleppo MAAR, what could be confirmed. The reply from Eimesh not only confirmed what Nuri had said but went on to question the plans of the plantation on technical grounds. He wrote that the site in question was in fact 'one of the good sites in our steppe in terms of plant cover'.[9] He went on to say that there 'are about 100 [tent and house dwelling] families living there all of the Abraz clan' a number of which had been 'moved from Ein al-Zarqah and Maraghah sites due to the establishment of reserves there'. The Director also noted that 'there are no alternative lands for the Abraz clan' and that there was 'a dispute between them and the neighbouring al-Haib clan'.

The plantation ultimately was never established, at least partly because of the complaints. But complaints are what they were, for local groups were not able to participate in the conception of the scheme – only respond once tractors appeared on the site. Indeed, the Minister of Agriculture had already signed off on the project, a point of no return in the establishment of other plantations. The Abraz incident threatened serious dislocation and local unrest; the governor over-ruled the Minister for Agriculture and Agrarian Reform in Damascus, and cancelled the work. The joint venture between the two provinces abandoned, the Aleppo authorities switched their focus to the establishment of a smaller plantation at Dabourah, a site held by clans of Ghanatsah, Hadidiyin (see Figure 12.2). In 1998 the land there was appropriated and planting begun. The tribal landholders at Aleppo's subsequent site for a plantation, Dabourah, were not so lucky. They, too, only found out about the 20,000 ha plantation after the authorities had already agreed and signed off on it. Leading Ghanatsah men took their case to the highest provincial levels, but even with their wealth and local influence, the plantation went ahead and the land was formally appropriated in 1998.

Conclusion

Increasing land scarcity, particularly along the fringes of the steppe where much of the land for plantations has been appropriated, has moved the customary land tenure system towards greater exclusivity and individualization of tenure. Much of the distance travelled by the customary system has been facilitated through tribal written treaty, usually with the authorities as guarantors. This was the case before and after 1958, the year tribal law and custom was officially abolished. Government development of steppe areas began in earnest in 1963 on an assumption that the region was not contributing its fair share to the national economy. The blame for this was placed squarely with a bankrupt and archaic system of tribal management. The authorities continued to participate in tribal resolution to see off local unrest, but when it has come to steppe development, particularly pasture improvement and management, ideology and latent ambivalence towards the moving tribes have precluded local participation in the decision-making process. The saga of plantation establishment in Aleppo province illustrates this point well. Those displaced went uncompensated, increasing local land scarcity went unnoticed by the authorities, and calamity in 1995 was only narrowly avoided. Despite this the establishment procedure in Aleppo province goes largely unchanged. Dabourah, the latest plantation, was already a foregone conclusion by the time local inhabitants realized what was happening.

There is much that could not be included in this chapter. At least two points should be mentioned. One is the problem the authorities are having in managing up-and-running plantations. Here there is everything from regular trespass, manipulation of the plantations by one tribal group to effectively take land from another, and a complete over-run of a plantation (Rae et al. 1997; Rae 1999; FAO 1999). The other problem concerns the Abraz case specifically. One way to relieve the pressure of rapidly growing land scarcity is by expanding land holdings elsewhere in the steppe, and this is exactly what Abraz did. In 1992, they were successful in securing, with government backing, a large share of Haib territory lying to the east of the artillery field. The action was led by the sheikhly faction of Abraz, Kanaknah, and the area to the east of the artillery field with their name on it covers much of what was taken. There had been fifteen households semi-settled at the site; with their area seriously truncated, twelve were forced to move their permanent winter camp to the outskirts of Aleppo.

The authority's inability to share responsibility for steppe development with the entrenched and sophisticated tribal system is the major undoing of most projects. The potential for synergy, cost effectiveness and sustainability with a collaborative approach to conception, development and application of interventions is significant. What has hampered progress to date is govern-ment resistance to devolving power, and less than a clear understanding on the evolution and current state of the customary land tenure regime. There is substantial opportunity for Syria to establish an enduring integrated and coherent management strategy for the steppe areas and lead the WANA countries in this respect.

Notes

1 Translation of an official party document, first published in *al-Munadil* (April 1966, 3: 13), taken from Van Dam (1996: 146).

2 For instance by stone piles, a furrow or some other visual sign such as crests of hills or a road.

3 Recent examples include: Minutes to the Resolution Meeting between Ghanatsah of the Hadidiyin and Sba'ah Btayinat over lands of al Del'a [east of Abu al-Naytel], Office for the Chief of Secret Police, Homs Province, 1800hrs on 14 December 1981; MAAR (1992) Agreement over lands at the Ja'ar site between Haib, Abraz, and Bu Shahab al-Din.

4 Amongst other considerations, the head office had information on the number of shrub seedlings available from the nurseries.

5 The committee should be composed of an agricultural economist, a geologist, and repre-sentatives from the provincial departments of the MAAR and the Steppe Directorate.

6 Prior to a complete ban in 1995, the authorities permitted licensed agriculture over limited areas of the steppe.

7 Letter from the Mudir, Directorate of Agriculture and Agrarian Reform to the Governor of Aleppo, No. 1942/16: 21.2.1995.

8 Letter to H.E. The Governor of Aleppo from Faysal al-Nuri and companions on behalf of the Abraz tribe: 13.2.1995.

9 Earlier in the paper some doubt was raised about the assumption that plantations were located only at degraded sites. The concern raised was that insufficient monitoring and inappro-priate theories on rangeland ecology, made it difficult to identify vulnerable areas. In this letter from the provincial director of MAAR, he has obviously been advised that the designated site might well not be appropriate. Letter from the Mudir, Directorate of Agriculture and Agrarian Reform to the Governor of Aleppo, No. 1942/16: 21.2.1995.

References

Barth, F. 1961. *Nomads of South Persia.* Oslo University Press.

Barzel, Y. 1989. *Economic Analysis of Property Rights.* Cambridge: Cambridge University Press.

Beck, L. 1981. 'Government Policy and Pastoral Land Use in Southwest Iran'. *Journal of Arid Environments,* 4.

Behnke, R. 1991. 'Economic Models of Pastoral Land Tenure'. *Proceedings of the Inter-national Rangeland Development Symposium.* Department of Range Science, College of Natural Resources, Logan, Utah, 1–11.

Délégation Générale de la France Combattante au Levant (D.G.F.C.L.) 1943. *Les Tribus nomades de l'état de Syrie.* Inspection des mouvances bédouines de l'état de Syrie, Les lettres françaises: Beyrouth, March.

FAO Land Tenure Service. 1999. Tenure Service Mission Report. 'Range Rehabili-tation and Establishment of a Wildlife Reserve, Syria'. FAO Italy: Government Co-operative Program GCP/SYR/009/ITA.

Heyd, U. 1973. *Studies in Old Ottoman Criminal Law.* Oxford: Clarendon Press.

Platteau, J. 1995. *The Evolutionary Theory of Land Rights as applied to Sub-Saharan Africa: A Critical Assessment.* Cahiers de la Faculté des Sciences Economiques et Sociales No. 145.

Rae, J. 1999. 'Rangeland Management in the Syrian Steppe: Tribe and State'. DPhilq thesis, Oxford University.

Rae, J., Arab, G., Jani, K. and Nordblom, T. 1997. *Socio-economics of Shrub Plantations,* M&M Project, ICARDA, Syria.

Rose, C. 1994. *Property and Persuasion: Essays on the History, Theory and Rhetoric of Ownership.* Boulder, Colorado: Westview Press.

Stewart, F., 1978 ca. 'Texts in Bedouin Law'. Unpublished personal copy.

—— 1986. 'Tribal Law in the Arab World: A Review of Literature'. *International Journal of Middle East Studies,* 19.

Van Dam, N. 1996. *The Struggle for Power in Syria.* London: I.B. Tauris.

Wilkinson, J. 1983. 'Traditional Concepts of Territory in south east Arabia'. *The Geographical Journal,* 149(3).

13

Animal Reintroduction Projects in the Middle East

CONSERVATION WITHOUT A HUMAN FACE

Dawn Chatty

Introduction

Conservation in the Arabian Peninsula, unlike Africa and elsewhere, does not have a long history. In other parts of the world, ideas and policies for the 'preservation of nature' and the conservation of plant and animal species were exported with the colonial administrations of, mainly, France and Great Britain. The Arabian Peninsula, however, was never a 'colony' of a Western power. Its neo-colonial period, which might have served to develop such an interest, was very short, and only lasted a few decades between the ends of the two World Wars. In addition, its mainly arid land mass was not suitable as a wooded reserve. Furthermore, it had few species of large mammals, making it unattractive for the development of wildlife reserves. Conservation and eco-tourism were therefore largely irrelevant in the Arabian Peninsula for most of the twentieth century. Only as the millennium began to draw to a close did a particular form of conservation – animal reintroduction – manifest itself in the region. Without the colonial baggage most other parts of the world had to carry, these conservation projects should have been able to avoid the mistakes and pitfalls that plagued similar efforts in other regions. That, sadly, has not been the case.

Projects to reintroduce *Oryx leucoryx*, the Arabian oryx, into Arabia were set up in the late 1970s and 1980s, first in Oman, then Saudi Arabia, Jordan, Israel and Syria. These projects, often couched in the contemporary developmental jargon of 'participation' and 'grass roots' support, in actuality continued to regard local human populations as obstacles to be overcome – either through monetary compensation or with special terms of local employment – instead of as partners in sustainable conservation and development. In this chapter, using the case study of an exceptionally 'successful' internationally-supported, wildlife reintroduction in Oman, I demonstrate that initial successes were later marred by serious setbacks which were grounded in the failure to draw the local population into the conception, planning and management of the project. That story needs to be told, and lessons drawn so that other animal reintroduction projects in the region can avoid making the same mistakes.

Conservation in Oman

The earliest expression of interest in conservation in Arabia only came in the middle of the twentieth century, when the alarming rate at which gazelle, oryx and other 'sporting' animals were being caught or killed became clear (Henderson 1974). In the southeast corner of Arabia, the Sultan of Oman, Saiid bin Taimur, issued a decree in 1964 banning the use of vehicles for hunting gazelles and oryx. Hunting parties from outside Oman were known to be taking advantage of the country' s wide open and indefensible borders to enter the country for sport. The Sultan therefore commanded the establishment of a 'gazelle patrol' to protect these graceful mammals in the central Omani desert which borders Saudi Arabia and the Trucial States (now the United Arab Emirates).

The Arabian oryx, which belongs to the family of horse-like antelopes, is closely related to the gemsbok of the Kalahari Desert, the beisa oryx of Somalia, the fringe-eared oryx of Kenya, and the scimitar-horned oryx of the Sahara. At one time, the Arabian oryx thrived throughout the arid lands of the Arabian Peninsula. During the latter part of the nineteenth century and the first decades of the twentieth century, its range gradually contracted. By 1917, the oryx survived mainly in two pockets: one, in the Great Nafud Desert in the north; and one in and around the Rub'-al-Khali. The northern population became extinct around 1950 (Stanley Price 1989: 37). In Oman, oryx were then still sighted throughout the Jiddat-il-Harasiis which borders the Rub'-al-Khali. By 1972, however, with no further sightings having been made for some time, the oryx was declared extinct in Oman and in the rest of Arabia.

Contrary to what some observers surmised, the indigenous human population bordering the Rub'al-Khali, the Harasiis tribe, which had shared its territory with the oryx for several centuries, had not played any significant role in its extinction. As Stanley Price makes clear, it was unlikely that Harasiis hunting pressure – in open country with only ancient rifles and camels to hide behind – could ever have eliminated the population (Stanley Price 1989: 42). It was the motorized hunting parties brandishing automatic weapons that succeeded in doing so, despite the indigenous tribe's wishes to preserve the animal in its vast, shared, arid environment.

In 1976 Sultan Qaboos bin Saiid put into effect a ban on the hunting or capture of all large mammals – specifically oryx and gazelle. This 'ban', however, was interpreted by some as being relevant to nationals only, and foreign hunting parties of wealthy neighbouring elite continued to enter the country. The Sultan was forced to issue a second decree, the following year, to clarify this ban. It unequivocally stated that no permission would be given to foreign hunting parties to enter and operate in Oman.

For several decades before the extermination of the last herd of wild oryx in 1972, a number of zoos around the world had been breeding a 'World Herd' in captivity for eventual reintroduction into the wild (Grimwood 1962). In 1974 Sultan Qaboos bin Saiid instructed his expatriate advisor on the conservation of the environment to explore the practicalities of restoring the oryx to Oman as part of its natural heritage. In 1977 and 1978 a consultant with the World Wildlife Fund and a veterinarian toured extensively through the interior of Oman with a Harasiis guide. They produced two unpublished reports for the International Union for the Conservation of Nature (IUCN), summarized by Jungius (1985). The reports concluded that the ideal habitat for the oryx reintroduction project was the Jiddat-il-Harasiis, in an area known as Yalooni, 'the best vegetated pan on the Jiddat, with resources of grazing, shrub and tree browse' (Stanley Price 1989: 60). It also recommended that the whole of the Jiddat-il-Harasiis should be proclaimed a wildlife reserve or sanctuary. These recommendations were accepted, and in 1980 the first oryx from the World Herd were flown back into the country and released into the main oryx enclosure at Yalooni.

Two years later, the first herd of ten animals was released into the wild. This herd slowly began to explore the desert and make the transition from supplementary feed to desert grasses and herbs. The ten Harasiis tribesmen who had been hired to track these animals gave them individual names and watched them each day, recording their movements and their behaviour as individuals and as a group. Two years on, in 1983, a second herd was released into the wild. By 1986, both herds were relying on fresh graze alone and supplemental feed was no longer necessary (Spalton et al. 1999). Between 1986 and 1989 the two herds grew and split off into a number of small groups. By the end of 1989, the total number of oryx in the Jiddat was recorded at 92 (Spalton 1992).

When the number of oryx grew to more than 100 in early 1990, the scientific managers of the project decided to change the monitoring procedure from an individual basis to sample population-base. The tribesmen continued to collect detailed information from a sample of known animals, and maintained low-level monitoring for the rest. The herd continues to grow, and the studies carried out by the scientists at the base camp and published in international journals indicated a success story of exceptional certainty (Stanley Price 1989; Spalton 1992, 1993; Beck et al. 1994; Kleiman et al. 1994; Spalton et al. 1999). By 1996, fourteen years after the first herd of ten animals had been released into the wild, and with a further thirty animals released in the intervening period, the total number of oryx in the wild was estimated at 400.

This conservation experiment, the reintroduction into the wild of an animal that had previously been hunted to extinction, had a grim underside. On 17 February 1996, Harasiis rangers found evidence that poachers in motor vehicles had chased a herd of oryx and captured two calves. Further evidence revealed that regularly over the coming months animals were being chased until they either fell from exhaustion or were knocked over by the poachers' vehicles. By October of 1996, 27 oryx were known to have been lost to poachers. This illegal hunting continued throughout 1997 and into 1998. By March 1998, the oryx population was estimated at 310 (Spalton et al. 1999). By August of the same year, poaching had so reduced the number of oryx, particularly the females who were the target of this illegal activity, that the project team decided the wild population could no longer be considered viable. Steps were taken to establish a captive herd from the 100 plus animals remaining in the wild. By late February 1999, the captive herd at Yalooni comprised 40 oryx (30 females, nine males and one calf) while eleven females and an estimated 85 males remained in the wild (Spalton et al. 1999: 174).

The astonishingly rapid unravelling of this internationally popular, well-funded conservation project has been laid at the feet of the institution of hunting. Spalton suggests that the original feasibility study for the Arabian oryx reintroduction by Jungius (1985) should, perhaps, not have concluded that the hunting threat which was the cause of the species' original decline had been eliminated. He suggests, instead, that the hunting threat had remained dormant, and that once the reintroduction project had successfully re-established a large, wild oryx population in the Jiddat, poachers exploited a demand outside the country for captive animals. Illegal capture emerged as the primary threat to wild oryx (Spalton et al. 1999: 175). This deduction echoes the kinds of conclusions commonly associated with African conservation projects. In the context of Oman, and more specifically the Jiddat-il-Harasiis, one must ask, what about the people? What about the nomadic pastoral Harasiis tribe who had shared their territory with the oryx for several centuries? What role did they play? What participation, if any, did they have in the reintroduction project, or more ominously in the illegal capture of oryx? What, if anything, might the tribesmen have been able to do to prevent the poaching, or at least reduce the extent of the activity? And, finally, what lessons can be learned for other reintroduction projects in Arabia?

Although there is an element of hindsight in what follows, it is essentially a catalogue of the mistakes and errors in judgement that eventually came to plague a project. For over a decade, I had observed and spoken out about the step-by-step development of an animal reintroduction project designed, planned and implemented with near total disregard for the indigenous human population who maintained a pastoral livelihood in this territory. The Harasiis, a tribe of nearly 3000 nomadic camel and goat-raising pastoralists inhabiting the Jiddat-il-Harasiis, where Yalooni is located, were greatly saddened by the extermination of the oryx with whom they had shared the same ecological niche for centuries. The Harasiis had been pushed into the remote, waterless plain of the Jiddat nearly 200 years earlier by stronger pastoral tribes (Chatty 1996: 81); the oryx, which had once graced the whole of the arid desert regions of south Arabia, had been pushed back until, by the middle of this century, it too was found only in the Jiddat (Stanley Price 1989: 37). The Harasiis had seen the progressive decline in numbers take place, but had been unable to stop the motorized hunting parties that descended upon them in their search for oryx herds.

The idea of setting up an oryx sanctuary in their traditional territory, however, had never been discussed with them, nor had they been consulted on the most suitable area to place such a sanctuary. The only Harasiis tribesman to be consulted, if the term could be used in this case, was a former guide to the expatriate advisor on the environment, during the advisor's days as an oil company liaison officer. This Harasiis tribesman was the guide to the IUCN exploratory party during the feasibility study mentioned earlier. He was not part of the political leadership of the tribe, although he and his brothers regularly petitioned the government for recognition as such. The decision to set up the project base at Wadi Yalooni was made in Muscat by the expatriate advisor and the international consultants on hand. Only when this decision had been approved by the Sultan were the tribal leaders summoned to Muscat and informed (see Chatty 1996: 136). For many years after, this meeting was talked about by the Harasiis with mixed feelings of remorse, and unease. The area which had been selected by the conservation officials was a most important reserve, or *hawta,* for the Harasiis. Wadi Yalooni was the one part of the Jiddat where they could rely on two seasons of growth whether or not there had been any rain. Its proximity to the Indian Ocean, and the presence of cold water offshore meant that dewfall and fog were common here (Fisher and Membery 1998). During periods of drought this part of the Jiddat was particularly important for the survival of their herds. On the other hand, they wished to see the oryx returned to the Jiddat, and they could not imagine that there would not be room for a reintroduced wild herd as well as their own livestock.

In 1980, at the time that the base camp for the oryx project was being set up and just before I commenced my own field work among the Harasiis, I was visited by the expatriate advisor to the oryx project. He wanted to know my intentions. I am not sure that I succeeded in reassuring him that I would not 'give the natives any funny ideas'. But when I asked him how he would deal with the impact which his project would inevitably have on the Harasiis tribe, he replied, 'I'm not interested in people, only plants and animals.' That statement accurately described the fundamental guiding principle of the project: the focus on plants and animals. It accounts for the astonishing success the project had in establishing a wild herd of nearly 400 oryx in the Jiddat, but it also points to its failure, the near destruction of the herd by illegal capture. With people left out of the formula, the oryx were reintroduced into a community that was not prepared or educated to protect them. The human element, that is the key factor for success, was not engaged at the beginning, nor twenty years later.[1]

The aims of the project, its goals, the implied restrictions on infrastructural development, and even the importance of their cooperation were never explained to the tribal community. Only gradually, as crises erupted, did the Harasiis learn what restrictions and infringements upon their lives this project placed. Once the project actually commenced, they were caught up with the spirit of the enterprise and were delighted not only at seeing the oryx returned to Oman, but also in the sudden creation, for a limited number of men, of 'dream' jobs as oryx rangers.[2]

The reintroduction project began life with a prolonged three-year 'honeymoon' period. During this time there were no conflicts between the indigenous population, the growing expatriate conservation management team, and other Omanis. Gradually, however, difficulties did appear. These arose as competition over grazing during prolonged drought between the herds of domestic goat and camel and the reintroduced oryx (Stanley Price 1989: 212–13), between the different lineages among the Harasiis tribe over access to employment and special benefits, and between the Harasiis and rival tribes who had been ignored in this conservation effort. The goodwill with which the project was initially accepted remained, particularly among the older generation that had grown up with the oryx. But the appearance of illegal hunting of gazelle on the Jiddat (first reported in 1986), and its yearly increasing level by rival tribesmen, and – some expatriates have said – disaffected youth, pointed to the serious flaws in the planning, design and implementation which top-down conservation projects all too often make.

As long as the Harasiis had no aspirations of their own, no desire to see an improvement in their access to water, no desire to have regular road grading, or infrastructural development in their traditional homeland, relations with the oryx reintroduction project and staff remained untroubled. But the Harasiis, like people everywhere, were opportunistic. Eventually they wished to improve their lives, and had no special desire to remain in a sort of pristine traditional state just for the sake of not changing. Slowly at first, and later with greater speed, the Harasiis came to realize what was being expected of them, and what constraints they were under. They came to appreciate that, in drought conditions, they were expected not to camp within the vicinity of an oryx herd, even when all other grazing areas had been depleted.[3] At about the same time, the tribe's long-standing campaign to have a water well dug by the Ministry of Water and Electricity in a promising area north of Yalooni, appeared to be blocked by the oryx reintroduction project managers. In addition, the Harasiis understood that their efforts to get the national petroleum company to regularly grade roads in the vicinity were also being thwarted.

All the while, the longstanding rivalry between the Harasiis tribe and its neighbours, the Jeneba, found new expression. A generations-old blood feud between the two tribes had been settled by the Sultan's representative in 1968, and relations between them cooled down. More numerous and better educated, and having had longer exposure to schooling, Jeneba tribesmen managed to get most of the skilled jobs available in the government tribal administrative centre (Haima), constructed in 1980, in the middle of the Jiddat-il-Harasiis. Some Jeneba wanted the better-paying jobs at the oryx project, but discovered that these positions were restricted to Harasiis tribesmen.

In 1996, the first oryx was reported to have been poached, and a Jeneba tribesman was suspected. Very rapidly over the next several years the incidence of poaching increased until, in 1998, out of an estimated wild herd of about 400 there were only 120 animals left in the Jiddat. Nearly all those caught or suspected of poaching were Jeneba tribesmen. This fact suggested, first, that intertribal rivalry was on the rise;[4] and, second, that the project has not maintained its popularity, especially with the Jeneba youth who grew up during the time when the oryx was extinct, so that it was never a part of their cultural tradition. To these rival tribesmen, and perhaps some disaffected, unemployed, Harasiis youth, the oryx sanctuary makes no sense other than to put wild animals first, before people and domesticated herds. They have no sense of ownership or participation in the animal sanctuary. They see no benefit to themselves, their families or their community. The opportunity to make some money by poaching thus becomes a difficult temptation to resist.

The steps being taken to declare the entire area a UNESCO World Heritage Site further alienated the indigenous population. A significant part of the Jiddat-il-Harasiis was identified for a national nature reserve in 1986 as a preliminary step in turning it into a UNESCO World Heritage Site. In 1994 this area of nearly 42,000 sq. km, with a nomadic pastoral population of 3000, was established by Royal Decree as the Arabian Oryx Sanctuary. The Harasiis tribe was not consulted nor was it educated as to the significance of this decree. Few, if any, Harasiis understood that the Jiddat would be divided into three land use zones: a core area with the strictest environmental protection; a buffer zone, with fairly strict protection, in which a limited number of activities would be permitted if they are compatible with conservation objectives; and a transition zone where most activities would be permitted unless clearly damaging to conservation objectives. Nor did they know that a land use and management plan had been prepared for the entire area.[5] The preliminary report, which has now been superseded by further study, clearly reveals that the Harasiis, who have occupied this arid land for the last two centuries, were, in the mid-1990s, still not being consulted or integrated into the conservation scheme in any way other than as passive participants.

The Harasiis, and also in their way the Jeneba, have already challenged the system. By working quietly and consistently for the past ten years the Harasiis have succeeded in having a reverse osmosis water plant built by the government in an area that is a buffer zone beside the sanctuary. This construction, much sought after by the Harasiis, contradicts the purpose of the conservation management land use zones set up by foreign experts for the Arabian Oryx Sanctuary. A similar situation is likely to occur in respect to local roads. The management plan intends that a careful network of local roads be established 'in consultation with the stakeholders in the area'. The priorities which must be observed include, in the following order, wildlife conservation, tourist access, mobility of government staff, and finally the *'legitimate movements of the indigenous pastoralists'* (emphasis mine). It is unlikely that the Harasiis, let alone the Jeneba tribe, will allow themselves to be the last consideration. Quietly and consistently, or with astonishing force, they will work to achieve what they feel is necessary for their communities' well-being. Whether it be petitioning government offices for more wells, reverse osmosis plants, or road building and grading efforts, or tolerating – if not encouraging – the capture and sale of oryx to wealthy private collectors in the UAE and Saudi Arabia, the pastoral tribes sharing the same and adjacent habitats as the reintroduced oryx will insist on being recognized.

It is in the interest of the state and the conservation authorities to try to bring the indigenous population into a truly participatory relationship with the project which goes beyond simply hiring a few men as rangers, or paying compensation. Such passive participation does not lead to a sense of integration or ownership. The long-term sustainability of conservation development in Oman requires a grass-roots, local-level approach which emphasizes co-ownership and management, capacity building, and institutional change management. With ever-increasing numbers of Harasiis youths attending the high school at Haima (the first class graduated in 1993) there is still the possibility that the local population (Harasiis and Jeneba alike) could be drawn gradually into the conservation project – through concerted education, curriculum development, and skilled employment – in a more significant capacity than the 'passive participation' (Pretty et al. 1994) of the past. Truly interactive participation and ownership may prove to be the only solution to the current malaise.

Conservation in Syria

The pastoral Bedouin tribes of Syria, Jordan and Saudi Arabia have struggled, for decades, with two opposing forces: one compels them to settle on the edges of the desert and engage in marginal agricultural production; the other forces them to move away to seek multi-resource livelihoods and pastoral subsistence across several national borders (Abu Jaber et al. 1978; Lancaster 1981; Chatty 1986, 1990). Establishment of the independent nation-state in the late 1940s and 1950s saw the culmination of several decades of sustained effort to control and break down pastoral tribal organization. Much of the tribal leadership was co-opted into the elite urban political scene, and land holdings once held in common were increasingly registered in the names of tribal leaders of important families and converted into farms.

In Syria, the 1960s was a period of strenuous government land reform, including the complete seizure of all common tribal land and the confiscation of the large tracts of land owned by tribal leaders. Following a three-year-long drought, in which over two million sheep died, the government instituted a programme to alleviate the problems caused by this ecological disaster. An internationally sponsored project was set up to revitalize the pastoral sector of the Syrian economy. Its foremost goal was to stabilize the mainly pastoral livestock population. This proved very difficult, mainly because the officials running the project did not understand Bedouin methods of animal husbandry.[6] In turn, the Bedouin had no trust in government, especially in light of the recent confiscation of grazing land, and the explosive expansion of agricultural development over nearly a third of the best rangelands of the *Badia* (Al-Sammane 1981: 32).[7]

After a number of years of poor project results, a handful of specialists launched a campaign to convince the agencies concerned with rangeland of the importance of studying the human factor. They argued that unless development programmes were in harmony with the customs and ways of life of the pastoral populations, the whole rangeland development scheme would fail. Bedouin as well as government cooperation was required in order to solve the problem. In 1967, Draz (1977) recommended that the government revive the Bedouin tradition of *hema* (i.e. returning control over range conservation and management of grazing lands back to the Bedouin). His recommendations for a return to a system of communal ownership appealed to the Syrian government's socialist orientation and the proposal was accepted. After several years of trial and error, a programme of cooperatives was implemented whereby block applications by tribal units for control over their former traditional grazing lands were generally granted by the government. Power and responsibility within a cooperative thus remained within a tribe, giving its members a participatory role in the programme. Today perhaps two-thirds of Syria's Bedouin population belongs to *hema* cooperatives and associated schemes, though government reports (Al-Sammane 1981) suggest that number is nearly 90 per cent. As membership has never been mandatory, but the individual choice of a tribesman within a lineage group, the majority of Syria's Bedouin are joining because they perceive a benefit from doing so. The benefit is both as an individual herd owner and as a tribesman in terms of access to managed grazing, preferential prices for feed, and some credit facilities.

Despite numerous ups and downs caused by changing legislation, and inadequate restraint on the spread of agriculture into the *Badia*, the current situation, which allows Bedouin a participatory voice in the running of cooperatives that were set up to accommodate traditional Bedouin land use patterns, is an improvement over the uncontrolled grazing of the 1950s, and the rigid government regulatory schemes of the 1960s. Flexibility and an acceptance of traditional Bedouin systems of exploitation and marketing have resulted in a national programme of some success at both the national and local levels.

Throughout the past thirty years of qualified success in the operation of *hema* cooperatives, the government has continued to experiment with protecting and conserving flora in the *Badia*. The rationale behind these measures and pilot projects has been an attempt to rehabilitate rangelands, protect threatened plant and shrub species, and stop the incursion of thorny bush. The hope has always been that the Bedouin would appreciate the benefit of fencing and exclusion and be inspired to do the same on traditional land holdings. Unfortunately this has not happened. Instead, the Bedouin express resentment at traditional common lands being confiscated for government experiments from which they perceive that they are deriving no benefit (Chatty 1995; Roeder 1996).

In 1992, Syria attended meetings of the Commission for Natural Parks and Protected Areas of the World Conservation Union (IUCN) in Sicily, and negotiated funding for a project to rehabilitate rangeland and to establish a wildlife reserve in the Palmyra *Badia*. This project was approved, and the Food and Agriculture Organization was drawn into the operation of the project as it appeared to have a development focus (improving food security). The project proposed to address three interrelated issues: diminishing grazing land, disappearing wildlife, and increasing requirements for supplemental feeding of domestic herds. It also proposed to incorporate some of the land holdings of three *hema* cooperatives into protected ranges, to set up restrictions on access by Bedouin and their domestic herds, and to run a programme to introduce new plant species. After two years of this three-year project, it expected to have obtained 'higher forage production from the *Al Badia* Rangelands to enable domesticated animals and wildlife to live in harmony on the land' (FAO 1995: 7). In the third year of the project, physical boundaries were to be established and 'the reserve will only be devoted to wildlife grazing' (FAO 1995: 7). In other words, at the close of the project, the Bedouin and their herds were to be excluded from an important area of rehabilitated rangeland.

The project is now in its second, three-year cycle and many of its goals have not yet been achieved. And although there is a recognition that the 'integration and effective collaboration of the beneficiaries to the programme' is required for sustainability, no visible effort has been made in the technical description of the project to incorporate the Bedouin in its planning, development, or implementation. Instead, or in addition, the project document specifies that the successes of similar schemes in Saudi Arabia and Jordan will be studied in order to increase the likelihood of success in Syria.[8] The indigenous Bedouin population, however, are only to be involved peripherally in the analysis of field data. Representatives from the grazing cooperatives are to be involved in the data recording process and in the discussion of results in order to 'develop their awareness on environmental protection' (FAO 1995: 11). The wildlife reserve, *Taliila*, which received eight oryx from Jordan's Shawmary reserve and sixteen gazelle from Saudi Arabia, will remain the home for these animals for the foreseeable future. The Bedouin have been excluded from any role in the planning and management of the reserve; even the four 'local' guards at the entrances of the reserve are drawn from the town of Palmyra.[9]

What is striking from this inventory of project 'facts' is the short memory of government. The lessons learned in the 1960s have simply been forgotten. Pastoralists cannot be separated from their animals or from their common grazing land. Furthermore, the underlying assumption of this project seems to be again turning back to the now stale assumption that it is pastoralists who are overgrazing, or overstocking, and that the solution is to reduce herd numbers and restrict their access to land in order to protect its carrying capacity. These assumptions are not only wrong (see, for example, Behnke et al.1993; Pimbert and Pretty 1995: 5), but simply provide a scapegoat for a problem rather than looking for sustainable solutions. Such a search requires the inclusion of the affected people. The Bedouin need to be part of the project. Their perceptions of the problems, their causes and their possible solutions need to be taken into account. Their needs for their own herds, their access to grazing land, water and supplemental feed need to be considered as well. For without accommodation of their needs, Bedouin will not support the project, rendering the international wildlife reintroduction effort unsustainable in the long term.

A quiet effort in the direction of mobilizing community resource management, of encouraging the formation of small 'user' groups, and of building capacity and managing institutional change is now underway. For the past two years workshops – initiated by the Food and Agriculture Organization – have been held at or near the site of the oryx reintroduction project. These have aimed at introducing the concepts of participation into more than just the vocabulary of project personnel. Workshops are bringing together government technicians, project personnel, extension teams and the Bedouin whose traditional grazing and watering rights have been compromised by the oryx and gazelle reintroduction project and associated government plantation and reseeding schemes. These workshops have moved, step by step, towards drawing all sides together to work for a common goal – maintaining the wildlife reserve while at the same time permitting limited resource use by the Bedouin and associated users. The end goal is to achieve further capacity building and truly participatory resource management.

Government efforts to rehabilitate the Syrian desert rangelands in the 1960s initially failed to meet their objectives. Only when the human element was integrated into project development was there some success (Draz 1977). Thirty years on, government and international development agencies were again proposing to rehabilitate parts of the desert and to establish a wildlife reserve – without any Bedouin consultation (FAO 1995; Roeder 1996). The lessons learned decades before appear to have been forgotten. Now, however, as the new millennium begins, the Syrian government and its international conservation partners are once again looking at the delicate balance which needs to be maintained between pastoralists, conservationists, and the environment. And through the medium of participatory resource management, sustainable conservation and development is being sought. It remains to be seen whether the delicate balance these pastoralists have managed to maintain with their environment is once again threatened by plans which do not take into account their experience, way of life, or indigenous knowledge.

Conservation in Jordan

The interest in conservation had a relatively early beginning in Jordan. In the 1960s, a group of Jordanians realized that the gazelle, which they enjoyed hunting, was becoming scarce. They turned from hunters to protectors and founded the Royal Society for the Conservation of Nature (RSCN), making it one of the oldest NGOs in the country. In time, the government of Jordan granted the RSCN the mandate to establish and manage protected areas and enforce wildlife protection laws.

In 1978–9 Jordan received oryx from the international 'World Herd' conservation effort. These animals were sent to Shawmary, the wildlife reserve in northern Jordan designated to hold and breed oryx for a return to the wild in the future. This reserve, an area of 22 sq. km, was originally fenced in 1958 for a desert farming experimental station. In 1975 it was taken over by the RSCN and the fencing was restored to keep domesticated livestock of the indigenous pastoral people out. As a captive breeding station, it has been particularly successful, and is now considered over-crowded. The possibility of successfully releasing oryx into the *Badia* of northern Jordan is negligible. The surrounding countryside is now too built up and furthermore there is no single human population ready to share the burden of protecting this reintroduced animal. The RSCN is considering releasing about 200 oryx from the Shawmary Reserve into the Wadi Rum where a Global Environment Facility (GEF) eco-tourism project, jointly managed and funded by the World Bank and the UNDP, is underway. However in light of the experience of Oman and other Arab states with wildlife release programmes, the RSCN is cautiously examining its approach. Furthermore, drawing on the lessons learned over the past five years in its own conservation project in Wadi Dana, the RSCN is hoping to integrate the indigenous Bedouin community of Wadi Rum into the early stages of planning and management of such a release programme.

In 1993 Wadi Dana was declared a protected area, a 'unique assemblage of relatively pristine habitats reflecting the environmental gradient across the Rift Valley' (Johnson and Abul Hawa 1999: 4). Several thousand people from nomadic pastoralists to settled farmers made their livelihoods from resources in this Wadi. None of the people living in and around the newly declared protected area had been consulted and most were openly hostile. The concept of 'protected area', for many, implied hunting and grazing prohibitions, and the withdrawal of traditional rights of use and exploitation of resources for their own social and economic needs. As Johnson and Abul Hawa (1999: 5) make clear in their study,

> There is little doubt that if the community interests had not been addressed, the general hostility towards the protected area would have remained … with the possibility of violent confrontation. Also, it would have been impossible to 'sell' the benefits of biodiversity conservation to such an alienated population, resulting in a lack of political and practical support for conservation initiatives locally and nationally. Furthermore, if no compensations were made for restricting traditional land uses, life would have been made even more difficult for these underprivileged and marginalised people.

The RSCN set out to find a way of dealing with the local people, not so much as a problem, but as an integral part of the protected area and a major factor in shaping the character of its ecosystem. Early attempts to integrate local people focused on developing alternative sources of income, such as fruit drying, jewellery making and extensive tourism services. These projects were of great success, adding much needed income to the villagers of Wadi Dana. They did not, however, make any mark on the Bedouin pastoralists who had been excluded from important grazing areas, and were particularly vulnerable. Early in the project's history, the Azazme had been asked not to bring their animals to graze in the protected area and they had been paid compensation (personnel communications, Antoine Swenne). This measure was only a limited and temporary success, as once the money was gone, the Azazme were left even more vulnerable and impoverished. The RSCN undertook an internal review to come to an understanding of the situation of all the communities within and on the borders of this protected area. This revealed that not enough effort had been made to understand the level of dependency of the Bedouin pastoralists on the protected area and also their tribal decision-making structure – particularly in terms of allocation of natural resources.

A socio-economic study was then commissioned by the RSCN which revealed that the tribal group most dependent upon the water and graze in the protected area was a Bedouin community, the Azazme, which had taken refuge in the Wadi in 1948 when their traditional lands were lost during the fighting to create the state of Israel. Now they were found in the most arid and marginal parts of Wadi Dana. As a result of this report, the management team in the project area decided to review some of their own preconceived notions, particularly concerning domesticated animals. Their previous exclusion of domesticated livestock from the project area was reconsidered. They decided instead, to allow the regulated grazing of goats in the protected areas, and to justify it on ecological grounds. Instead of regarding the goats as 'public enemy number one', they decided to find a way of managing them and still achieve a reduction in grazing pressure. As a result, the planned tightening of a 'no graze' regulation was deferred and an innovative goat fattening scheme was introduced. The Azazme agreed to a graduated stock reduction in goat numbers over ten years as the worth of each individual animal rose. These community efforts and others are all contributing to a growing trust and sense of confidence among the local community and the project staff, making further innovative schemes more likely to succeed in the future.

In 1998, the RSCN was commissioned to manage one of Jordan's most beautiful nature sites and tourist attractions, Wadi Rum. It is also home to a large number of Bedouin who have managed the tourism operations in the area for many years without any interference, basing themselves at the fort made famous by T.E. Lawrence. Over 250 Bedouin four-wheel drive vehicles are registered with the Ministry of Tourism to drive visitors around Wadi Rum. It is now the RSCN's job to develop management strategies which ensure 'the protection and restoration of the natural qualities [of Wadi Rum] and the continuing development of tourism' (Johnson and Abul Hawa 1999: 10). Drawing on the fundamental lessons learned at Wadi Dana, the RSCN is determined to ensure that the local Bedouin are involved in the development of the site from the very inception of the project. Beginning with a full socio-economic study, the project management has identified the seven major tribal users and also drawn up a consensual sketch of the traditional land use rights in the newly designated protected area. The leaders of each tribal group have been invited to take part in selecting members for a steer-

move away. Some refused. They simply could not understand that the survival of their herds of goats was less important than wild oryx.

4. The estimated number of oryx poached in 1996 is drawn from a number of informants both on the Jiddat itself and in the capital, Muscat. The former manager of the oryx project station at Yalooni, Roddy Jones, pointed out to me that the pattern of poaching in the Jiddat is suggestive of traditional tribal raiding. The Jeneba obviously see the 'endangered species' gazelle and oryx as 'belonging' to the Harasiis. So the act of poaching is an expression of economic and political rivalry.

5. The report *Preliminary Land Use and Management Plan: The Arabian Oryx Sanctuary* was commissioned by the Ministry of Regional Municipalities and Environment in 1995. It has since been superseded by further studies, but a management plan has yet to be finalized. This fact, as well as the alarming drop in oryx numbers in the wild in Oman, has been noted with concern by the Report of the Rapporteur of the 23rd session of Bureau of the World Heritage Committee, UNESCO Headquarters, Paris, July 1999.

6. Bedouin animal husbandry is based on risk minimalization rather than the more common western market profit motivation. See Shoup 1990: 200.

7. The Bedouin 'dry farmed' cereal crops during years of good rain, but the large-scale cultivation in this arid zone had never occurred before.

8. The Jordanian Dana Project did not originally integrate the indigenous population into the planning and implementation of the project. Despite recommendations to the contrary by a number of social scientists and ecologists (Antoine Swenne, personal communication) it relied on a combination of passive participation (limited employments wardens) and a programme of monetary compensation to buy off the indigenous Bedouin and secure their promise not to use the grazing areas earmarked solely for protected wildlife. Since then, lessons have been learned and a more progressive approach is being applied to other conservation projects in Jordan. Information on the Saudi oryx and gazelle project at Mahazat As-Said Reserve has been limited to brief public relations information in the IUCN Bulletin (no. 3, 1993) and the occasional article in *Oryx*. It is very unlikely that there has been any indigenous pastoralist participation in the planning or implementation of this wildlife reserve which could be regarded as a 'scientific research station' rather than a project aiming at long-term conservation sustainability.

9. During a consultation visit in 1997, I engaged in a discussion on the hiring of local Bedouin for the reserve as a way of beginning to integrate them into the project. The British wildlife expert at the time rebuked my suggestion, saying that 'Bedouin would not work for the salaries I am offering'. The sums concerned were minimal – a matter of $20 or $30 a month. The significance of local, indigenous participation for the long-term success of the project, however, seemed to have been lost on the wildlife expert.

References

Abu Jaber, K., et al. 1978. *The Bedouin of Jordan: A People in Transition.* Amman: Royal Scientific Society.

Al-Sammane, H. 1981. *Al- Birnamij al-Suri li-Tahsin al-Mara'i wa Tarbiyat al-Aghnam* (Syrian programme for the improvement of range and sheep production). Damascus: Ministry of Agriculture and Agrarian Reform.

Beck, B., Rapaport, L., Stanley Price, M. and Wilson, A. 1994. 'Reintroduction of captive-born Animals'. In P. Olney, G. Mace and A. Feistner (eds). *Creative Conservation: Interactive Management of Wild and Captive Animals*, pp. 287–303 London: Chapman and Hall.

Bell, H. 1987. 'Conservation with a Human Face: Conflict and reconciliation in African land use planning'. In D. Anderson and R. Grove (eds). *Conservation in Africa: People, Policies and Practice*, pp. 79–101. Cambridge: Cambridge University press.

Behnke, R., Scoones, I. and Kerven, C. (eds). 1993. *Range Ecology at Disequilibrium: New Models of Natural Variability and Pastoral Adaptation in African Savannas.* London: Overseas Development Institute.

Chatty, D. 1986. *From Camel to Truck.* New York: Vantage Press.

—— 1990. 'The Current Situation of the Bedouin in Syria, Jordan and Saudi Arabia and their Prospects for the Future'. In C. Salzman and J. Galaty (eds). *Nomads in a Changing World*, pp. 123–37. Naples: Istituto Universitario Orientale, Series Minor.

—— 1995. 'Hired Shepherds: The Marginalization and Impoverishment of Pastoralists in Jordan and Syria'. Amman: CARDNE.

—— 1996. *Mobile Pastoralists: Development Planning and Social Change in Oman.* New York: Columbia University Press.

Draz, O. 1977. *Role of Range Management and Fodder Production.* Beirut: UNDP Regional Office for Western Asia.

FAO. 1995. *Rangeland Rehabilitation and Establishment of a Wildlife Reserve in Palmyra Badia (Al-Taliba).* Rome: Document no. GCP/SYR/003.

Fisher, M. and Membery, D. 1998. 'Climate'. In S. Ghazanfar and M. Fisher (eds). *Vegetation of the Arabian Peninsula.* Dordrecht: Kluwer Academic.

Grimwood, I. 1962. 'Operation Oryx'. In *Oryx*, 6: 308–34.

Henderson, D. 1974. 'The Arabian Oryx: A desert tragedy'. *National Parks and Conservation Magazine*, 48(5): 15–21.

International Union for the Conservation of Nature (IUCN) 1994. *Guidelines for Protected Area Management Categories.* Commission on National Parks and Protected Areas. Gland: IUCN.

—— 1993. *The World Conservation Union Bulletin*, 3/93: 10–12.

Johnson, C. and Abul Hawa, T. 1999. 'Local Participation in Jordanian Protected Areas: Learning from our Mistakes'. Paper given at the Conference, *Displacement, Forced Settlement, and Conservation,* St. Anne's College, 9–11 September.

Jungius, H. 1985. 'The Arabian Oryx: Its distribution and former habitat in Oman and its reintroduction'. *Journal of Oman Studies*, 8: 49–64.

Kleiman, D., Stanley Price, M. and Beck, B. 1994. 'Criteria for Reintroductions'. In P. Olney, G. Mace and A. Feistner (eds). *Creative Conservation: Interactive Management of Wild and Captive Animals.* London: Chapman and Hall.

Lancaster, W. 1981. *The Rwala Bedouin Today.* Cambridge: Cambridge University Press.

Leybourne, M. et al. 1993. 'Changes in Migration and Feeding Patterns Among Semi-Nomadic Pastoralists in Northern Syria'. *Pastoral Development Network Paper 34a.* London: Overseas Development Institute.

McCabe, J.T., et al. 1992. 'Can Conservation and Development be coupled among Pastoral People? An Examination of the Maasai of the Ngorongoro Conservation area, Tanzania'. *Human Organization*, 51(4): 353–66.

Oman, 1995. *Preliminary Land Use and Management Plan: The Arabian Oryx Sanctuary.* Muscat: Ministry of Regional Municipalities and Environment.

Ostrowski, S., Bedin, E., Lenain, D. and Abuzindada, A. 1998. 'Ten Years of Arabian Oryx Conservation Breeding in Saudi Arabia – Achievements and regional perspectives'. *Oryx*, 32(3): 209–22.

Pimbert, M., and Pretty, J. 1995. *Parks, People and Professionals: Putting Participation into Protected Area Management.* Geneva: United Nations Research Institute for Social Development (UNRISD). Discussion Paper 57.

Pretty, J. et al. 1994. *A Trainer's Guide to Participatory Learning and Interaction.* IIED Training series No. 2. London: IIED.

Roeder, H. 1996. 'Socio-economic Study of the Bishri Mountains'. Cologne: *Deutsche Gesellschaft für Technische Zusammenarbeit* (GTZ).

Shoup, J. 1990. 'Middle Eastern Sheep Pastoralism and the Hima System'. In J. Galaty and D. Johnson (eds). *The World of Pastoralism: Herding Systems in Comparative Perspective.* London and New York: Guilford Press.

Spalton, A. 1992. 'The Arabian Oryx (*Oryx leucoryx*) Re-introduction Project in Oman: 10 years on'. *Ungulates,* 91: 342–6. Paris: SFEPM.

—— 1993. 'A brief History of the Reintroduction of the Arabian Oryx (*Oryx leucoryx*) into Oman 1980–92'. *International Zoo Yearbook,* 32: 81–90.

Spalton, A., Lawrence, M. and Brend, S. 1999. 'Arabian Oryx Reintroduction in Oman: Successes and setbacks'. *Oryx,* 32(2): 168–75.

Stanley Price, M. 1989. *Animal Re-introductions: The Arabian Oryx in Oman.* Cambridge Studies in Applied Ecology and Resource Management. Cambridge: Cambridge University Press.

14

Environmental Conservation and Indigenous Culture in a Greek Island Community

THE DISPUTE OVER THE SEA TURTLES

Dimitrios Theodossopoulos

Introduction

'What good can a turtle do to a human? Why do we have to pay so much attention to them? Here in Vassilikos the turtle has caused great harm to our community. It went against the interests of the people.'

In this short narrative quotation, the 'people' (*oi anthropoi*) harmed by the 'turtle' are the Vassilikiots, the inhabitants of Vassilikos, a community of farmers and tourist entrepreneurs on the Greek island of Zakynthos in southwestern Greece. The 'turtle' (*e helona*, here emphatically used in the generalizing singular) stands for the Mediterranean loggerhead turtle,[1] a species of sea turtle threatened with extinction. Engaged in a bitter dispute over the politics of turtle conservation, Vassilikiots do not have the most sympathetic attitude towards this particular species. Parts of their land have already become a natural reserve for the reproduction of the turtles, while the formal establishment of a Marine National Park in the wider area prohibits some Vassilikiots from building on their landed property and engaging in tourism-related enterprises. Unsurprisingly, those measures have incited considerable local protest against environmental conservation.

Written from an anthropological perspective, this chapter focuses on the indigenous resistance to turtle conservation in Vassilikos. Anthropology,

Milton (1996: 22–5, 213–14) maintains, facilitates the study of environmentalism in two fundamental ways: either by deciphering aspects of human interaction with the physical world or by focusing directly on the discourses and practices of environmental groups as a cultural phenomenon in its own right. Nevertheless, one can identify a third major anthropological contribution: the application of the study of the human–environmental relationship as an exegesis of environmental disputes. Tensions arising between indigenous communities or actors, who are directly affected by the application of environmental policy at the local level, and the advocates of that policy (environmental groups, state representatives or policy makers) can be effectively explained by a culturally informed analysis of the indigenous approaches to the environment. The discipline's sensitivity towards indigenous points of view, enables anthropologists to document the voices of those who understand the protection of the environment in terms that differ significantly from the ones established by those who manage and implement environment protection.

Thus, anthropology is in a unique position to fill the record of existing knowledge on the human dimension of environmental disputes, and in particular 'of local or indigenous knowledge or practices in the area of wildlife management and control' (Knight 2000: 5). My ethnographic account in this chapter contributes towards this purpose, as it attempts to explain Vassilikiots' opposition to environmental conservation in terms of social priorities and cultural values. Vassilikiots juxtapose their own understanding of the natural world with the environmentalists' practices and ideals. They stress their own 'household-focused', 'anthropocentric' priorities in their relationship with their immediate environment, which is perceived by them as the field of daily work, toil and continual hard labour. The conservationists, on the other hand, prioritizing an 'ecocentric' understanding of the environment over practical or economic concerns, formulate an approach on environmental management, which excludes the inhabitants of Vassilikos from direct access or control of their land.

The individuals of Vassilikos who protest against the establishment of the National Park are not only those who have been directly affected by it. They are the great majority of the people of Vassilikos, those related to the protesters by ties of kinship or obligation, and those who simply fail to comprehend the priorities of the environmentalists. They view the prioritization of the turtles over the resident population as a perversion of the 'natural order of things', 'a great injustice' inflicted on their relatives or neighbours, who are simple farmers wishing to advance the living standards of their families. The gradual development of small-scale, family-run tourism enterprises is locally perceived as the 'only vital' option in escaping from poverty and rendering 'life in the countryside worth living for'. Thus, the explicit ban on tourism development within the territory of the National Park has deprived some of the local residents of what is perceived as their only realistic chance for achieving 'progress or prosperity' (*prokopi*) and 'a better life' (*mia kalyteri zoï*). Deprived of the possibility of fulfilling the economic potential of their land, they are destined to face the future from a particularly disadvantaged position.

They are forced to relocate their dream of developing tourism outside the territory of the National Park, on lands owned or controlled by other islanders, or simply defy the authorities and fight against the conservation regulations.

In fact, the great majority of Vassilikiots have chosen to follow the path of direct confrontation with environmental conservation. In defiance of the turtle protection regulations, some local individuals continue operating tourist enterprises – family tavernas, refreshment kiosks, sun-umbrella and canoe renting – within the vicinity of the park. They bitterly engage the environmentalists, ignore the warnings of the police and deter the reluctant agents of the state from demolishing their illegal constructions in the turtle protection zone. These are, however, only ephemeral victories, confined to defying the application of the law at the local level, but not the law itself (Theodossopoulos 2000: 73). The conservation regulations state explicitly that, sooner or later, all tourism development in the vicinity of the park should cease and the existing facilities be abandoned and demolished.

In the ethnography that follows, I will describe in detail the circumstances of the conservation dispute in Vassilikos, and devote special attention to the cultural worldview that informs the indigenous resistance to conservation. Starting from a short description of Vassilikiot social history and economy in the recent past, I will briefly examine the introduction of tourism into the community and the obstacles created by conservation in the development of tourism. Then I will focus on the resistance of the local inhabitants to the conservation restrictions, their anti-environmental discourse and their expression of resentment towards the agents of environmental conservation. In the third and critical section of this chapter, I will describe in detail the cultural dimension of the Vassilikiots' perceptions of the natural world. The local protagonists enact through their daily toil or 'struggle' (*agonas*) an embodied and constantly realized relationship with their immediate environment which, in turn, justifies and informs their claim that they are – contrary to the environmentalists – the rightful guardians of order on their land.

Vassilikos, Tourism and Conservation

In the early part of the twentieth century most inhabitants of Vassilikos were landless tenants (*sembroi*)[2] living and working on the estates of landlords (*afentades*). They had to cultivate the land according to highly exploitative tenancy agreements, a particular set of regulations known in the local dialect as *sembremata*.[3] According to those rules, the greater part of the agricultural produce was allocated to the landlords, while the landless tenants were content to receive what was merely necessary for their sustenance. Driven out by poverty, many young Vassilikiot men had to migrate to other parts of Greece or even abroad. Vassilikos, less attuned to social changes happening elsewhere on Zakynthos since the end of the nineteenth century, remained a predominantly rural area where the overwhelming majority of adult men and women devoted their daily work and toil to subsistence farming.

Most Vassilikiot families acquired land of their own in the years following the Second World War. In narrative recollections of the recent past, the process of land acquisition is presented as slow and painful, the product of years of hard manual labour realized by two generations of Vassilikiots. It marked the beginning of a new era of relative independence from wealthy landlords and the start of 'a better life' (*mia kalyteri zoi*) for the inhabitants of Vassilikos. Soon most Vassilikiots secured the ownership of some land and broader changes in the island's economy were about to bring to the community the long awaited prosperity. While most of the local men and women devoted their energies to farming, utilizing any available resource on their newly acquired land, suddenly great numbers of tourists started appearing on their island.

In the last three decades, Zakynthos has been radically transformed into a major Mediterranean tourist destination. In the 1970s and 1980s, tourist resorts with massive complexes of auxiliary facilities were erected on several parts of the island to accommodate the growing numbers of tourists arriving *en masse* on charter flights from Britain, Germany and Scandinavia. Vassilikiots responded to the new economic advantages of tourism only slowly and gradually. Lacking the capital for considerable investments and relying essentially on close family co-operation, they developed tourism on a small-scale household basis (cf. Galani-Moutafi 1993: 250; Zarkia 1996: 156). Nowadays, some families run restaurants – there are over forty in Vassilikos – others own small all-purpose shops or mini-markets. The overwhelming majority of the local households maintain tourist apartments or room rentals, while some entrepreneurial individuals make reasonable profits by means of renting canoes and sun-umbrellas at the beach, selling soft drinks in temporary cafeterias or, even, constructing summertime bars by the sea.

In Vassilikos the introduction of tourism did not radically sever the relationship of the local inhabitants with agriculture and animal husbandry (Theodossopoulos 1997b: 253–4; 1999: 613). Most Vassilikiots continue to define themselves as 'farmers' (*agrotes*), despite the fact that they earn most of their yearly income from tourism related activities. They retain animals on their farms, maintain their olive groves and devote considerable attention to farming chores and duties. The tourist market readily consumes locally produced farming products, such as salad vegetables, fruit and cheese. Quite often the same families that produce the goods own the restaurants or the mini-markets where the goods are sold. Tourist enterprises are often part of the immediate household environment and work invested in farming directly reflects on the quality of the service provided to tourists. As Vassilikiots maintain, 'having rooms for rent in an olive grove requires both the rooms and the olive trees to be well cared for'.

Furthermore, the aura of tradition and authenticity associated with farming lifestyles is treated by both tourist operators and Vassilikiots alike as an additional tourist attraction. Unlike other, heavily developed parts of the island, Vassilikos obtained a reputation as an ideal location for quality tourism. It attracts tourists who trek through the countryside, interact with the local inhabitants and make the most of what is loosely defined by both visitors and hosts as 'Vassilikos' beautiful nature'. In this respect, Vassilikiots' environment and the aesthetics of 'nature' are not merely visually consumed by the tourists (Urry 1995), but also – as I will demonstrate in the following sections – strategically and discursively manipulated by indigenous residents and environmentalists in the context of local environmental politics.

As most Vassilikiots openly admit, tourism has greatly improved their lives. It put a definite end to the migration of local men or, as the older folks put it, 'it kept the young people (*tous neous*) at home'. It also enhanced the position of women in the family, since it provided them with more opportunities to directly contribute to their households' economies (cf. Galani-Moutafi 1993, 1994; Kenna 1993: 85–6; Theodossopoulos 1999). The most senior men and women interpret the new prosperity (*prokopi*) achieved through tourism as their fair reward for the difficulties and hardship endured in the past. The younger Vassilikiots understand their engagement with the new tourist enterprises as the entrance ticket to modernity and a more comfortable lifestyle. They argue that all people who are born in Vassilikos 'deserve' (*axizoun*) to profit from tourism and, thus, are expected to 'improve or fix their lives' (*na ftiaxoun tin zoi tous*) accordingly.

The entrepreneurial freedom of some Vassilikiots, however, is seriously threatened by the establishment of a Marine National Park in the vicinity. The park includes parts of the coastal environment, and imposes limits and prohibitions on tourist development on the land adjacent to it. The primary objective of the environmentalists who have lobbied for its establishment is the protection of the loggerhead turtle. The local beaches are an important breeding site of this ancient species and those who have studied its behaviour maintain that for the egg-laying of the turtles to take place, noise and light pollution on the surrounding land should be kept to a minimum. Consequently, the presence of curious tourists, noisy bars or restaurants and large tourist resorts is explicitly prohibited. Even the reflected light stemming from proximate developments at night, does not merely discourage the egg-laying adult turtles, but also disorients the hatchlings on their first attempt to leave their nest and venture to the sea. The environmentalists demonstrate with 'hard' scientific data that the prerequisites for turtle breeding and the development of tourism are mutually exclusive (Margaritoulis et al. 1991; Cape 1991; Arapis 1992).

The Greek state, which is in general supportive towards the development of tourism, under the continuous pressure of environmental groups, had no alternative but to promote the creation of the National Park, in a reluctant effort to conform to the pro-environmental regulations of the European Union. In 1984, a Presidential decree prohibited any construction on the land proximate to the turtle breeding sites and, in the subsequent years, additional laws established the legal grounds for the creation of the National Park. Surprisingly enough, the Greek government, reluctant to pay the cost of compensation to the Vassilikiot landowners affected by the conservation legislation, avoided any further measures towards the formal and final establishment of the park. Thus, while on the one hand some Vassilikiot families were prohibited from developing their land, on the other any claims for com- pensation were effectively ignored.

Vassilikiots, under the encouragement of local politicians and some state officials, were allowed to believe that in the future a favourable solution to their problem would be achieved. But eventually they realized that they were being deprived of the most vital asset of their land, the potential to develop tourism. While their neighbours in other parts of the island made significant profits, they saw themselves as being 'left behind' (*na menoun piso*) in the pur- suit of 'progress' (*prokopi*). After several years of complaints, they entered the most determined phase of their resistance to environmental conservation. They declared the environmentalists their enemies or people who were 'unwelcome on their land' (*anepoithimitoi*). During the late eighties and early nineties they engaged in a bitter conflict with the representatives of environ- mental organizations, and occasionally, with the state authorities. Further- more, as I will describe below, they continued building on the prohibited area. Defying the law, the Vassilikiots have dramatically shown their resistance in terms of masculine demonstrations of self-assertion that well-deserve to be described as stances of 'performative excellence' (Herzfeld 1985: 16).

Resisting Environmental Conservation

It was a cold winter afternoon during the early days of my fieldwork and the small all-purpose shop, which also served as a coffee-house of a sort, was packed with Vassilikiot men. They were relaxing after a hard day's work in the fields, watching the television news, drinking and commenting on current national and international politics. Taking advantage of the cordial and relaxed atmosphere, I was deeply immersed in a conversation with a couple of local men, when suddenly an intense, vehement booing united every- body's attention. The television was reporting on the actions of Greenpeace in some part of the planet, and all men present in the room instantly reacted with rage at the mere sound of the name of that particular environmental group. 'They are dangerous (*epikyndinoi*), malicious (*mohtiroi*) people', one of my interlocutors commented, 'they did great harm (*megalo kako*) to the peo- ple here in Vassilikos'.

At the time this incident took place,[4] Vassilikiots already had a good insight into the objectives and protest-oriented profile of Greenpeace. Some of its professional members had visited the island twice, deliberately instigat- ing some negative local reaction, which was consequently used by the media to attract attention to the cause of the turtles. Other environmental groups had more constant and constructive involvement in the turtle conservation dispute on Zakynthos. 'The Sea Turtle Protection Society' (STPS), a Greek environmental NGO, had been working towards the protection and the study of the loggerhead turtle since the 1980s. Its members were predomi- nantly Greek biology students and graduates – although foreign volunteers were recruited on a seasonal basis – and their admirable devotion to environmental conservation had already caused considerable aggravation in Vassilikos. In the late 1980s, the STPS established a permanent presence on the island, supported by a semi-organized nucleus of Zakynthian environmentalists. The society also found an additional, more powerful and well-known ally, the World Wildlife Fund (WWF), which included on its agenda, apart from the turtles, another endangered species of local fauna, the Mediterranean monk seal.[5]

By the early nineties, STPS, WWF and Greenpeace had already achieved a remarkable level of co-operation for the sake of turtle protection. Having successfully got the media on their side, they advocated that the future of the turtle was hindered by tourist development in Zakynthos and, in particular, by the activities of a minority of 'selfish' and 'profit oriented' local farmers who pursued their interests against the welfare of the local environment. The overwhelming majority of the Greek members of the above-mentioned NGOs had a middle-class background and resided permanently in urban centres (cf. Cotgrove 1982: 19, 34, 52, 93; Lowe and Goyder 1983: 10–11; Harries-Jones 1993: 46; Berglund 1998: 37). They had little experience with rural lifestyles and created little opportunities for maintaining long-term contact with the local population in the countryside of Zakynthos. Their views about the environment represented predominantly radical versions of environmentalism (Milton 1996: 74–8, 205–6; see also, Cotgrove 1982; Hays 1987; Norton 1991; O'Riordan 1976; Worster 1977), which 'exclude human beings from the natural realm' (Knight 2000: 11). However, despite occasional disagreements on abstract ideological matters, at the practical level of policy making and advocacy, theoretical arguments were put aside for the greater benefit of the turtle (cf. Norton 1991; Milton 1996: 78).

The gradual increased professionalization and efficiency of the environmental NGOs based in Greece, was associated with WWF's decision to buy some of the land adjacent to the turtle breeding ground in Vassilikos, in an effort to ensure that no further development would take place on it. The owners of this land accepted the money raised by the WWF as compensation, discouraged by waiting for so many years for a positive resolution to their problem. They said: 'what's the purpose of keeping land, if we are not allowed to have any control over it?' Most of their neighbours, however, have remained uncompensated for their conserved property. 'We will never sell our land', they argue, 'the ecologists have no right to evict us from our property'. By creatively articulating notions of unwritten justice with a cultural symbolism on the significance of the land and their aspirations for material and social progress (Theodossopoulos 2000), they formulate an original anti-conservation discourse with the sole purpose of defying the conservation legislation. Here are some of their poignant remarks:

> I can't sell my land. I can't see the land of my father being sold to foreigners and especially to the ecologists … I want to keep my land and make something nice on it. This is the land I worked. This is the land I ploughed with a wooden plough. [It is] here that I struggled (*edo palepsa*), [it is] here that I'll grow to be an old man (*edo tha geraso*).

This is my land, the land my father secured with sweat and effort. What do you expect me to do? Offer it as a present to the ecologists? (*Na tin hariso stous ecologous?*) These are serious matters (*sovara pragmata*), my dear (*matia mou*). We are talking about people's land (*ti gi ton anthropon*) ... the land of our fathers, the land we work, the land we sweat ... people's legal land ...

I will never sell my land. Look at this man [: a particular name is stated]. He sold his land to the WWF and now comes to my place to fish and moor his fishing boat.

But the resistance of the Vassilikiots to the conservation legislation is not confined to words. Since the early nineties, they have taken the matter 'into their own hands' (*sta heria tous*) by openly challenging the imposition of the National Park on their lives. Unable to physically attack the state, they turned their attention onto the visible representatives of environmentalism on Zakynthos, the people they collectively call with the generalizing term 'the ecologists'. After threatening or, even, physically attacking the defenders of the turtle on several occasions, they developed a frightening reputation among the circles of environmentalists. As some of the latter explained, one had to be really intrepid to conduct scientific measurements on the turtles in Vassilikos. During some seasons of intense conflict in the early nineties, the most heroic 'ecologists' could only approach the local turtle beaches from boats – since the Vassilikiots themselves controlled the main village road – and after hastily taking some measurements they had to flee to the safety of other parts of the island.

Men in Vassilikos, but also the women, appear content with the fierce reputation they have created among 'the ecologists'. 'It is good that we showed our power (*dynami*)', they explain, 'now the ecologists are scared of us ... they do not bother us so much any more'. Confident of their solidarity and secure within the immediate territory of their community, they apply their own practical solutions to the problem. Some of them build illegal constructions within the conservation zone, others rent sun umbrellas on the egg-laying turtle beaches, a few more run restaurants or bars on strictly prohibited ground. When the environmentalists launch their complaints the villagers ignore them or, even, scare the most annoying and persistent 'ecologists' with threats of violence. During the infrequent visits of the police, the Vassilikiots evoke their own versions of what constitutes justice and legitimacy. They emphasize their poverty, their family obligations and the legal titles of their conserved land. Here are a couple of examples:

We are poor farmers. My father and grandfather bought this land with their sweat ... The ecologists promised compensations. We have been waiting and waiting ... We are still waiting ... Now, we have lost our patience ... We have to protect the interests of our children.

This land which we possess today belongs to us. It was bought by our grandfathers and our parents in 1955. They didn't snatch this land from anybody else. Nobody gave this land to us for free. This land is the outcome of the labour and sweat of three generations, who lived and toiled all their lives, having as their only dream to possess this land, their land ...

In their reluctant attempts to impose the conservation restrictions, the state authorities are faced with determined crowds, which often include women and elderly people, and have always withdrawn unsuccessfully. Civil servants of lower rank, like the drivers of bulldozers attempting to demolish illegal constructions, are even more easily intimidated by masculine displays of rage and despair. Stories about Vassilikiots waving hunting guns at individuals who attempted to implement the conservation prohibitions, were commonplace during my fieldwork days. I still remember the trembling voice of an old man when he narrated to me how a relative of his defended an illegal tourist enterprise: ' "This is my land", my cousin said to the driver and the civil servants from the town. "This is my land and you'd better go away ..." ' Then lowering his voice, he whispered slowly to me: 'My cousin was so angry that he could even have murdered someone. The people from the town saw that he was "determined" (*apophasismenos*) and left.'

Relating to the Natural World

Over the course of the turtle dispute in Vassilikos, 'outsiders', journalists and environmentalists, attempted several times to explain the reasons behind the local protest. Their explanation, however, remained within the confines of the apparent and assumed economic interest of the Vassilikiots who were accordingly portrayed as immoral individualists strictly concerned with the maximization of their own profit at the expense of the local ecosystem. This kind of reductionist explanation, espoused primarily by the environmentalists, narrowly focuses on the economic utility of the disputed land and completely ignores the cultural matrix – or the 'cultural reason' (Sahlins 1976) – that informs the indigenous resistance to the environmentalists' ideals (Theodossopoulos 1997b). To fully understand the Vassilikiots' lack of appreciation for environmental conservation, some light must be shed on indigenous cultural views of the natural world. This is why my analysis in this section will focus on a thorough exploration of the cultural perceptions that inform the Vassilikiot relationship with the physical environment and the non-human beings living in it.

As I have already emphasized, the people who object to conservation in Vassilikos are not merely the ones who are directly affected by it. The overwhelming majority of Vassilikiots feel culturally obliged to empathize with their affected neighbours and relatives, while simultaneously, they fail to appreciate the objectives and moral justifications for the environmental conservation of their land. Their relationship with the local environment is constantly informed by a pragmatic, 'working' relationship with it. Far from being perceived as in need of protection, the natural world is understood as the physical and conceptual site where a farmer has to 'sweat' and 'struggle' in a daily repetitive battle to utilize and develop the land's resources. This work-oriented disposition towards the land is articulated in metaphors that symbolically represent daily life in the countryside as a 'struggle' (*agonas*) and discursively communicate a confrontational, contesting relationship with the physical environment. As du Boulay has noted:

> ... the winning of bread from the rocky fields is, as the villagers say, 'an agonising struggle' (*agonia*). For the greater part of the year nature, if not actually hostile to man, is at least relatively intractable. Day after day the farmer wears himself out in clearing, burning, ploughing, double-ploughing, sowing, hoeing, weeding; all through the year there are risks from hail, floods, drought, locusts, diseases ... (du Boulay 1974: 56).

Pruning-shears, sickles, scythes and even fire are some of the means summoned by Vassilikiots in their attempt to control 'wild vegetation' (*agriada*). Even after the introduction of modern power saws and other mechanical devices, most of the work is still completed with traditional equipment and constant manual labour. In fact, cutting undesirable vegetation well deserves to be accounted as 'struggle', since most of the weeds or thorn bushes exhibit a remarkable ability to resist extermination. During their repetitive struggle with the wild flora and while wiping away their sweat, Vassilikiots engage in rhetorical exclamatory observations of this type:

> I cut you and then I cut you again ... but there you are damned thing ... how many times do I have to pull out your roots?

> Look at those useless weeds (*paliohorta*) ... look at them! They come out of nowhere ... it was only a month ago when I destroyed them (*otan ta halasa*). Here they are! They never stop growing ... this struggle will never end (*autos o agonas dhen tha teleiosei pote*).

This constant and repetitive confrontation with the wild elements of the environment is carried out, not merely in the cultivated fields, but also in land proximate to tourist facilities where the pressure of keeping the vegetation under control is much greater. The Vassilikiot farmers, who recently became tourist entrepreneurs, apply to all parts of their property their own farming standard of what constitutes a properly managed environment. According to their aesthetic criteria, 'a well-cared piece of land' is one subjected to human labour and control to such a degree that the human input or 'sweat' (*ydrotas*) is easily perceived and appreciated by neighbours and tourists alike.

Vassilikiots' emphasis on the significance of human labour invested in the land is attuned with Christian cosmology, which explicitly places 'man' (*ton anthropo*) in a hierarchical position with reference to the environment and the other non-human creatures that live in it (cf. Papagaroufali 1996). Within this anthropocentric framework, human needs are prioritized over the needs of other beings, while the human actor is expected to assume responsibility – and responsibility here is manifested as both control and care – over animals and the natural environment. Metaphors drawn from the Bible, and in particular from Genesis constitute a rich source of ideological support to the Vassilikiots' perceptions. Religious 'discourses of truth' (cf. Foucault 1984) are thus often used by the local farmers in their attempts to articulate their own relationship with physical nature and legitimize its primacy over other worldviews, such as that maintained by the environmentalists. This is how one Vassilikiot put it:

> When God expelled Adam from Paradise, God gave him this land and said: this land is for you, to work it hard and earn your bread from it. This is what we do all the time. We work the land. We care for the animals and everything living on it.

Work in Vassilikos is locally conceptualized as a purposeful, constant 'struggle' to keep the environment 'in order' (*se taxi*). An environment 'in order' is, in turn, a productive field upon which human labour is realized for the benefit of the rural household. Within the farm environment, the notion of 'order' is the most central concept pertaining to the organization and management of the natural surroundings. Order is defined, created, imposed and safeguarded by both men and women farmers and it is realized as the conceptual and material outcome of hard work and 'struggle'. Domestic animals, vegetable gardens and animal shelters, even tourist apartments, all have to meet the prerequisite of order set by the farmers.

Keeping domestic animals in order and caring for them is an indispensable part of a Vassilikiot farmer's daily responsibility. Du Boulay (1974) has very neatly described domestic animals as the lower members of the rural household. Like human household members, animals are subject to reciprocal obligations and privileges. This reciprocal character of the human–animal relationship is treated by Vassilikiots as one further manifestation of the ideal order within the farm. 'Caring for the animals well' effectively means ensuring that one's domestic animals are not in any physical pain or danger, are well-fed, well-sheltered and, of course, 'kept in order'. Part of caring for the animals is also teaching them their place in the order of the farm, and punishing them when they fail to respect it. Vassilikiots maintain that animals, like children, can 'learn' their place and position within the farm and they constantly try to teach them, by use of punishment or reward, how to remain within given specifications of 'order'. If animals are left undisciplined and unattended, Vassilikiot farmers argue, they will most certainly regress to a much undesirable state of anarchy.

Slaughtering animals is likewise interpreted as an inevitable stage in the reciprocal relationship between human and animal members of the rural household. In this respect, the death of some domestic animals is conceptualized as part of the context of order and care on the farm. 'What will we give the animals to eat if we don't kill some of them?' the Vassilikiots argue, 'how are we supposed to maintain our house(holds)?' The human–animal relationship on the farm is thus constructed upon the simple rule of reciprocity that envelops the very meaning of animal husbandry as this is understood in Vassilikos. According to this rule, domestic animals receive 'care' from their owners and are expected in return to respect the 'order' of the farm and the sacrifices this order entails. Vassilikiots poetically summarize this as follows:

> Then you have the 'ecologists' who come to you and say why do you kill your own animals ... they say we don't care for animals ... We care for those animals

every day … We raise them … the ecologists get meat from the supermarkets. Then they come here to tell us how to run our farms …

Look at this lamb. We care for this lamb. We provide for it. The lamb has a good life because we provide for it. When the time comes the lamb will have to die. It had a life, a good life … The lamb will die and give life to the rest of the animals … The lamb could not exist without the human provider (*horis ton anthropo*).

The rules of reciprocity I have just outlined, however, apply insofar as the animals in question exist within the context of 'care and order' established on the farm. Hence, Vassilikiots do not feel any obligation towards wild animals that are seen as existing beyond and above the farmers' control, and are subsequently classified according to their potential 'usefulness' (*hrisimotita*) or harm (*zimia*). For example, most species of wild birds are conceptualized as useful, since as game they satisfy the Vassilikiots' celebrated passion for hunting. The local hunters shoot edible[6] birds at any given opportunity, that is, whenever they spot them while carrying a gun. Hunting is in turn considered in Vassilikos as an important activity representative of, and indispensable to, the local culture and tradition. Like the prohibitions raised by the turtle conservation, seasonal hunting regulations are dealt with by the assertive Vassilikiot hunters in an uncompromising manner: 'Hunting is a tradition for us', they say, 'an important part of our lives. The ecologists will never succeed in making us give it up. We will be hunting on our land for ever, like we always did.'

Finally, Vassilikiots' attitude towards 'harmful' (*vlavera*) wild animals is one of unequivocal animosity. In this category are classified all wild creatures that 'cause harm' (*kanoun zimia*) to the household–farm continuum, either by destroying crops or fishing nets, or worse by preying on domestic animals. Vassilikiots are sworn enemies of these pernicious beings and seek to destroy them by guns, traps, poison or whatever other means possible. Their hostile sentiment towards them is a further manifestation of the local farmers' constant struggle to defend care and order on the farm against the uncontrollable aspects of the physical environment. Vassilikiots explicitly express their anger (*thymos*) and sorrow (*lypi*) when domestic animals fall victim to wild predators. The attack by wild animals is treated as a violent interruption to the process of order and care established through hard work and considerable effort. The 'harmed' domestic animals are in turn, referred to as 'being lost' (*hanontai*) and the care invested in them as equally 'lost' (*hameni*) and irretrievable.

Conclusion

The people of Vassilikos are very angry about the ecologists. In the beginning a few of them came. We gave them hospitality. We welcomed them on our land. They said they were counting the turtles … Then they kept on coming. More and more of them, every summer. They said we couldn't build on our land. We couldn't do this or that … All this because of the turtle [narrated by a sixty-year-old Vassilikiot, who is a farmer and tourist entrepreneur].

During the twenty-year long dispute over the conservation of the loggerhead turtles on Zakynthos, the inhabitants of Vassilikos had been repeatedly voicing their objections to the application of environmental legislation to their land. At the practical, observable level of environmental politics, some Vassilikiot families have become victims of a socially inconsiderate environmental policy. Uncompensated and deprived of control over their land, those unfortunate Vassilikiots are not merely deprived of the possibility of developing tourism, which is the most vital productive resource of their land. They are forced either to pursue opportunities with tourism in other parts of the island, where they do not own land suitable for such an undertaking, or to ignore the conservation measures and engage in a painful confrontation with the authorities.

At the level of media attention and pro-environmental public discourse, the indigenous Vassilikiot approach to the natural world is conspicuously overlooked. In the turtle politics of Zakynthos, environmental conservation is prioritized over those who 'work and care for the land', while the future of the turtle – an animal species that exists outside the culturally defined parameters of care, order and reciprocity – has overshadowed the future of the local human caretakers. Instead of being in a position to manage their immediate environment as rightful heirs of a well-established culturally informed anthropocentric tradition, the Vassilikiots affected by conservation have become outcasts on their own land.

To justify their resistance to turtle conservation, Vassilikiots draw upon their own cultural approach that informs their relationship to the natural world. They understand their immediate environment in terms of their own constant work or 'struggle' (*agonas*) with it – an agonistic attitude well documented in the regional literature (Friedl 1962: 75; du Boulay 1974: 56; Kenna 1990: 149–50; Hart 1992: 65; Dubisch 1995: 215). This confrontational spirit towards the land is part of a constantly enacted and embodied relationship manifested in the context of daily work. According to the same cultural logic, physical 'nature' (*e physi*), far from being thought of as fragile, threatened or in need of protection, is locally understood as powerful and resistant to human intervention and control. In order to extract the required resources from it or make the existing ones productive, Vassilikiots know very well that they have to invest considerable toil (*mohtho*) or hard labour.

This general cultural theme of controlling the productivity of the land through constant input of human action upon it is further enacted in Vassilikiot performances of struggle in the environment of the farm. As I have already noted in the previous section, the human actor is perceived as responsible for the imposition of order on the farm and the undertaking of care for domestic animals. The latter are treated as lower members of the household compound and are expected to reciprocate the care they have received from their caretakers (cf. du Boulay 1974). Wild animals, by contrast, are perceived to exist outside the humanly managed context of care and order and are thus conceptualized as either potentially useful resources for the household, or as harmful entities that disrupt and destroy what the farmer has built with sweat and labour. To safeguard order on the farm, the human protagonists are expected

he Dispute over the Sea Turtles | 257

to assume a central position in it. This role of human guardianship over non-human beings and the environment is further validated by the anthropocentric orientation of Christianity in relation to the natural world (cf. White 1968; Worster 1977; Morris 1981; Thomas 1983; Serpell 1986; Ritvo 1987; Ingold 1988, 1994; Tapper 1988; Willis 1990; Davies 1994), which constitutes a rich source of moral and ideological support for the Vassilikiots' claim that they are indeed responsible for managing their immediate environment.

In the context of the turtle dispute in Vassilikos, the great majority of Vassilikiots have failed to sympathize with or understand the priorities of the environmentalists, largely because they do not apprehend their own immediate environment as being ecologically threatened or in need of protection. They look at their olive groves and perceive their humanly managed land as beautiful and well cared for: 'this beauty is the result of our constant labour and care' they explain. In contrast to the inhabitants of cities like Athens, which is where most of the environmentalists reside, Vassilikiots maintain a living and embodied relationship with the natural world as opposed to the conservationists who appear as merely 'talking' about it. 'The ecologists talk theory (*theoria*), we talk action (*praxi*)', Vassilikiots argue, and add: 'The ecologists and the journalists who talk about the protection of nature don't live here. What do these people know about protecting this land? What do they know about living in the countryside and caring for the land?'

Despite its rhetorical fervour, the Vassilikiot anti-conservationist discourse remains largely unsuccessful in contexts other than the immediate locality of Vassilikos. As other ethnographers have noted, far away from home Greek local actors tend to appear intimidated by the formality and impersonality of the law and bureaucracy (Campbell 1964; Loizos 1975; Herzfeld 1992). When it comes to direct representation in the process of making laws, Vassilikiots remain effectively muted (Theodossopoulos 2000: 73). Consequently, they confine their resistance to obstructing the application of the undesirable legislation at the local level. Vassilikiot men and women, frustrated by the restrictions of environmental conservation, have responded to state inaction with their own initiatives. To demonstrate their determination not to be parted from their land, they consistently and deliberately ignore the conservation regulations, build illegal constructions in the conservation area or engage in tourist related enterprises on the turtle breeding grounds.

On the other hand, the state authorities, intimidated by collective performances of indigenous resistance, have failed to impose the conservation restrictions. The leniency of the Greek government in the application of the law is related to the state's inability to provide the affected landowners with compensation. Under the pretext that more environmental planning should be completed and critically examined, delay and postponement in finalizing the conservation measures was instituted as the unofficial policy of the state. As Herzfeld has noted about a conservation dispute in Crete, when the need for precise documentation is involved, 'nothing is more permanent ... than the temporary' (1991: 251). In the meantime, several Vassilikiot families affected by turtle conservation have been deprived of the most vital productive resource of their land, the locally perceived right to develop tourism. Thus, environmental conservation, which is considered by most Vassilikiots as meaningless or unnecessary, has deprived some local households of their long-awaited claim to progress or prosperity (*prokopi*).

The ethnography presented in this chapter suggests that the indigenous resistance to environmentalism in Vassilikos cannot be adequately explained without considering the indigenous approach of dealing with the animate and inanimate physical world. Vassilikiots' understanding of the environment is rooted in their distinct, lived, and constantly enacted relationship with it. This is a relationship built upon the axiom of reciprocity; 'in accordance with the logic of gift exchange', Bourdieu maintains, 'nature brings her fruits only to those who bring her their toil as a tribute' (1990: 116). Those Vassilikiots who feel evicted from their land for the sake of turtle conservation, persistently articulate their resistance in terms of a cultural strategy which is meaningful to them. They establish an intimate connection with the land by investing their toil in it and realizing its productive resources. Tourism is only one of them. Although Vassilikiots know very well that their voice will probably never be heard in the social *milieu* of the environmentalists and the environmental planners, they keep on arguing:

> We don't want the ecologists on our land. They only cause trouble. They did harm (*zimia*) to several people here. They try to tell us what to do with our property. What to do in our own fields. The land we work with our own hands …
>
> We didn't go to their place to tell them how to run their own homes. If the ecologists care for the turtles, then why don't they take them to their own property? Why don't they take care of them on their own land?

Notes

1. Loggerhead sea turtle, *Caretta caretta.*

2. The term *sembroi,* translated here as landless tenants, carries in Zakynthos feudal connotations comparable to the English term 'serfs'. This meaning is not applicable to other parts of rural Greece, but it is rather indicative of the island's feudal past.

3. Different kinds of *sembremata* arrangements regulated different forms of cultivation and animal husbandry. For more information on particular standardized *sembremata* arrangements and the actual negotiation between labourers and landlords, see Theodossopoulos (1997a: 33–8, 72–8, 108–9). For an examination of similar arrangements on Corfu, see Couroucli (1985: 87–94).

4. My fieldwork in Vassilikos, Zakynthos, was carried out from 1991 to 1993 in a period lasting over eighteenth months.

5. Mediterranean monk seal, *Monachus monachus.*

6. For more information on the criteria which constitute edibility in the indigenous system of animal classification, see Theodossopoulos (1997a: 144–5).

References

Arapis, T. (ed.) 1992. Thalassio Parko Zakynthou: proypotheseis and protaseis gia to shediasmo, tin idrysi kai leitourgeia tou. Independent study, supported by WWF-Greece, Greenpeace-Greece, Sea Turtle Protection Society of Greece.

Berglund, E.K. 1998. *Knowing Nature, knowing Science: An Ethnography of Environmental Activism.* Cambridge: The White Horse Press.

Bourdieu, P. 1990. *The Logic of Practice.* Stanford: Stanford University Press.

Campbell, J.K. 1964. *Honour, Family and Patronage: A Study of the Institutions and Moral Values in a Greek Mountain Community.* Oxford: Oxford University Press.

Cape, C. 1991. 'Turtles or Tourists? Zakynthos: How tourism could assist the conservation of the loggerhead turtle'. MSc Dissertation in Conservation, University College London.

Cotgrove, S. 1982. *Catastrophe or Cornucopia.* Chichester: John Wiley & Sons.

Couroucli, M. 1985. *Les oliviers du lignage: une Grèce de tradition Venitienne.* Paris: Maisonneuve et Larose.

Davies, D. 1994. Christianity. In J. Holm (ed.) *Attitudes to Nature.* London: Pinter Publishers.

Dubisch, J. 1995. *In a Different Place: Pilgrimage, Gender, and Politics of a Greek Island Shrine.* Princeton: Princeton University Press.

Du Boulay, J. 1974. *Portrait of a Greek Mountain Village.* Oxford: Clarendon Press.

Foucault, M. 1984. 'Truth and Power'. In *The Foucault reader* (ed.) P. Rabinow. London: Penguin.

Friedl, E. 1962. *Vassilika: A Village in Modern Greece.* New York: Holt Rinehart and Winston.

Galani-Moutafi, V. 1993. 'From Agriculture to Tourism: Property, labour, gender and kinship in a Greek island village' (part one). *Journal of Modern Greek Studies*, 11: 241–70.

—— 1994. 'From Agriculture to Tourism: Property, labour, gender and kinship in a Greek island village' (part two). *Journal of Modern Greek Studies*, 12: 113–31.

Handman, M.E. 1987. *Via kai poniria: antres kai gynaikes s' ena elliniko horio.* Athens: Ekdoseis Kastanioti.

Harries-Jones, P. 1993. 'Between Science and Shamanism: The advocacy of environmentalism in Toronto'. In K. Milton (ed.). *Environmentalism: The View from Anthropology*, pp. 43–58. London: Routledge.

Hart, L.K. 1992. *Time, Religion, and Social Experience in Rural Greece.* Lanham: Rowman & Littlefield Publishers.

Hays, S.P. 1987. *Beauty, Health and Permanence: Environmental Politics in the United States 1955–1985.* Cambridge: Cambirdge University Press.

Herzfeld, M. 1985. *The Poetics of Manhood: Contest and Identity in a Cretan Mountain Village.* Princeton: Princeton University Press.

—— 1991. *A Place in History: Social and Monumental Time in a Cretan Town.* Princeton: Princeton University Press.

—— 1992. *The Social Production of Indifference: Exploring the Symbolic Roots of Western Bureaucracy.* Chicago: The University of Chicago Press.

Ingold, T. (ed.) 1988. *What is an Animal?* London: Unwin Hyman.

—— 1994. 'From Trust to Domination: An alternative history of human–animal relations'. In A. Manning and J. Serpell (eds). *Animals and Human Society.* London: Routledge.

Kenna, M.E. 1990. 'Family, Economy and Community on a Greek Island'. In C.C. Harris (ed.). *Family, Economy and Community.* Cardiff: University of Wales Press.

—— 1993. 'Return Migrants and Tourism Development: An example from the Cyclades'. *Journal of Modern Greek Studies*, 11: 60–74.

Knight, J. 2000. 'Introduction'. In *Natural Enemies: People–wildlife Conflict in Anthropological Perspective.* London: Routledge.

Lison-Tolosana, C. 1966. *Belmonte de los Caballeros: A Sociological Study of a Spanish Town*. Oxford: Clarendon Press.

Loizos, P. 1975. *The Greek Gift: Politics in a Greek Cypriot Village*. Oxford: Basil Blackwell.

Lowe, P. and Goyder. J. 1983. *Environmental Groups in Politics*. London: Allen & Unwin.

Margaritoulis, D., Dimopoulos, D. and Kornaraki, E. 1991. 'Monitoring and Conservation of *Caretta caretta* on Zakynthos'. Report submitted to the EEC (Medspa-90–1/GR/28/GR/05) and WWF (project 3825).

Milton, K. 1996. *Environmentalism and Cultural Theory: Exploring the Role of Anthropology in Environmental Discourse*. London: Routledge.

Morris, B. 1981. 'Changing Views of Nature'. *The Ecologist*, 11: 130–7.

Norton, B.G. 1991. *Toward Unity among Environmentalists*. Oxford: Oxford University Press.

O'Riordan, T. 1976. *Environmentalism*. London: Pion Limited.

Papagaroufali, E. 1996. 'Xenotransplantation and Transgenesis: Im-moral Stories about Human–animal Relations in the West'. In P. Descola & G. Palsson (eds). *Nature and Society: Anthropological Perspectives*. London: Routledge.

Ritvo, H. 1987. *The Animal Estate*. Cambridge, MA: Harvard University Press (Penguin 1990).

Sahlins, M. 1976. *Culture and Practical Reason*. Chicago: University of Chicago Press.

Serpell, J. 1986. *In the Company of Animals*. Oxford: Blackwell.

Tapper, R.L. 1988. 'Animality, Humanity, Morality, Society'. In T. Ingold (ed.). *What is an Animal?* London: Unwin Hyman.

Theodossopoulos, D. 1997a. 'What use is the Turtle?: Cultural perceptions of land, work, animals and "ecologists" in a Greek farming community'. Unpublished PhD Thesis, University of London.

—— 1997b. 'Turtles, Farmers and Ecologists: The cultural reason behind a community's resistance to environmental conservation'. *Journal of Mediterranean Studies*. 7(2): 250–67.

—— 1999. 'The Pace of the Work and the Logic of the Harvest: Women, labour and the olive harvest in a Greek island community'. *J. Roy. Anthrop. Inst.* (N.S.), 5(4): 611–26.

—— 2000. 'The Land People Work and the Land the Ecologists Want: Indigenous land valorisation in a rural Greek community threatened by conservation law'. In A. Abramson and D. Theodossopoulos (eds). *Land, Law and Environment: Mythical Land, Legal Boundaries*, pp. 59–77. London: Pluto.

Thomas, K. 1983. *Man and the Natural World: Changing Attitudes in England 1500–1800*. London: Penguin Books.

Urry, J. 1995. *Consuming Places*. London: Routledge.

White, L. 1968. *Machina ex Deo: Essays on the Dynamism of Western Culture*. Massachusetts: MIT Press.

Willis, R. (ed.) 1990. *Signifying Animals*. London: Unwin Hyman.

Worster, D. 1977. *Nature's Economy: A History of Ecological Ideas*. Cambridge: Cambridge University Press.

Zarkia, C. 1996. 'Philoxenia: Receiving tourists – but no guests – on a Greek island'. In J. Boissevain (ed.). *Coping with Tourists: European Reactions to Mass Tourism*. Oxford: Berghahn.

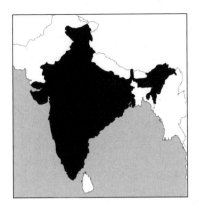

15

Displacement and Forced Settlement

GYPSIES IN TAMILNADU

Daniel Meshack and *Chris Griffin[1]*

Introduction

This chapter examines the displacement in Tamilnadu of former Outcaste forest-dwellers known as Narikuravas, Vagri or Kurrivikaran, and the problems they face. Since they are neither agriculturalists nor a 'service caste' (see Mines 1984), or for that matter classified 'tribal' or 'indigenous' with an historical claim to stewardship of, or access to, particular country, they fall outside the social space usually examined by anthropologists. Rather, they are commercial nomads, peripatetics or Gypsies[2] who (like Gypsies elsewhere) have traditionally lived physically apart from surrounding populations and with little sense of identity or attachment to one particular locality (Werth 1993).

This raises the conceptual issue of whether 'displacement' and 'forced settlement' are appropriate terms to apply to such inveterate wanderers? And we will say they are. We will argue that 'displacement' here refers not to induced dislocation from time-honoured places per se but instead to displacement from a specific ecological niche without geographic boundary – namely, 'the forest'. Furthermore, we maintain that because such dislocation has compelled many Narikuravas to opt for more permanent settlement than previously, the term 'enforcement' is also appropriate.

Our first task is to view the construction of Narikurava identity from three perspectives: their own and that of foreign scholars who claim to know them well; other Tamils; and Tamilnadu intellectuals and advocates for Narikuravas who are themselves Dalits or 'Untouchable'. After that we will describe Narikurava culture and social organization, and show how deforestation, displacement and forced settlement have affected them, and end with a brief overview of the resultant problems and how these might be addressed. We adopt a broadly comparative approach by drawing attention to parallels in the circumstances of peripatetics elsewhere.

People and Forests

Tamilnadu has a population of 56 million, of which 19 per cent comprises *Scheduled Castes* consisting of unlisted castes, converts and Dalits, and 1.03 per cent *Scheduled Tribes.* In principle, the remaining 80 per cent are caste Hindus of the three 'clean' *varna,* Brahmin, Kshatriya and Vaishya, and 'unclean' Shudra. However, some say in reality the latter count for 31 per cent, Brahmins for less than 4, and Vaishya and Kshatriya for none (Saraswathi 1974). It is only when other high castes are included with Brahmins, that the upper strata – classed *Forward Classes* or Communities [FCs] in contrast to *Backward Classes* [BCs,] made up of Shudras – amount to 49 per cent.

Dalits who stand outside the caste-varna ideology because of ritual 'uncleanliness' are normatively protected by the Constitution from discrimination and have reserved quotas in the public service, but pragmatically most face daily stigma and discrimination. Meanwhile, somewhere between BCs and *Scheduled Castes* are those designated *Most Backward Classes* [MBCs] which in Tamilnadu include Narikuravas.

According to Prabhaker (undated), India's Outcastes were people historically 'co-opted' into Hindu society where 'untouchability' was forced on them. Hence comes the term Fifth Caste, though others are *Panchamas* and *Adidravidas.* Most are landless and many are bonded labourers, and they make up most of Tamilnadu's poor.

Officially distinguished from caste Hindus and *Adidravadas* are the many groups of indigenous people known collectively as *Scheduled Tribes* or *Adivasis* whose social and cultural origins pre-date the advent of Hinduism. Hunters and collectors, swidden-cultivators, and agriculturists, *Adivasis* occupy the hills and forest, and though sometimes less stigmatized than *Adidravadas* and BCs are usually economically worse off (Shah 1975). To our knowledge as a category they exclude commercial nomads.

Thurston (1965, first published 1909) included Kurrivikaran in his magnum opus on South Indian tribes and castes, while others have listed them among the Criminal Tribes. Werth remarks on their universal reputation for 'dirtiness' (1996: 6), and officials in Tamilnadu classify them as MBC. Yet for all the stigma and perhaps *because* of general low self-esteem, Narikuravas consider themselves superior to Dalits. We estimate there are about 239 Narikurava 'settlements' or social clusters in Tamilnadu, of which 35 per cent are semi-permanent. This amounts to 7920 families or 47,520 individuals – a figure close to the one reckoned by Werth (1993).

According to the National Remote Sensing Agency (NRSA), between 1972 and 1982 India lost 1.3 million hectares of forest annually, eight times the 0.15 million hectares officially reckoned by the Forests Department (Fernandes et al. 1988: 3). In the same decade 'closed' or dense forest decreased from 14.12 per cent of total land surface to 10.96, while 'open' or degraded forest increased marginally from 2.67 to 3.06 per cent. In comparison to other states, Tamilnadu's record was by no means the worst; even so, the 1952 National Forest Policy which reckoned optimum forest cover at 33 per cent estimated the national figure at 22 per cent, and Tamilnadu's at seventeen. Moreover, state income from logging and forestry which in 1980–81 totalled 170 million rupees, had by 1995–6 multiplied seventeen times to around 3 billion rupees, far exceeding income rises in agriculture and fishing. To comprehend what this and the state reservation of 1,914,000 hectares of forest, out of a total 2,200,000 (Fernandes et al. 1988), means for Narikuravas, we must view the larger picture.

To start with, the 'quality of life' in 'Western India' is generally acknowledged as being higher than that in 'Eastern India' – defined as states lying east of a line drawn between Delhi and Chennai (formerly Madras) – and in Tamilnadu this is explained by several factors. They include: the successful control of birth rates, improved agricultural productivity, a large and expanding industrial sector, and the development of a 'very well-balanced urban pyramid' (Rothermund 1993: 191) which as well as Chennai (3.8 million) includes Coimbatore, Madurai, Tiruchirappalli and Salem, and many smaller cities and towns. However, as a result, vast areas of forest have been cleared, some simply for city firewood (Baker and Washbrook 1975), and thus the enhanced quality of life for some has come at a cost for others – including Narikuravas.

Little is known about Narikuravas before the mid to late nineteenth century. Most of our knowledge comes from Thurston (1965), from Werth (1993, 1996), and from what we can deduce from studies of other castes and tribes, and economic histories. Following the English East India Company's appearance in the 1700s, political control of Peninsular India fell to three governments, the Native States of Cochin, Native States of Travancore, and the Company Madras Presidency, where in each case indigenous claims to forest were ignored in favour of their own or local zamindars.

The Forest Department established in Madras in 1865 remained ineffectual until 1892 when the British administration (established in 1858) introduced its Forest Act. In the interim, agricultural castes continued to graze animals and gather firewood in the forest as per tradition. And in Andhra, these professional graziers included Lambardis (Baker and Washbrook 1975: 99–100), of whom more presently. It was not until a new Chief Conservator was appointed early the following century that the idea of common access to forest resources was challenged and customary practices began to be curtailed. Morris summed it up like this:

> In effect the forest department replaced the agents of the Raja, except in the case of the Zamindar forests of Madras, which had the status of private lands right up to the Estates Abolition Act of 1947. Not that government was unaware of the rights of the hill tribes, but they were clearly subordinated to the fundamental premise that the forests were its property, and its policy represented the interests of overseas capital and of the high caste and propertied groups in the plains (1982: 40).

Consistent with what Narikuravas tell us we can therefore deduce that during the first half of the twentieth century, Narikuravas' traditional patterns of hunting, trapping and peripateticism were affected by market changes, causing many to be displaced and forced into urban migration and less mobile living.

Identity

For present purposes Narikurava identity can be viewed three ways: (1) as they see themselves or as social scientists – especially anthropologists – *say* they do; (2) as Tamils generally do; and (3) as Tamil intellectuals and advocates for Narikuravas do. And bearing in mind one of us is an anthropologist, and the other a Dalit advocate for Narikuravas, we will try to keep our views as clearly distinguished as possible from others in these categories.

Self-Identity

According to Werth (1996), one of the few outsiders to know them well, the Narikuravas probably originate from Gujerat and Rajasthan,[3] but now live throughout Peninsular and some of Central India.

Although outsiders refer to them as Narikuravas, they call themselves Vagri, and see themselves as culturally distinct by virtue of language, religious practice, and social organization into twenty-two exogamous patrilineal sub-clans or *jathi,* each with its own female deity to which is offered blood-sacrifice (see Werth 1993, 1996). Within *jathis* individual conjugal and extended households have their own shrines and deity images. In this sense, both from an insider's and anthropologist's perspective, Vagri have a strong sense of *collective* identity and a marked sense of *family* and *individual* identity. Dependent on context this tension is reminiscent of other Gypsy societies (Berland 1982).

Quite the most important element of Vagri identity rests on people's association with the 'forest': *chahadi* in Vagrivale, *kaadu* in Tamil, *jangal* in Hindi. Thus they call themselves *Jangal jathis* and *Kaadu Raja* (Lords of the Forest), and make a point of camping on the side of fields lying closest to the forest when they work as watchmen for agricultural castes. Moreover, viewed etically, this association with the 'forest' or 'uncultivated' side of Nature (as opposed to 'fields' and Culture), and the pride Vagri take in being united with it, is reciprocated by agriculturalists who for their part view the forest as redolent of all that is 'primitive' and 'dangerous'.

Some Narikuravas tell us they are the distant cousins of Lambardis. Thurston says (1965: 182) they claimed their original male ancestor was brother to the first male ancestor of the Lambardis and to the founder of the Dommaras – a caste or tribe of acrobats. Lambardis were originally a north-eastern tribe of long-distance carriers who by the eighteenth century transported grain and other goods by bullock all over India, as well as providing cattle, grain, and manpower to various armies (Thurston 1965; Moreau 1995). Later they turned to pastoralism and agriculture in the Deccan, to grazing in Andhra (Baker and Washbrook 1975), as did others in North Arcot. Indeed today, the Narikuravas (who believe themselves superior) trade with Lambardis in Tiruvannamalai.

According to Werth (1996) *Lambardi* and *Sukkali* are South Indian terms for those in the north known as Banjaras who call themselves *God* or *Godmati.* This is congruent with Thurston's remark that in the 1891 Madras Census, Lambardis and Sugalis were 'practically the same' (1965: 210).

Another community of peripatetics in Tamilnadu, known generally as Tombar but who call themselves Banjanyo and tell outsiders they are *Gujerato* (Werth 1996), are of interest, given that *Gujerato* is also the name of a Narikurava sub-clan. Furthermore, *Tombar* is cognate with *Dombar* who, according to Shashi (1994: 154), are a forest-dwelling 'gypsy' tribe of hunters-and-collectors known also as *Domb,* whom the British once classed a 'criminal tribe'.

Clearly one can only speculate on how the Tombar or Banjanyo may at one time have been one with the Lambardis and Banjaras – whose name Moreau (1995) claims meant 'inhabitant of the wood'. Some close connection is likely, as with the Narikuravas who claim distant kinship with Lambardis – and Dommaras (Thurston 1965) – and trade with them in North Arcot.

Lewis (1991) describes Dombs in Orissa as 'gypsy-like' people who specialize in usury and bear training. These are presumably the same people whom Fawcett a century earlier called 'an outcast jungle people, who inhabit the forests on the high-lands fifty to eighty or a hundred miles from the east coast, about Vizagapatam' (in Thurston 1965: 174). Probably Dravidian, certainly moneylenders, labourers, scavengers and traders, Thurston considered them an offshoot of the north western Dom.

Fraser (1995), an authority on European Gypsies, believes Dom may be ancestors of the European Rom, and originally have been Dravidians. He also acknowledges – while personally disagreeing for linguistic reasons – that other specialists identify Banjari or 'Lamani-speaking' Banjaras as the Roma's ancestors.

Thurston (1965) likened Dombs to the Panos in the Khond country of Orissa and suggested the labels were interchangeable, while Bailey identified Panos in Orissa as 'outcastes' (1960: 124). Meantime, in Madras, *Dom, Dombara, Pano,* and *Paidi* are all synonyms for a Scheduled Caste (Saraswathi

1974), and it is this consensus about 'untouchability' which prompts us to inquire how other Tamils see Narikuravas.

General Tamil Views

Other Tamils generally perceive Narikuravas as culturally distinct and socially removed from themselves on several counts: appearance, unclean occupations, forms of shelter and habitat, spatial mobility, and 'unclean' culinary practices. Put together they add up to a near sub-human type of Outcaste.

Appearance

Dress and decoration function as universal markers of identity but nowhere more so than in India. Many Vagri men wear red turbans or head bands over hair drawn into a topknot. Men of middle years and upwards sprout handlebar moustaches which come into contact with the blood of sacrificial animals drunk ritually and so take on a hallowed status like the turbaned head itself – physical locus of deity visitations. Breech clouts, worn underneath *lungis* or wraparounds, if glimpsed by passers-by (as is possible with pavement dwellers) are taken as further evidence of the 'primitive'.

Women traditionally favour a maroon mid-calf length skirt sometimes with a decorated hem; the maroon and red represent blood-sacrifice, though how many outsiders realize it we cannot say. The skirt strikes a bold contrast with the sari or tunic and trousers worn by other Tamilnadu women and is adapted to mobility, but saris given as gifts by Christians are increasingly being worn. The skirt is also functional in other ways: like Gypsies in America (Sutherland 1975) and Britain (Okely 1983) Narikuravas women can purposely pollute men by the touch of their hem or hoisting the garment up to expose the genitals. Men, for their part are less powerful than menstruating women when it comes to pollution, but can inflict harm by the well aimed toss of a loin cloth. A bodice, worn with a drape over one shoulder (or both) tie-able at the waist for carrying an infant, virtually completes the woman's outfit. Her remaining decoration consists of nose-studs, earrings, toe-rings, and (like men) bead necklaces and tattooed hands and arms.

Unclean and Low Status Occupations

Narikurava and *Kuruvikkaran* are ethnonyms for jackal hunters and bird catchers. Vagri not only trap and eat birds, and sell them to others to eat, but where conditions allow they hunt and eat jackal (*canis aureus*), rabbit, jungle cat, Indian fox, and squirrel; all of which marks them out as 'unclean'. Yet this does not deter affluent caste Hindus from seeking out the charms they fashion from the 'horn' or skull bone of jackal – known as *narikombu.*

Narikuravas beg, sing, dance, make and hawk bead necklaces, sell fishhooks and heavy-duty needles (*oosi*) made on small anvils, lead performing monkeys, raise pigs, recycle waste, hawk cosmetics, scavenge food, sell catapults to tourists, and vend their women's hair (cut for religious occasions) to tribes making hair-pieces.

Shelter

Early last century the Vagri lived in thatch shelters transportable on ox-carts (Thurston 1965). Today their shelters vary according to location, mobility, wealth and preference. In Arani, a hundred kilometres from Chennai, residents of one piece of waste-ground occupy low thatch shelters just big enough for two. Others arrange themselves as domestic groups on mats surrounded by their chattels. On a road outside a village close to Arani, a larger group occupies several one-room houses big enough to stand in, made from plaited coconut leaves. Barred from the village wells they draw water from a pond. In the same area in the dry season we met several small extended travelling family groups camped shelterless by the road.

On the outskirts of Chennai, in a legal permanent encampment of twenty households, the poorest families live beneath tarpaulins or in shacks, while two better off families occupy mud and tiled roofed houses complete with television. On a similar site in central Chennai, the city fathers, after years of lobbying from READA (Rural Education and Development Association, Chennai), have built tiny brick and tile homes for about sixty families. Yet around the corner on a major road their relatives camp on the pavement, where the women must suffer the sexual predations of passers-by while their men work scavenging at night.

Habitat

Like Gypsies elsewhere, most Narikuravas inhabit the outskirts of villages, towns and cities, which taken in conjunction with other factors lends them an identity. As for this choice of habitat, some say they freely chose it and others insist they were forced into it by deforestation and the advent of forest regulations. Whatever the actual or perceived reasons, these marginal habitats only reinforce the Narikuravas' reputation as outcastes and dependent 'primitives'. This is not to say they were previously self-sufficient; peripatetics never are. Instead, as hunters and collectors involved in trade and passing service they have long operated in what Morris (1982), speaking of the Hill Pandaram, calls a 'dual economy'. That is to say, a subsistence economy involving a significant measure of social autonomy, combining trade with others.

Spatial Mobility

Being regarded as nomadic is such a negative marker of identity that even former peripatetics like Irular snake-catchers and Korova fortune-tellers and basket-makers today look down on Narikuravas, as do most Dalits. It reminds us that for much the same reason officials and agricultural castes hold the Hill Pandaram in contempt (Morris 1982: 45), yet just why nomads come in for this is something we will return to.

Culinary Practice

Being reputed consumers of flesh – cow, jackal, pig, jungle cat, and bird life – qualifies Narikuravas as unclean. Being known to eat discarded food, reheat meals already mixed with spices, drink *arrack*, and fight among them-selves[4] only adds to this status and to their reputation as bluffers, tricksters, and sometime poultry thieves.

In sum, general Tamil perceptions of Narikuravas are a result of a conflu-ence of factors: notions of the 'primitive' associated with the forest, their social separateness, marginal habitats, appearance, polluted occupations, unclean culinary habits, simplicity of shelter, and ideas of nomadism. Yet considering how India has long been a place of 'travelling specialists' (Singer 1955 cited by Misra 1971), and how spatial mobility has sometimes been used to explain regional and national integration (Marriott 1955; Beals 1974: 12) – one might say 'identity'– it is appropriate we try to situate Narikuravas within that larger picture.

Intellectual Views

The small amount of work on the Narikuravas has been done by foreigners like Moffat (1979) and Werth (1993, 1996, 2000), while Indian social scientists have somewhat ignored the subject and exceptions to the rule tend not to be academicians. Instead, like one of us, they are academy trained activists in the Church of South India who also happen to be Dalits. Sometimes evan-gelical, but not always, their views – like Meshack's – are informed by soci-ology, and focus primarily on human rights and development.

A manual metaphor for Dalits' own position was given to us by Devasa-hayam, a Professor of Dalit theology at Gurukul Seminary, Chennai, before he became Bishop of Madras in 1999. Raising a hand to display four fingers 'naturally' aligned and a thumb protruding to one side was, he suggested, an apt model of the caste-system. Many Dalits, including Devasahayam, owe their upward social mobility to a non-Hindu education, and their consequent rejection of Hindu 'hegemony' and 'Sanskritization' of Dravidian culture leads them to regard themselves as no less 'indigenous' than *Adivasis*. Thus M.K. Azariah, the Bishop of Madras before Devasahayam, speaks of 'the two *Adi* societies' declaring: 'to say Dalits in India are those who are the poor and the oppressed is a gross understatement about people. They are a "No" [sic] people made strangers on their own native soil, deprived of their properties as well as their personal human dignity and basic human rights' (Azariah undated: 89–90).

From this perspective if *Adidravadas* are 'strangers' like *Adivasis* then Narikuravas are twice over, which matches Moffatt's description of them as 'without any reservation Untouchables to the Untouchables'[5] (1979: 144). Therefore in turning for a moment from Dalits as 'indigenous',[6] to the idea of 'strangers', and particularly Simmel's (1964) typical conception of the 'stranger' as a trader who is physically 'close' but socially 'distant' to others, who is expected to move on, but in fact remains, we have a construct as applicable to Narikuravas as once it was to Jews and Gypsies in Europe.

Identity: Conclusion

Having examined Narikuravas' identity from three perspectives, the question that arises is why other Tamils view them so negatively, given India's long history of peripateticism and distance travelling.

Anthropologists like Marriott (1955) have attempted to show how different modes of communication led to the development of regional cultures and the idea of Indian civilization as a whole, and in so doing have stressed, *inter alia,* the role of trade. Others, including Morris (1982), have discussed the role of middlemen in the domestic and overseas market for tribal forest produce. It has been said that most sociological studies examine agricultural castes and villages to the neglect of those non-agriculturalists who have an important role in creating supra-local forms of social organization. Such criticism led Mines (1984) to study the Kaikkoolars of Tamilnadu, a caste of artisan-merchants dealing in textiles, and Rudner (1994) to examine the activities of Nakarratars, a merchant caste who in the seventeenth century transported salt long distance. Other communities of long distance carriers were the Banjaras, whom as we mentioned earlier, traded salt (Moreau 1995), and the Lambardis who dealt in cattle and mixed merchandise. Nevertheless, when it comes to Gypsies, the detailed studies are scarce.

It is sometimes claimed that Indian villages were economically self-sufficient until the mid-nineteenth century. Bailey cautiously admits this may have been the case in Orissa (1960: 125), and Epstein (1967) implies it for some South Indian villages. Nevertheless, the century 1850–1950 witnessed both an expansion in public education, colonial governance, markets and industry opening up new economic pathways for caste-Hindus, and the advent of affirmative action via the Constitution for *Scheduled Castes* and *Tribes.* For commercial nomads not classified as either, however, these opportunities have proved elusive.

In agrarian societies where individual and collective reputations are so often based on people's perceived connections with the land, Gypsies can constitute an ambiguous category at best. Similarly, in industrial societies, urban dwellers can both admire and resent Gypsies' physical mobility and their unwillingness to become permanently tied into wage labour. In Ireland, for example, nomadic tinkers were treated as outcasts for decades, in fact even itinerant labourers and craftsmen were held in low repute (Evans 1957) albeit not as low as tinkers. Internal migrants too have sometimes been marked as strangers lacking roots or culture. In China, the Hakka who for centuries emigrated southwards from north-east China, were simply not *allowed* to acquire a sense of place (Blake 1981). According to Leong (1997), in eighteenth-century Dongguan they were thought of as vagrants, dogs, and sub-human, and to their Punti neighbours especially they long remained 'kind of semi-barbarians, living in poverty and filth' (p. 74) and 'beyond the Chinese racial and cultural pale' (p. 80).

Until quite recently 'boat-people' of the Hong Kong territories and South China coast were deemed 'barbarians' and 'uncivilized' because officials found them difficult to 'classify' or pigeon-hole. It was only when Barbara

Ward, the geographer, described their symbiosis with land-people that such ideas were challenged, and their status improved (Blake 1981; Acton 1985). Finally, the *real* lives of nomads and other travellers are often misunderstood or purposely ignored by officials in particular, and misrepresented by more sedentary peoples generally.

Social Organization, Values and Beliefs

Narikuravas in Chennai and North Arcot are organized into two ranked patrilineal, non-local, exogamous clans and several ranked non-local *jathis* or sub-clans. Each clan has a female deity to whom its leader periodically sacrifices a male buffalo or goat; each *jathi* has its own form of the deity; and each household owns its particular effigy and shrine (Werth unpublished). Buffalo sacrificers seek marriage partners among goat sacrificers and vice versa. This way Narikuravas not only maintain clear symbolic boundaries with non-Narikuravas – since their rituals play no integrative role with other Hindus (ibid.) – but they also maintain internal ones.

A normative residential marriage rule stipulates initial neo-localism but peripatetics have to be economically adaptable which in turn is helped by the possession of a wide network of kin and affines. Settlements or encampments are therefore as much 'sites of travel' – where the fission and fusion of family clusters occurs – as they are 'sites of dwelling' (Clifford 1992).

Between 1986 and 1996 Narikuravas settlements in Tamilnadu increased in number from 182 to 249, though whether this was the result of displacement or simply further evidence of fissioning is uncertain. What Narikuravas themselves say is that much of their displacement in the twentieth century resulted from a tightening of Forest Department regulations – often in the name of conservation – on the trapping and sale of animal products to middlemen supplying national and international markets. The fact that they were illiterate and rarely understood the changes in regulations also made them vulnerable to arrest and arbitrary punishment – another reason for moving to town.

India's Forest Acts have long been subject to official abuse. Fernandes et al. (1988) tell how functionaries in Orissa favoured the interests of commercial middle men more than those of the tribal groups which they were charged to protect. And in Bastar, Madhya Pradesh, where officials' exploitation of tribespeople dates back to 1894, the exploitation continued even after an inquiry into gross malpractice between 1955 and 1975 resulted in Departmental re-organization (Anderson and Huber 1988).

An observation by an Indian anthropologist, though not intended to refer to the peripatetics here, nevertheless captures the essential failure of communications between locals and forest officials which has been a major feature of that nexus:

> Every position, every way of seeing can be as true or as untrue as any other, in so far as they are derived from different ways of life. But when they come

together in the frame of day to day living they generate questions and they initiate discourses and dialogues over differences and inequality. The social spectrum across which these positions are arranged has at one end people marginalised in different sectors of work; at the other end are people enclosed in the routine of the official dominant modes of work. From the standpoint of the latter, to be with forest dwellers, to go and live with them, to know and to learn from their experiences and social life, is a backward movement in time (Savyasaachi 1998: 99).

Werth (2000) documented five *jathis* – Gujerato, Pavar, Dabhi, Mevado, and Selyo. We found in one rural settlement of North Arcot alone, individuals who could name up to twenty-two *jathis*, eleven for the Jambalavala clan which sacrifices buffaloes to Kali, and eleven in the Nandevala which offers goats to Meenatchci. In another settlement of the district people were able to name twenty-four *jathis*, eight Jambalavala and sixteen Nandevala,[7] with Werth's Pavar and Dabhi both included among the Jambalavala. In a third settlement, in Chennai, we found three actual *jathi* –Pavar, Gujerato and Navthal. Nowhere did we hear of Mevado and Selyo.

In the two rural settlements five *jathis* were named in both places and there was a general consensus as to which clan they belonged. Sekru, Goindo, and Ramo were Jambalavala or 'buffalo', and Heero and Kheto were Nandevala or 'goat'. Incuding the Chennai settlement, we noted forty-three *jathi* names and found three or four present in each settlement.

In the legal permanent settlements set up with READA's help, *jathi* and clan leaders along with the heads of various settlement associations (*sangams*) form a general committee (*panchayat*) under an elected head – usually a man but occasionally a woman – who can resolve internal conflicts and communicate with officials.

The nuclear family is the main unit of social organization in Narikuravas society which though itself patriarchal, openly acknowledges a woman's right to a satisfying marriage. Infant betrothal is the norm and children who are betrothed are referred to as 'husband' and 'wife'. Polygyny too is normal but in practice limited. Marriage partners are often sought in *jathis* where a family already has affines. A female 'given' by one *jathi* is thus balanced by one 'taken' in return. In this way inter-generational alliances may be forged between families. Couples seeking to avoid an arranged marriage, simply elope.

When a death occurs among forest dwellers the body is buried and the relatives move on. In the new, more permanent settlements, relatives burn the deceased's clothing and shelter before burial and then vacate the area for twelve days in the case of a dead man, and eleven if a woman. This is reminiscent of British and Irish Traveller practice where once a dead person's tent or wagon was burnt, but now the caravan is sold to strangers.

It is not at all surprising that Narikuravas should pride themselves on their knowledge of fauna and flora. Some keep pigs and fowls but their proper expertise lies with the small mammals like jungle cat, Indian 'fox' (*vulpes Bengalensis*), jackal (*canis aureus*) and squirrel which they trap, net or shoot with ancient and illegal muzzle-loading rifles. Birds, including waterfowl, are lured with whistles or mimicry, and netted, shot, or killed with catapults. Some families keep a pet cat, dog, or squirrel, or a trained monkey which continues to live as part of the family long after its performing days are over.

The Narikuravas are also knowledgeable about natural medicine which possibly may explain the alleged good health of many of them and why their remedies are sought by outsiders. In fact, the medicines of one man in a small country town were so well known that the local western trained medical practitioners acted collectively to stop him. What is disturbing about this is not just one poor family's loss of income, or the potential loss of knowledge important to Narikuravas' health generally, but the fact that it has to some extent been accelerated by displacement and forced settlement done in the name of conservation, deforestation and 'development'.

Problems

In this final section we will sketch the major problems faced by Narikuravas and tentatively suggest how they might be addressed.

1. *Status.* According to READA the economic opportunities necessary for Vagri development will only occur when their status as a *Most Backward Class* is changed to *Scheduled Tribe.* This may seem problematic given that 'tribals' or *Adivasis* are 'indigenous' by virtue of ancient claims to custody of *particular* tracts of country, but in so far as these labels are official constructs representative of particular positions, and in as much as Narikuravas are essentially a tribe or caste of nomad whose dual economy has long entailed change and adaptation, it is a human rights issue that their status as a 'tribe' be acknowledged.

2. *Land.* Access to clean water, shelter, primary health care, and basic education is virtually non-existent, and officials frequently blame the Narikuravas, their nomadism and big families for it. They believe Narikuravas Gypsies should settle down and limit their offspring and, of course, some already do so. We suggest however that more thought be given by the authorities to making land available in rural and urban areas for habitats adapted to a peripatetic niche of settlement combined with periodic travelling. Where READA has already succeeded in this direction the benefits are evident, but more steps need to be made towards taking commercial nomadism seriously, and if urban Narikuravas are to acquire new survival skills.

3. *Legal Aid.* This is essential if Narikuravas are to improve their conditions of life; moreover without basic literacy they will fall prey to inept lawyers. Thus far READA has managed to acquire some legal assistance through social networking, and in the medium run its role as mediator will remain important, but in the longer term, Narikuravas must have the educational and social skills to deal effectively with and pay for lawyers, and beyond that have people of their own trained in social work and law.

4. *Shelter.* For want of shelter many Narikuravas women live in fear of rape and sexual molestation by strangers who share that stereotype of 'wild' and promiscuous beauties that some non-Gypsy men have of Gypsy women in Europe. For this reason some among them deliberately affect a dirty, dishevelled, appearance as a deterrent, although it does not stop long-distance bus and lorry drivers carrying soap and toiletries as means of sexual inducement.

A lack of proper shelter also means families have to shelter from the rain in bus stations and other public covered areas. By contrast, where land has been made available and shelter erected in safe habitats (as in central Chennai) lives have improved immeasurably, even if small regard was given to their lifestyle and culture. Proper consultation is essential.

5. *Health.* Anecdotal evidence suggests Narikuravas health is better than would be expected, but hard data on morbidity and mortality are lacking, and there are grounds for concern. Many Vagri women have been sterilized without explanation, more-or-less compulsorily. There is also an urgent need for sex health education, and especially on the matter of HIV/AIDS, considering these people's mobility and the women's vulnerability to the sexual advances of strangers and long distance drivers. Moreover, concern is even greater because most major funding agencies have now shifted their HIV/AIDS programmes out of South India to the North.

Class and caste-centrist medical and social work professionals need to learn from and value Vagri knowledge of the mind, body, spirit nexus and acknowledge their traditional medicines. Much also has to be done to assess the needs of the handicapped – including children.

6. *Literacy.* In a world constructed and represented by mass media, based in large part on the written word, and where dealing with money-lenders, police, and lawyers demands confident daily self-assertiveness, basic literacy is essential. READA has achieved an important beginning but the majority of individuals and communities remain disadvantaged.

7. *Education and Training.* There is a pre-school in one Chennai settlement, and a young woman in another one on the outskirts has completed secondary education thanks to her mother's economic enterprise. Many adults, including this woman and her mother, have learned negotiating skills through role-play workshops organized by READA. In settlements far from the forest, these and basic literacy skills have resulted in some people relinquishing begging and scavenging in favour of more profitable work. Even so, plain hunger is a problem which some can only deal with by lying down to sleep.

Prasad (1980) indicates that tribal people are not 'against' education when it is sensitive to culture. Hostility only sets in when teachers' attitudes and the curriculum alienate, when 'primitive' becomes a put-down word for tribal or minority children taught unsympathetically alongside non-tribal children, and when children are nomadic with traditional vocations in acrobatics, puppetry, smithing, and begging. For these last he proposed the use of mobile schools, an arrangement found among many other nomadic groups, and it is the one we believe might have a role in Tamilnadu.

Notes

1 Both authors wish to thank the Centre for Refugee Studies, Oxford University, for facilitating Daniel Meshack's visit to the Oxford Conference.

2 'Commercial nomads', 'peripatetics' and Gypsies are synonyms for nomads living primarily from the sale of goods and services (see Rao 1987); Misra (1971) prefers 'symbiotic' nomads. As an 'ideal type' the concept distinguishes this kind of nomad from others, but we must stress that today not all Gypsies are nomadic; that nomadism itself is a matter of degree; and that our occasional use of the word 'itinerant' should not be taken to mean 'nomad' or 'nomadic'. By 'nomad' we mean one whose spatial mobility is embedded in the traditional life of a conjugal or domestic *group* that is itself mobile and connected over several generations with similar groups. An 'itinerant' either travels alone or in a group (domestic or otherwise) which lacks such connections.

3 The authors are very grateful to Lukas Werth (who worked mainly in Namakkal district, near Salem) for providing them with copies of his work, some still unpublished. Our own data were collected in Chennai and North Arcot but, like Werth's, draws on wider experience in Tamilnadu.

4 Thurston reckoned one reason for the low status of Dommaras, Dombs, Domban, or Dombas –who 'know absolutely nothing of hunting' (1965: 175) – was the fact that they ate cats, dogs, and crows which died naturally. He also noted they made and sold 'spurious jackal horns' (p. 185), a Narikuravas speciality. Moffat (1979) independently observed Narikuravas ate crows.

5 For a discussion of 'Untouchable', see Charsley (1996).

6 In 1987 *Adivasis* claimed 'indigenous' status before a UN Working Group on Indigenous Populations.

7 As this community (and the others) claimed eleven *jathis* for both clans, yet named twenty-four, we may have to revisit this datum. What we can say fairly confidently is that throughout Tamilnadu twenty-two is the total figure usually given.

References

Acton, T.A. 1985. 'The Social Construction of the Ethnic Identity of Commercial Nomadic Groups'. In J. Grumet (ed.). *Papers from the Fourth and Fifth Annual Meetings, Gypsy Lore Society, North American Chapter.* New York: Gypsy Lore Society, Publication no. 2.

Anderson, P.S. and Huber, W. 1988. *The Hour of the Fox.* Seattle: University of Washington.

Azariah, M. undated (originally 1989). 'Doing Theology in India Today', in A.P. Nirmal (ed.). *A Reader in Dalit Theology*, pp. 85–92. Madras: Gurukul Lutheran Theological College & Research Institute.

Bailey, F. G. 1960 (originally 1955). 'An Oriya Hill Village: I'. In M.R. Srinivas (ed.). *India's Villages*, 2nd edn. Bombay: Asia Publishing House.

Baker, C. J. and Washbrook, D.A. 1975. *South India: Political Institutions and Political Change 1880–1940.* Delhi: The Macmillan Company of India.

Beals, A. 1974. *Village Life in South India.* Chicago: Aldine Publishing Company.

Berland, J.C. 1982. *No Five Fingers are Alike.* Cambridge, Mass. and London: Harvard University Press.

Blake, F.C. 1981 *Ethnic Groups and Social Change in a Chinese Market Town*. Honolulu: University Press of Hawaii.

Charsley, S. 1996. '"Untouchable": What's in a Name?'. *The Journal of the Royal Anthropological Institute*, 2(1): 1–23.

Clifford, J. 1992. 'Traveling Cultures'. In L. Grossberg, C. Nelson and P.A. Treichler (eds). *Cultural Studies*. New York: Routledge.

Epstein, S. 1967. 'Productive Efficiency and Customary Systems of Reward in Rural South India'. In R. Firth (ed.). *Themes in Economic Anthropology*. ASA Monographs 6, London: Tavistock Publications.

Evans, E. E. 1957. *Irish Folk Ways*. London: Routledge & Kegan Paul.

Fernandes, W., Menon, G. and Viegas, P. 1988. *Forests, Environment and Tribal Economy*. New Delhi: Indian Social Institute.

Fraser, A. 1995. *The Gypsies*. Oxford, UK and Cambridge, USA: Blackwell.

Leong, Sow-Theng. 1997. *Migration and Ethnicity in Chinese History: Hakkas, Pengmin and their Neighbours*. Stanford: Stanford University Press.

Lewis, N. 1991. *A Goddess in the Stones*. London: Picador.

Marriott, McKim. 1955. 'Little Communities in an Indigenous Civilization'. In McKim Marriott (ed.). *Village India*. Chicago: University of Chicago Press.

Mines, M. 1984. *The Warrior Merchants*. Cambridge: Cambridge University Press.

Misra, P.K., Rajalakshmi, C.R. and Verghese, I. 1971. *Nomads in the* [sic] *Mysore City*. New Delhi: Government of India.

Moffat, M. 1979. *An Untouchable Community in South India*. Princeton: Princeton University Press.

Moreau, R. 1995. *The Rom*. Toronto: Key Porter Books.

Morris, B. 1982. *Forest Traders*. London and New Jersey: Athlone Press and Humanities Press.

Okely, J. 1983. *The Traveller-Gypsies*. Cambridge: Cambridge University Press.

Prabhaker, M.E., undated (c.1990) 'Developing a Common Ideology for Dalits of Christian and Other Faiths'. In Arvind P. Nirmal (ed.). *Towards a Common Dalit Theology*, pp. 53–79. Madras: Gurukul Lutheran Theological College and Research Institute.

Prasad, S. 1980. 'Education of Scheduled Tribes and Nomads'. In. L.P. Vidyarthi and B.N. Sahay (eds). *Applied Anthropology and Development in India*. New Delhi: National Publishing House.

Rao, A. (ed.). 1987. *The Other Nomads*. Köln: Böhlau Verlag,

Rothermund, D. 1993. *An Economic History of India from Pre-Colonial Times to 1991*. London: Routledge.

Rudner, D.W. 1994. *Caste and Capitalism in Colonial India*. Cambridge: Cambridge University Press.

Saraswathi, S. 1974. *Minorities in Madras State: Group Interests in Modern Politics*. Delhi: Impex India.

Savyasaachi, 1998. 'Unlearning Fieldwork: The flight of an Arctic tern'. In M. Thapan (ed.). *Anthropological Journeys: Reflections on Fieldwork*. New Delhi: Orient Longman.

Shah, G. 1975. *Politics of Scheduled Castes and Tribes*. Bombay: Vora and Co.

Shashi, S. S. P. 1994. *Encyclopaedia of Indian Tribes*. vol. 2. New Delhi: Ammol Publications.

Simmel, G. 1964. 'The Stranger'. In K. Wolff (ed.). *The Sociology of Georg Simmel*. New York: The Free Press.

Singer, M. 1955. 'The Cultural Pattern of Indian Civilization'. *Far Eastern Quarterly*, xv(4): 23–36.

Sutherland, A. 1975. *Gypsies: the Hidden Americans*. Prospect Heights: Waveland Press.

Thurston, E. 1965 (originally 1909). *Castes and Tribes of Southern India*. vols, 1–7, Madras: Johnson Reprint Corporation.

Werth, L. 1993. 'The Vagri of South India and their Ancestors'. *Journal of the Indian Anthropological Society*, 28: 275–84.

—— 1996. 'Forgotten People: the Vagri and Other Peripatetic Communities of South Asia', Conference on Asian Minorities, University of Minnesota, 1996.

—— 2000. 'Kinship, Creation and Procreation among the Vagri of South India'. In A. Rao and M. Böck. *Theories of Creation and Procreation*. New York/Oxford: Berghahn.

—— Unpublished. 'The "Untouchables of the Untouchables"? Myth, Ritual, and Society of the Vagri of Tamil Nadu'.

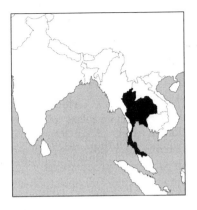

16

Karen and the Land in Between

PUBLIC AND PRIVATE ENCLOSURE OF FORESTS IN THAILAND

Jin Sato

Introduction[1]

Thailand has historically been a sparsely populated country with an abundance of open land. With an increasing scarcity of land, however, a legal framework limiting the rights to land use has gradually developed. To halt further 'encroachment' on state land, the government has initiated projects for land allotment, purportedly for the landless poor. It has also invested in tree planting and expanding protected areas for the environment, particularly since the late 1960s. Despite these initiatives, illegal logging has continued, forest cover has declined, and socio-economic conditions in rural areas have not improved as intended. Thailand is now considered to be one of the worst forest managers among tropical countries (Poffenberger 1990).

Practitioners and scholars in Thailand appear to be divided into two groups over the link between forests and local people. The first regards local people as a threat to forests and gives supreme priority to ecological preservation in the interest of the state as a whole. From this perspective, forests should be 'protected' from local people. The second emphasizes the rights of local people to stay where they are and attempts to find a balance between meeting local needs and achieving conservation objectives. The two points of view even divide various government agencies and NGOs. This contrast became more salient in the process of drafting a new community forestry bill, which has yet to be enacted. This proposed bill contains pioneering legislation that permits eligible local people, as communities, to use and manage forests. The central point of dispute in this drafting has been whether to allow local people to use and manage those forests inside protected areas, and whether the capacity of local people to manage forests is credible from the viewpoint of the government.

Scholars of common property have emphasized the question of 'who has access to what'. I believe one should go further and ask why certain people *end up* using certain types of resources and not others. My observations in Thailand suggest that forests that are collectively controlled by local communities are rare, and those that generate direct and significant economic benefits to the local people are even rarer.[2] As we shall see later, the Karen hill people, who have resided in the forests of western Thailand for over 200 years, have been evicted from their homeland which has now been designated a world heritage site. They have been squeezed into a small plot of land in the buffer zone surrounding their home forest. Unable to allow for an adequate fallow period, Karens were forced to abandon their rotational shifting cultivation for the sake of 'global' environmental concerns. Having deprived them of their own system of resource use, it is futile to debate whether these villagers now have the capacity to manage a forest. Seeking the sustainable use of forest land with exclusive focus on the village-level mechanics of collective action, therefore, will not suffice.

In this chapter, I will first examine the history of forest policies in Thailand. Second, I will discuss how the Karen, one of the hill peoples, have been struggling to survive between the forces of capitalistic development and forest conservation. Two Karen villages were selected to conduct a detailed study of forest use and dependency, the type of critical data often missing in defining the problem. I argue that the state's effort to reduce the Karen's forest dependency or even evict them from the forests does not lead to the stated objective of conservation.

Earlier Policies: Forest Clearance and Agricultural Expansion

Until the first half of the twentieth century, Thailand was a 'kingdom of forests'. For a long time, forests were perceived as dark dangerous jungle where brutal animals resided; they were seldom considered economically valuable or something to be owned. Even in the late nineteenth century, European loggers had to rely on the hill tribes for most of their labour since the lowland Thais feared going into forests (Falkus 1990). Similarly, from the government's point of view, forest land was considered unproductive and something to be developed. In the legal code of the Ayutaya period (fourteenth to eighteenth centuries) encroachment into forested areas by villagers was described as '*bukburk*' (pioneering) rather than '*bukruk*' (encroachment) which is the term that is currently used to denote the same practices (Chamarik et al. 1993). This indicates that forest clearance was officially encouraged rather than discouraged for a long period of time.

Thailand has increased its agricultural production through territorial expansion and has relied very little on intensification (De Konnick and Dery 1998: 15). The amount of land area allocated for agriculture has doubled between 1910 and 1940 and tripled again between 1940 and 1970. Although some planned resettlement programmes have been initiated by the government, most of the territorial expansion occurred as a result of spontaneous settlement of villagers themselves (Uhlig 1984).[3] Thailand has supported its successful economic growth up until the 1980s with the massive loss of its natural capital. Diminution of forest areas has been clearly detected by satellite images particularly since the 1970s. Official statistics reveal that the 38 per cent forest cover in 1970 has been reduced to 26 per cent in 1995. Even the often underestimated official figures show that Thailand has lost 50 per cent of its forest in the past thirty years (Figure 16.1).

The rapid loss of forest resources was damaging to the government, because timber exports were one of the main sources of foreign exchange. In 1975, the government banned the export of raw timber, and, in 1977, Thailand became a timber importing country in net. Thailand was forced to abandon forestry based on natural stands. In 1988, when large-scale flooding allegedly caused by excessive logging killed hundreds of people in the southern region, the media and NGOs created a mass movement against commercial logging, which in 1989, resulted in an official policy to halt commercial logging in state-owned public land.

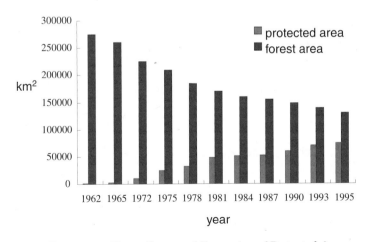

Figure 16.1 Decrease in Forest Cover and Expansion of Protected Areas

Note: Protected area represents the added area of national parks and wildlife sanctuaries (Source: RFD [1996])

State Protection of Forests from Villagers

During the rapid expansion of private agricultural lands, the Thai government paid little attention to unoccupied places (except for those that contained valuable goods such as teak). These places, primarily forested, were 'open access' lands. Public lands were in fact private lands for the government and there was no policy to allocate those lands to the landless population. Explicit designation of 'public land,' not surprisingly, meant exclusive 'state land' reflecting the interests of the government (Sayamon 1995). For the farmers, public land chiefly meant future reserves for farm land, and, unless it was a watershed, the economic function of public land was not considered critical to their production. People have settled in to those open access forest lands only in situations of disaster, land shortage due to population growth, or to escape a plague.[4]

Conflicts between farmers and officials have stimulated governments to strengthen their protection measures. Some scholars have described this process as internal 'territorialization' (Peluso and Vandergeest 1995). Instead of establishing external boundaries at the national borders, areas of exclusive control by the state are created within the borders. Areas with valuable natural resources are usually the first places to be 'protected'. Many of the forests already had people residing prior to their designation as state forests, and the delay of demarcation invited more migrants from other areas that intensified land related conflicts.[5]

Diminishing forest resources and the expansion of protected areas altered the nature of forest administration in the country. Forest reserves in the traditional sense were patches of forests waiting to be harvested. The mandate of the RFD was therefore to manage these reserves in an economically profitable way. However, the depletion of forest lands and the inevitable banning of commercial logging in 1989 created an identity crisis for the RFD, and it was forced to transform its mandate from production to protection. The protection of forests was justified not only on the basis of environmental objectives, but also as a way to attract foreign assistance. An increase in the availability of economic assistance and the expected income from tourism are the basic motivations behind the expansion of protected areas (Ghimire 1994).

Battles Over the 'Land in Between': Land Reform and Forest Conservation

Competition for land has intensified not only between farmers and the state, but even among the different government departments, particularly in the land where property demarcations are still vague. These areas are typically called the degraded forests (*paa suam soom*). Degraded forests were formerly covered with forest but have now been transformed into agricultural land. Most of this land is legally owned by the state and has become the target for future plantation sites by the RFD, or is handed over to the Agricultural Land Reform Office (ALRO) to be distributed to the landless poor. The status of this land, therefore, is ambiguous and controversial in many ways.[6] The precise number of farmers actually living on these lands is uncertain, and estimates vary between 1.5 million to 8 million (about 13 per cent of the total population) (Lynch and Alcorn 1991; Christensen and Rabibhadana 1994).

To counter the problem of people illegally residing on state-owned land, the government had implemented certain 'participatory' projects that would give various degrees of land rights to farmers. Among the projects that took place in the forest reserves, the most notable for the purpose of this chapter, is the land reform programme implemented by the ALRO. This is a scheme whereby the government obtains unused land from landlords or state-owned lands and distributes it to landless farmers. In addition, the ALRO will support the development of infrastructure (e.g., pond digging, road construction) along with the provision of land certificates. The land to be obtained was originally sought from large-scale landlords. However, it soon became apparent that obtaining land from powerful landlords would be politically difficult, and even if this were possible, this group could not supply as much land as was needed. From the mid-1980s, the government has shifted its attention to forest reserves, officially under the control of RFD, as a potential source of land to be given to the landless. This move, naturally, has motivated some capital investors to encroach further into forest land, and as a result, an attempted 'reform' has in some cases intensified the concentration of land with the elites (Charasdamrong 1997).

The growing political power of NGOs working against forced relocation, as well as the basic lack of land onto which to move these people, have created certain tensions inside the state-owned forest land. In response to protests by the 'Assembly of the Poor,' the Chavarit government issued a cabinet resolution on 22 April, 1997. The resolution acknowledged the land rights of farmers to remain inside protected areas provided they had been there before the protected area designation, and were willing to cooperate with the state in forest conservation. Yet the next administration, led by Chuan, attempted to attack the opposition party, which was supported by poor farmers, by revoking the previous resolution. In June 1998, they passed a new resolution, which stipulated that the RFD should have the right to evict all those who are inside the state-owned forests (Ekachai 1998). Regardless of these policy shifts, the basic underlying fact is that there is no land available for those who have been evicted. The fact that the RFD has been forced to give away 30 per cent of its territory to the ALRO clearly manifests the consequence of forest mismanagement by the state. The government is now forced to recognize the presence of 'illegal encroachers' on state land by issuing them land certificates to curtail further encroachment (see Table 16.1).

Figure 16.2 illustrates simply the changing connections between exclusive state lands represented by area 'A,' and private farm lands represented by area 'C'. 'B' represents the ambiguous area, or the land in between the two that is owned by the state yet utilized by local villagers. The 'land in between'

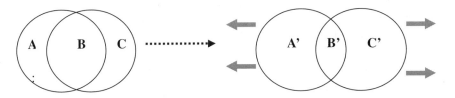

Figure 16.2 Shifting Property Relations and the Land in Between

Note: When land was abundant and population scarce, the public lands used exclusively by the state were limited to areas for mining, military bases, and teak harvesting (A). Private lands with exclusive property rights were also limited because the majority of the land was, in principle, open-access, and guaranteed by usufruct rights based on customary laws (C). A large portion of land, therefore, belonged to the state yet was *de facto* privately accessible (B). As the economy became more market-oriented, private assets gained importance and the legal structures that protected private rights expanded. This was reflected in the expansion of private farm land (C'). The expansion of area C' was strongly supported by development policies and foreign aid. The large-scale introduction of commercial crops in C' lands including, sugar cane, coffee, tapioca, maize, and eucalyptus played a major role in soil erosion. The State, on the other hand, also strengthened its exclusive control over 'public lands,' the area where forest land was most prevalent (A').

areas have served not only as economic alternatives but also as a political haven for disaffected groups fleeing political repression, particularly when communist movements were active in the mid-1970s (Hirsch 1990).[7]

My focus is on what is happening in the B' areas. A typical B' area can be found in the fringes of national parks and other types of protected areas now often referred to as 'buffer zones' (Figure 16.3). In recent literature on protected area management, fringe areas of the core conservation zones are given particular importance as places to meet the basic needs of the local population as well as providing an additional ecological layer (Sayer 1991).

Buffer zones attract much attention not only because of their relation to the core protection area, but also because they happen to fall under two contradicting national policies in Thailand: land reform and forest conservation. As mentioned earlier, the competition between the Agricultural Land Reform Office (ALRO) and the RFD in the buffer zone has been the most acute recently. From the RFD's point of view, forest reserves in the buffer zones are the first targets that should be included in the expanding protected area. Unfortunately, there are no clear scientific criteria to determine lands that should be given to the farmers through the ALRO, and lands that should remain under the RFD's control. For the ALRO, deforestation is not the primary issue; in a sense, it is a necessary part of its solution to resolve the landless problem.

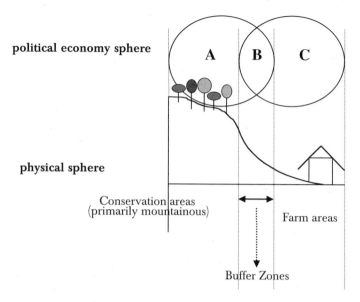

Figure 16.3 Land in Between

People in Between

The Karen under State Enclosure

There are nine official categories of hill people in Thailand (Bhruksasri 1987), among whom the Karen population is estimated to be about 600,000 (Hayami 1996). In the eighteenth and nineteenth centuries, the Pwo Karen in western Thailand were considered allies by the Thai king. They served the king as scouts in anticipation of the Burmese invasions from the west in addition to providing valuable forest products such as ivory, cinnamon, and cotton (Gravers 1994). The British also valued the Karen's knowledge of timber and mobilizing elephants for their harvesting in the mountains and jungles of the frontiers. King Chulalongkorn's administrative reform in the early twentieth century, however, marginalized the Karen's previous position in the Thai polity (Jorgensen 1996). During the 1960s, a communist insurgency extended to the Thai border, and the economically marginalized hill people rose to centre stage once again. In the 1970s, many of the Karen joined communist groups which made them politically dangerous 'tribes' in Thai society.

The Huai Kha Khaeng (HKK) wildlife sanctuary extends over the Uthai Thani and Tak provinces of western Thailand. It is a central component of the kingdom's integrated protected area, which covers 14,000 sq. km (Yellow Stone National Park is about 5,500 sq. km), including five wildlife sanctuaries, eight national parks, and three would-be national parks. Cubitt and Stewart-Cox note that 'for a country that is widely known and often criticized for its deforestation, this is astonishing. It would take a grueling hike for about a month to get from north to south of [HKK]' (Cubitt and Stewart-Cox 1995: 124).[8] The HKK forest was designated a wildlife sanctuary under Thai law in 1972 and later, in 1992, became the first natural UNESCO world heritage site in Thailand.

For a long time, the Karen along with other hill people have been identified as 'forest eaters' (Gravers 1994). As early as 1923, a commentator of Karen economy noted that:

> In some places the same clearing is cultivated three years in succession, but in others a new clearing is made every year. In the past, large tracts of forest have been destroyed in this way, and even now the Forest Department has taken the matter up, a good deal of destruction takes place in valuable forests (Andersen 1923: 55).

More recently, the Thailand Development Research Institute (TDRI) argued that 'for the Royal Thai Government, the hill tribes pose a series of profound political, social and ecological problems. Much highland deforestation ... can be laid directly at their door' (TDRI 1987: 80). The people once blamed for their backwardness and opposition to modernization and development are now being blamed from an environmental standpoint. The proposed strategy drawn from this assumption is, naturally, to move people away from the precious forests.

> As described by Mr. Chatchawan Phitsamkham, the superintendent of the Huai Kha Khaeng Wildlife Sanctuary, people have been attracted to the forest and its resources like ants to a lump of sugar. The ICAD (Integrated Conservation and Development) activities we propose are intended to turn people *away* from the protected areas, to attract them to 'new and sweeter lumps of sugar' outside the forests (World Bank 1993: 11, emphasis in original).

Having been evicted from the HKK and now squeezed into the 'buffer zone' of an internationally recognized protected area, the Karen are not only denied their traditional farming techniques, but also modern chemical agriculture for fear they will damage the forest. This is a 'late developer's trap' – you cannot go back but neither can you move forward in the same way that 'modern' farmers have. Michael Dove summarizes the dilemma by saying 'the challenge is to achieve the benefits achieved by past paths while *not* following them' (Dove 1994: 1069, emphasis in original).

Rotational shifting cultivation of upland rice was the central economic activity of the Karen. Their production portfolio consisted of yam, taro, cotton, tobacco, chilies, and various vegetables in the same plot of approximately 5 *rai* (1 *rai* = 0.16 hectare) per household. Banana trees and sugarcane often surrounded swidden agriculture plots (Gravers 1994). Metal products (such as knives), guns, salt, and certain cloth had to be purchased from the local market in a distant town in exchange for their agricultural products. Although little is known as to what extent the Karen impacted the forests in the past, at the minimum, their long-term continuous residence in the same area demonstrates a successful implementation of the swidden system. The rotation cycle of a fallow period was between ten to fourteen years. A villager once told me with frustration, 'if our farming system is such a bad thing, the Huai Kha Khaeng forest would never have become a world heritage site; it would have disappeared by now.'

Despite the blame placed on hill people and local villagers living close to rich forests, there are almost no data on the nature of their forest use. This is surprising because states often gather data that will enable them to effectively execute projects that will address those problems. The persistent absence of empirical data on villagers' actual forest use tempts us to think that the state does not perceive villagers as real threats, but only uses them as pawns to deflect public attention from corruption, large-scale infrastructure development, and illegal logging by public officials themselves.[9]

What follows is an analysis of data that has been completely missing from the debate over whether to allow the Karen to live in the buffer zone, and to assess the legitimacy of the state's project in dealing with local peoples. I have focused almost solely on the material connection between the Karen and forests because this, over and above cultural attachments and non-material linkages people have to their forests, seems to be the central concern for the policy makers.

Changes in the Karen Economy

I studied the Karen communities adjacent to the HKK forest complex from March 1996 to March 1997. There are more than twenty villages within the 5 km distance from the HKK boundary. Four of these villages were the Karen communities with the longest histories in the area, who are also the most dependent on forest resources. Among the four villages, I selected two as intensive study sites for the following reasons: (1) the populations of both villages were about the same and reasonably large to reflect various types of livelihood strategies within each village; (2) although their distance from forested areas was the same, village A has inferior access to roads and markets, which enabled me to measure the influence of the market economy in relation to their forest use; (3) both villages have similar histories of migration and initial property holdings (they were both 'kicked out' of the HKK in the '70s). Table 16.2 shows basic data on the villages.

The 1970s were a critical turning point for the Karen economy. A Thai logging company built the first road that penetrated into the villages. The road significantly pushed the subsistence economy into the market economy. Most notably, middlemen could now reach the formerly inaccessible villagers, which encouraged the Karen to plant cash crops. Mono-cropping of maize increased the weeds, and ploughing was needed to prepare the land. Repetitive use of the same land made the soil harder and more difficult for animals to plough, so that farmers had to rent tractors from Thai moneylenders in the lowlands. The introduction of tractors reduced the opportunity to harvest different crops throughout the season. Whereas traditionally there had been a time-lag between the planting of rice and the maturing of other crops such

Table 16.2 Basic Data on the Two Research Villages (1995)
(1 baht = 3¢; 1 rai = 0.16 ha)

	Khongsao Village (A)	Ban Mai 2 Village (B)
Population	224	230
Average annual income[a]	18,270 baht	33,026 baht
Self-sufficiency of rice	64%	39%
Frequency of labour exchange[b]	3.7 days per year	3.5 days per year
Average size of land under cultivation	6.3 rai	10.4 rai
Access to the market	difficult access by car	accessible by car
Village access to electricity	no	yes

Notes

[a] The high average income figure is because high salaried people (school teachers, clinic doctors) are included. Without these non-farming people, Village A's average income would be around 10,000 baht and that of village B around 20,000 baht.

[b] 'Labour exchange' is represented by measuring the average of total labour days exchanged among villagers during the planting season of July and August.

as vegetables and potatoes, the need to plough mechanically forced the Karen to give up intercropping.

Meanwhile, highland development policies initiated by the state from the early 1960s did not have their intended effects. Attempts to introduce paddy rice techniques as a way to push the Karen out of shifting cultivation largely failed, with few farmers adopting them.[10] Coffee plantation projects in the 1980s also failed because of inadequate market connections. At the same time, forcing people to move out of the core area may have had an unintentional negative impact on the forests. The RFD considers forest fire as one of the major threats to the HKK today, with the sanctuary being damaged by fire every year. Ironically, because there are no people resident in the area, there is nobody to detect and counter forest fires until the flames are large enough to attract the attention of the forest fire unit station in the park.[11]

The responses of the Karen to the ever-tightening government regulation of forests (such as the abolition of shifting cultivation) and land (the imposition of a private property system) divide the Karen roughly into two groups: those who have intensified their cash crop production to escape from poverty, often at the expense of their subsistence production; and those who could not afford to follow this strategy and were forced to depend on other means, such as working on other farmers' residual land, borrowing money from local moneylenders and relying on collecting forest products. Many of them are unable to work in the cities to earn cash because they lack citizenship, personal connections, language skills or the confidence.

The ALRO is now responsible for implementing the integrated conservation and development project (ICDP) in the buffer zone area. Their latest approach to both conserving the HKK and responding to the needs of the farmers, particularly those falling under the second category above, is to try to shift people's economic orientation away from the forests by promoting income-generating activities in and outside the buffer zone. This new approach, although recognizing the basic right of people to live in the area, still assumes that local people are the main abusers of forests. Poverty, together with a purported lack of knowledge about how to farm appropriately, are believed to be driving people to unsustainable resource use (PEM consult 1996). The next section will examine the validity of these assumptions.

Analysis: Karen Forest Use and Dependency

Before discussing the kinds of interventions that might be conducive to conserving forests, it is important to understand how people interact with them on a daily basis. Unfortunately, the exact nature of forest dependency by local residents has seldom been investigated, and no site-specific information exists on this subject. Previous attempts to measure forest dependency in other areas of Thailand have tended to suffer from one or more of the following shortcomings: (1) The span of time is often limited, and year-long variability is not taken into account; (2) when potential seasonal biases are accounted for, the size of the samples is often too small; (3) where the samples are large enough, the selection of households often ignores the economic stratification within villages, which might strongly affect levels of forest dependency; (4) data collection often relies on villagers' memories or the recording of daily forest resource consumption by the villagers themselves, but such data are not reliable when one is dealing with the use of 'illegal' resources.

In an attempt to avoid these shortcomings, my own measurement of forest dependency of local villagers is based on different wealth ranks within villages; this makes comparisons possible and questions the common demonization of poverty per se as the central cause of excessive resource exploitation. I developed two indicators to capture forest dependency.

(1) *Income dependency* is the percentage and amount of income people obtain from selling forest products. It is important to note that we measure the *relative* proportion of income derived from forests.[12] When one villager derives all his income from forests whereas the other derives only 50 per cent, we can say that the former is more dependent on forests even though they may have an identical income in absolute terms.[13]

(2) *Livelihood dependency* is measured by the variety and intensity of forest products consumed as daily food contained in each meal. Direct use of forest resources requires labour investments and must, therefore, reflect the importance attached to the activity by villagers. I did not measure the amount of firewood or charcoal people consumed in the area because there seemed to be little variation among the households.

For income dependency, information was collected from all households through informal interviews and cross-checking. The choice of indicator for measuring livelihood dependency was difficult, but I eventually decided to observe the consumption of meat, for three reasons. Firstly, meat consumption reflects the households' connections both to the market and to the forest; meat was likely to be income elastic and can be expected to reveal differences in consumption between rich and poor. Secondly, people prefer to eat meat if they have a choice; it is an important source of protein, though they do not get to eat it often. The obstacles to meat eating include the labour required to hunt, forest guards, and/or the cash required to buy meat at the local market. Thus, meat consumption will likely reflect the general economic well-being of a household. Thirdly, from the viewpoint of the forest guards, animals are a primary forest resource to be protected. Forest animals are thus contested resources and consumption of them reflects the desperation of villagers under adverse circumstances.

Using a wealth-ranking exercise, twenty sample households were selected from each of the two villages (forty in total). With the help of two assistants from each village, I observed thirty randomly selected meals to observe the frequency of forest meat consumption in the following seasons: between 20 July and 20 August (the busiest time of the year for planting seeds and preparing the soil), another thirty meals in October (a rainy month with no major work, yet economically most difficult), and an additional twenty meals from February to March (during the dry season and after the harvest, so people are generally better off at this time). These sampling periods reflected various seasonal conditions that may affect villagers' access to the forest. The Karen in this area normally take two meals per day with little difference in the content of each meal. This provided the total of eighty samples for each household. Village assistants were permitted to choose which meal to observe each day.[14]

From the ranking exercise and discussions with villagers, I was able to extract criteria for wealth (and poverty) that people implicitly use to evaluate each other. Some of the common criteria were size and neatness of house, amount of land, number of children (fewer is better in this case), debt, income, family labour power, and self-sufficiency in rice. These elements were all combined in complex ways to come up with the total ranking of households. From these indicators, I learned that the amount of land under cultivation is strongly correlated with income dependency on forests.[15] My hypothesis is that because the busiest time for planting and weeding overlaps with the bamboo harvest season, those with larger land holdings cannot afford to allocate labour for obtaining forest resources; differences in dependency are not necessarily due to wealth. The results of the data collection based on this method are presented in Tables 16.3 and 16.4.

Some observations can be made from these data. In terms of income dependency based on the amount of land, less wealthy families tend to depend more on forest products (i.e., bamboo shoots). This is not surprising given the fact that their private workable land is limited. Forest products are important not only in terms of supplementing income but also in equalizing the flow of income throughout a year (bamboo shoots generate income during the months of July and August when no other income sources exist). In terms of livelihood dependency, no clear disparities were found between the

Table 16.3 Income Dependency Based on Amount of Land under Cultivation

Area under cultivation (rai)	Village A (limited access to the market)		Village B (better access to the market)	
	Annual income from forest (avg. baht)	% share of income from forest (avg. baht)	Annual income from forest (avg.baht)	% share of income from forest (avg. baht)
More than 20 rai	3500 (n = 1)	14%	1512 (n = 12)	4%
15 to 19 rai	1780 (n = 10)	11%	2833 (n = 6)	8%
10 to 14 rai	2000 (n = 3)	22%	2916 (n = 6)	17%
Less than 9 rai	1664 (n = 22)	34%	4552 (n = 19)	35%

Notes:

a) The price of bamboo shoots in village A is lower than that of village B by 20 per cent because of bad road access.

b) Area under cultivation is different from area under occupation, because many farmers do not have enough capital to fully cultivate their land. The figure includes only cash crops.

c) Those who have a regular salary (e.g., school teachers) are excluded from the sample. It was interesting to find that even a relatively large family did not split up its labour to maximize its income from the two sources (i.e., forest and agriculture). It was natural for them to work together in the same place as a family.

Table 16.4 Livelihood Dependency Based on Wealth Rank (one-year weighted average)

Wealth Rank	Village A	Village B
A (sufficient)	39%	33%
B	60%	33%
C (average)	39%	44%
D	56%	37%
E (poor)	52%	46%

Note: Figures are calculated based on the weighted average of frequency in each season. For wealth rank B, the sample was n = 1, so it is likely to be more biased than others.

higher and the lower-ranked. However, access to the market and good roads seems to differentiate the level of dependence between the two villages. Furthermore, the year-long study showed that, regardless of wealth and market access, much of the villagers' meals are composed not of meat but of rice, chili, and some vegetables. Meat consumption depends on seasons, the willingness to go into the forest, and on mere luck. Much of the villagers' protein intake comes from fish and small animals that are not the central target for conservation. There is thus no justification for characterizing the Karen as poachers of wild animals. At the same time, however, the frequency of meat intake, although lower than expected, should not give the impression that forest resources are unimportant to Karen lives. For many, especially the poor, the forest remains their only source of livelihood.

The dependency on forests observed among the majority of the population in the research site, set against the strict policing of the forests by the RFD, indicates that the Karen have limited opportunities to generate income and secure food. But this does not justify blaming them for the destruction of forest. Hunting activities are often carried out not in the forest, but in the Karen's fields during harvesting, when wild animals such as pigs come out to feed on the crops. There is almost no selling of animals in and outside the village. Timber use is also very limited; most houses are constructed with bamboo. Encroaching farmland into the conservation areas is too risky and too easily detected by the guards. On the other hand, some Karen do take advantage of the ambiguous demarcation of the sanctuary and look for forest products around the border. In many cases, they know where to look for the products while escaping the eyes of the forest guards.

Shifting cultivation has almost disappeared in this area, mostly because, as a result of state conservation policies, there is no place to shift to. Furthermore, shifting cultivation is ill-suited to the prevailing land registration and economic system where rewards are given to those who cultivate the same land every year and generate cash. In this sense, the nature of the Karen's forest dependency has changed dramatically from farming in their forests to collecting resources from the state forest. The separation of farm land and forest land is becoming increasingly sharp. Ironically, Karen living on the fringes of the HKK are now the target of development projects aimed at reducing the impacts of their *non-traditional* agriculture, which they only recently adopted after being forced to abandon their traditional shifting cultivation.

Concluding Remarks

The living conditions of marginal hill people on the edge of biologically rich forests illustrate a concentration of stress from various forces. The twin facts that hill peoples are concentrated in the northern and western forested areas and that they traditionally practise shifting cultivation have encouraged most government officials to see them as the principal cause of deforestation (Rigg and Stott 1998: 108). However, one can also argue that if the forested areas survived for so long with hill people living in and around them, the Karen cannot be *the* cause of environmental destruction. The future connection of the Karen with the forests has to be examined in the larger context of increasing privatization of lands and intensifying calls for preservation of biodiversity.

To sustainably manage forest resources, the social mechanisms to organize links between the government, village communities, and individuals become crucial. One cannot focus exclusively on a single component of a system while ignoring the rest. With regard to the state, we must ask what kinds of resources are given to the people, what forces are affecting a particular allocation, *before* asking how those resources can be sustainably managed in a participatory fashion. With regard to an individual or household, we must pay attention not only to forest resources under collective use per se, but also to private prerequisites (such as adequate farm land) that enable people to effectively take part in communal activities.

Notes

1 This chapter substantially relies on the author's previous article published in *Development and Change* (Sato 2000).

2 In the northeast of Thailand, for example, a field survey conducted by a team of NGOs and universities reported that among the eight community forests examined, only one was identified as rich and productive, whereas the rest were degraded (Danthanin et al. 1993: 122).

3 Uhlig notes that forest clearance by farmers was not necessarily caused by population growth, but that the pressure of debt repayment created by the rapid incorporation into the market economy was important as a motivating factor (Uhlig 1984).

4 The legal definition of forests relies on the Forestry Act (1941): 'Forest is a land without occupants in accordance with the land law' (*Praraachabanyat Paamai Po. So. 2484, Matraa 4 (1)* [Forestry Act of 1941, Article 4, Section 1]). Those who have legal ownership of land in the countryside, however, are still a minority, and this passive definition of forest certainly helped the government to legally confiscate some of the ambiguous land into state property.

5 Incessant encroachment of farmers into forest areas, despite the increasingly rigid regulations, occurred partly because of government inaction to the farmers' settlements. Particularly in the 1970s, forests were hiding places for anti-government groups and communists. Forest clearance, therefore, was conducive to uncovering and suppressing the movements of these people, while conciliating the farmers by not punishing them for their illegal forest clearance (Flaherty and Jengjalern 1995).

6 Lohmann estimates this area to be about 5,600 sq. km which is approximately 10 per cent of the total land area (Lohmann 1996).

7 Migration may be due to population growth accompanied by a reluctance to subdivide land, debt foreclosure, loss of tenancy rights, insufficient capital to purchase land on the part of young families, or forced displacement caused by infrastructural schemes and plantation projects (Hirsch 1990: 36).

8 HKK and the adjacent Tung Yai Naresuan world heritage sites consist of a complex mosaic of evergreen and deciduous forests. They lie at the meeting point of four biogeographic zones and derive elements of their flora from the west and north (Himalayan) and south (Sundaic) regions. At least 120 mammals, of which five are endemic, 401 bird species, 41 species of amphibians, and 1207 species of fresh water fish have been recorded (Nakasathien and Stewart-Cox 1990).

9 For example, the RFD began to count the number of people living inside the conservation forests only in the 1990s.

10 The main reasons why farmers did not adopt this high-yielding technique were twofold: (1) the difficulty in obtaining flat land to retain water, and (2) the difficulty of mixing a variety of crops to satisfy their subsistence needs.

11 In 1994 alone, 68,271 *rai* (10,923 hectares) reportedly burned down inside the sanctuary (Noikorn 1998). While I was conducting my fieldwork, a villager informed me that fire fighters themselves often set fires in the forests because their daily wages double when they are actively fighting fires.

12 Measuring income was more challenging than expected. Villagers, in general, do not know exactly how much they earned last year. The total income from maize, which was the central source of income for the majority of villagers, was much easier to estimate because they sell this

crop once a year. In contrast, bamboo shoots are sold on a day-to-day basis in smaller units. Based on accounts of villagers, I cross checked the information with the middlemen who had some statistics of purchases to come up with the best estimate.

13 Unlike an ordinary economic survey that often relies on income (which is often unavailable), or the amount of land holding, wealth ranking will allow us to observe how villagers *themselves* view wealth and poverty in their community. It will also minimize the snap-shot bias of relying on a single quantifiable measure.

14 Forest animals that they often hunt and eat are mostly 'illegal' in a strict sense. To increase the reliability of the data, I waited about a month before selecting capable villagers to help me with data collection. Based on my discussions with them, I developed a matrix that they could easily use to document observations. I asked my assistants to use different symbols for the meat that comes from the forest and from the market to obtain a rough idea of the relative proportion of food that originates from each source.

15 Initially, I analysed income dependency based on wealth categories. All households were more or less dependent on selling bamboo shoots at a similar level. I needed to look further into the composition of 'wealth' and decided to use the size of land as a criterion to differentiate villagers.

References

Andersen, J. P. 1923. 'Some notes about the Karens in Siam'. *Journal of Siam Society*, 17: 51–8.

Bruksasri, W. 1987. 'Minorities and Politics: Hill-Tribe Development and Integration Strategies', paper in the *Proceedings of the International Conference on Thai Studies*. The Australian National University, Canberra (3–6 July), pp. 245–53.

Chamarik, S. et al. 1993. *Paa chumchon nai pratheet thai: paa phonkeet roon kap phaap ruam khong paa chumchon nai pratheet thai* (Community Forests in Thailand: Direction for Development, vol.1, Overview of the Tropical Rainforest and Community Forests in Thailand). Bangkok: Sathaban chumchon thong thin phatthana. (in Thai).

Charasdamrong, P. 1997. 'The Land of No Return'. *Bangkok Post*, 16 February.

Christensen, S. and Rabibhadana, A. 1994. 'Exit, Voice and the Depletion of Open Access Resources: The political bases of property rights in Thailand'. *Law & Society Review*, 28(3): 639–55.

Cubitt, G. and Stewart-Cox, B. 1995. *Wild Thailand*. Bangkok: Asia Books.

Danthanin, M. et al. 1993. *Paa chumchon nai prathet thai: neao thand kan phattana lem 3: pa chumchon phak tawan ok chiang nua* (Community Forests in Thailand: Direction for Development, Vol. 3, Community Forests in the Northeast), Bangkok: Sathaban chumchon thong thin phatthana. (in Thai).

De Koninck, R. and Dery, S. 1998. 'Agricultural Expansion as a Tool of Population Redistribution in Southeast Asia'. *Journal of Southeast Asian Studies*, 28(1): 1–26.

Dove, M.R. 1994. 'North-South Relations, Global Warming, and the Global System'. *Chemosphere* 29(5): 1063–77.

Ekachai, S. 1998. 'The Fight is Far from a Result'. *Bangkok Post*, 2 July.

Falkus, M. 1990. 'Economic History and Environment in Southeast Asia'. *Asian Studies Review*, 14: 65–79.

Flaherty, M. and Jengjalern, A. 1995. 'Differences in Assessments of Forest Adequacy among Women in Northern Thailand'. *The Journal of Developing Areas*, 29: 237–54.

Ghimire, K. 1994. 'Parks and People: Livelihood issues in national parks management in Thailand and Madagascar'. *Development and Change*, 25: 195–229.

Gienty, D. 1967. *Thailand's Forest Development and Its Effect on Rural Peoples.* Bangkok, Thailand: United States Operations Mission.

Gravers, M. 1994. 'The Pwo Karen Ethnic Minority in the Thai Nation: Destructive "Hill Tribe" or Utopian Conservationists?' In *Asian Minorities: Three Papers on Minorities in Thailand and China.* Copenhagen Discussion Papers No. 23. Denmark: University of Copenhagen.

Gunatilake, H. M. 1998. 'The Role of Rural Development in Protecting Tropical Rainforests: Evidence from Sri Lanka'. *Journal of Environmental Management,* 53: 273–92.

Hafner, J. and Apichatvullop, J. 1990. 'Farming the Forest: Managing people and trees in reserved forests in Thailand'. *Geoforum,* 21(3): 331–46.

Hayami, Y. 1996. 'Karen Tradition According to Christ or Buddha: The implications of multiple reinterpretations for a minority group in Thailand'. *Journal of Southeast Asian Studies,* 27(2): 334–49.

Hirsch, P. 1988. 'Spontaneous Land Settlement and Deforestation in Thailand,' in J. Dargavel et al. (eds). *Changing Tropical Forests: Historical Perspectives on Today's Challenges in Asia, Australia and Oceania.* Canberra, Australia: Centre for Resource and Environmental Studies, pp. 359–376.

—— 1990. *Development Dilemmas in Rural Thailand.* Singapore: Oxford University Press.

IUCN. 1996. *A Review of the Thai Forestry Sector Master Plan.* IUCN.

Jorgensen, A. B. 1996. 'Elephants or People: The Debate on the Huai Kha Khaeng and Thung Yai Naresuan World Heritage Site'. Paper presented at the 48th Annual Meeting of the Association for Asian Studies, Honolulu, Hawaii, 11–14 April.

Kasetsart University, Faculty of Forestry. 1987. *Assessment of National Parks, Wildlife Sanctuaries and Other Preserves Development in Thailand. Final Report.* Bangkok: Kasetsart University.

Kemp, J. 1981. 'Legal and Informal Land Tenures in Thailand'. *Modern Asian Studies,* 15: 1–23.

Lohmann, L. 1996. 'Freedom to Plant: Indonesia and Thailand in a globalizing pulp and paper industry'. In M. Parnwell and R. Bryant (eds). *Environmental Change in South-East Asia: People, Politics and Sustainable Development.* London: Routledge.

Lynch, O. and Alcorn, J. 1991. *Empowering Local Forest Managers: Towards more Effective Recognition of Tenurial Rights, Claims and Management Capacities among People Occupying 'Public' Forest Reserves in the Kingdom of Thailand.* World Resources Institute.

Mehl, C. 1990. 'The Promise of Social Forestry: Evolution and Sustainability'. In S. Tongpan et al. *Deforestation and Poverty: Can Commercial and Social Forestry Break the Vicious Circle?* TDRI Research Report No.2, Chonburi, Thailand: TDRI.

MIDAS Agronomics. 1991. *Study of Conservation Forest Area Demarcation, Protection, and Occupancy in Thailand.* Vol. I, Appendix II, p. 1. Prepared for the World Bank.

Nakasathien, S. and Stewart-Cox, B. 1990. *Nomination of the Thung Yai-Huai Kha Khaeng Wildlife Sanctuary to be a UNESCO World Heritage Site.* Bangkok: Wildlife Conservation Division, Royal Forest Department.

Noikorn, U. 1998. 'Forest Fires Rage at Sanctuary: Insufficient staff seen as main problem'. *Bangkok Post,* 11 March.

Peluso, N. and Vandergeest, P. 1995. 'Territorialization and State Power in Thailand'. *Theory and Society,* 24: 385–426.

PEM consult. 1996. *Project Document: Huai Kha Khaeng Complex- Integrated Conservation and Development Project.* DANCED/Ministry of Environment and Energy.

Poffenberger, M. (ed.). 1990. *Keepers of the Forest: Land Management Alternatives in Southeast Asia.* West Hertford, Connecticut: Kumarian Press.

Pragton, K. and Thomas, D. 1990. 'Evolving Management Systems in Thailand.' In M. Poffenberger (ed.). *Keepers of the Forest: Land Management Alternatives in Southeast Asia* , pp. 167–86. West Hertford, Connecticut: Kumarian Press.

Ramithanon, S. et al. 1993. *Paa Chum Chon nai Prathet Thai: Neo thang kan Patthanaa.* Lem thi 1. (Bangkok: Sathaban Chumchon thong thin phtatthana, 1993) (in Thai)Åi Community Forests in Thailand: Directions of Development, Vol.1Åj.

Rigg, J. and Stott, P. 1998. 'Forest Tales: Politics, policy making, and the environment in Thailand'. in U. Desai (ed.). *Ecological Policy and Politics in Developing Countries: Economic Growth, Democracy, and Environment,* pp. 87–120. State University of New York Press.

RFD (Royal Forest Department). 1996. *Forestry Statistics of Thailand.* Bangkok: Royal Forest Department.

—— 1971. *Prawat Krom Paamai 2439–2514* (History of the Forest Department 1896–1971) Bangkok: Royal Forest Department. (in Thai).

Sato, J. 2000. 'People in Between: Conversion and conservation of forest lands in Thailand'. *Development and Change.* 37(1): 155–77.

Sayamon, K. 1995. *Panhaa tangkotmaii lae nayobaai khong kaanchai amnaatrat thii kiyokap kaan thiidin nai kheet paa* (in Thai) (Policies and Legal issues relating to the Excercise of State Power in the Forest Reserve Land, Masters Thesis, Faculty of Law, Chulalonkorn University, 1995).

Sayer, J. 1991. *Rainforest Buffer Zones: Guidelines for Protected Area Managers.* IUCN.

TDRI (Thailand Development Research Institute). 1990. *Land and Forest: Projecting Demand and Managing Encroachment.* Bangkok: TDRI

—— 1987. *Thailand: Natural Resources Profile.* Bangkok: TDRI.

Uhlig, H. 1984. *Spontaneous and Planned Settlement in Southeast Asia.* Hamburg: Institute of Asian Affairs.

Vandergeest, P. 1996a. 'Mapping Nature: Territorialization of forest rights in Thailand'. *Society and Natural Resources.* 9: 159–75.

—— 1996b. 'Property Rights in Protected Areas: Obstacles to community involvement as solution in Thailand'. *Environmental Conservation,* 23(3): 259–68.

World Bank. 1993. *Conservation Forest Area Protection, Management, and Development Projects.* Vol. 4. Prepared by MIDAS Agronomics Company, Bangkok, Thailand.

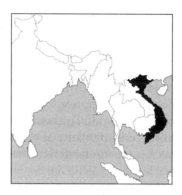

17

Lost Worlds and Local People

PROTECTED AREAS DEVELOPMENT
IN VIET NAM

Pamela McElwee

Introduction

In the last ten years, the Socialist Republic of Viet Nam has been proclaimed a 'biodiversity hotspot,' mainly due to the discovery of several new mammals previously 'unknown' to science. As a result, this small country has been the site of a concerted effort on the part of conservation organizations and international development agencies to improve environmental protection. In particular, the rise in deforestation beginning with the end of the Franco–Viet Nam war in 1945, through the American war from 1960 to 1975, and continuing after reunification of North and South in 1975, has been characterized as the most pressing environmental issue facing the country. One solution to this problem of deforestation has been to develop state-managed protected areas.

These parks are, for the most part, designed with input from Western conservation organizations and development agencies (cf. Wege et al. 1999). Furthermore, on paper, the goals are to protect them according to Western conservation ideals. While older protected areas in Viet Nam were often demarcated on an ad hoc basis, and included a wide range of disturbed and less disturbed habitat without much thought to system-wide representation, new parks have been proposed on the basis of biodiversity protection and in areas 'pristine' enough that they need enforcement against human encroachment. Mammals and birds are usually taken as the best indicators of biodiversity and 'undisturbed' habitat, and these species are used as the justification for new parks and as reasons to strictly protect the older ones.

This should not be surprising. Most countries in the world now have some sort of protected area system that encompasses a range of species in attempts to protect biodiversity, however defined by national governments. But Viet Nam's attempts to replicate a protected area system that one could find in America faces extreme difficulties. Most importantly, Viet Nam is one of the most densely populated nations on earth. This means that there is really no area in the entire country that can accurately be described as pristine or undisturbed; years of warfare and a growing, poverty-stricken population have seen to that. Most parks in Viet Nam are surrounded by resident populations of tens of thousands, and in some cases, hundreds of thousands of people. Yet rarely are park management plans formulated that specifically address these populations. This chapter attempts to answer the question why inflexible concepts of biodiversity conservation and national parks free of human use are being adopted in Vietnam. What is the rationale, and what will be the effect?

The History of Protected Areas in Viet Nam

Historically, interest in protected areas by the government of Viet Nam can only be traced back to the early 1960s and the personal influence of Ho Chi Minh, an advocate of natural resources conservation. When Viet Nam created the first national park in 1962, President Ho Chi Minh personally dedicated Cuc Phuong National Park, one hundred miles south of Hanoi.[1] He said then, 'Forests are gold. If we know how to conserve and use them well, they will be very precious.' Since then, his phrase *Rung La Vang* (forests are gold) has become a slogan for various state conservation plans.

Cuc Phuong was Viet Nam's first established national park, and since Cuc Phuong, ten more national parks have been created.[2] In addition to national parks, there are thirty-two 'Cultural, Historical and Environmental Reserves' (CHERs) and sixty-five nature reserves. Many of these reserves were created by the state in two administrative orders in 1977 and 1986 (Cao Van Sung 1995), and another round of parks has been added in the last five years since the Biodiversity Action Plan (BAP) for Viet Nam was adopted by the government in 1995.

Before the BAP, one million hectares (ha) of land were in these three protected categories, and the government said it intended to double that amount to two million ha before the year 2000. It has now met that goal and has an area of 2.1 million ha in some form of 'protected' designation (Viet Nam News Agency 1998). The distribution is 254,807 ha in national parks, 1,719,408 ha in nature reserve areas, and 145,359 ha in CHERs (Ministry of Agriculture 1997b). The whole system is classified by the Ministry of Agriculture as the 'Special-Use Forest System' (*rung dac diem*), to be distinguished from the production forests (*rung san xuat*) and protection forests (*rung phong ho)*, which are watershed areas. As the name of the system indicates, environmental plans are highly biased towards forest protection and reforestation. Although many scientists have pushed for marine reserves, coastal mangrove reserves, wetlands reserves, and protected fishing areas, the majority of the current preserves in Viet Nam are terrestrial, non-coastal forested areas (Cao Van Sung 1995).

'National parks' usually consist of a strictly protected inner core in which almost all anthropogenic activities are banned. Outer cores allow for such activities as regenerating and replanting schemes, some 'low-impact' resource use, and recreational activities. Buffer zones that allow for regulated production activities are supposed to be managed either by or with the help of park officials, but are outside most parks' officially demarcated boundaries.

The second designation of biodiversity zones is in 'nature reserves.' Nature reserves are protected areas primarily designed for conservation and scientific research; tourism is not encouraged. The administration of these zones falls mainly to each of the provinces, but boundaries are to be developed with the Ministry of Forestry. Provincial and district officers of the Forest Protection Department (*Kiem Lam*) are usually the ones in charge of day-to-day management. Finally, 'cultural, historical and environmental reserves' (CHERs) contain historical relics and scenes with aesthetic interest. Examples include a former jailhouse for anti-French revolutionaries. Tourism to these areas is encouraged, and although these CHERs are administered by the Ministry of Forestry, most are considered to contain little of scientific or ecological value (Tran Lien Phong 1995).

Looking at the protected area system that has developed to date, it is easy to see that it is biased towards forested, mountainous ecosystems. These are also the areas where the largest numbers of ethnic minorities in Viet Nam live (15 per cent of Viet Nam's population are non-Vietnamese, belonging to 53 distinct ethnic groups). The current system of parks and reserves predominates in these ethnic minority areas; of the nature reserves with the strictest restrictions on land uses, almost all abut areas inhabited by ethnic minorities. Other sensitive ecosystems populated and used by lowlanders, such as coastal mangrove forests or coral reefs, have not been emphasized in these biodiversity strategies. The state has even admitted that 'most of the special use forests have been allocated in remote or mountainous areas under difficult geographical conditions and less developed socio-economies' (Ministry of Agriculture 1997b).

However, rather than recognizing forest use rights of these minorities, many of whom are extremely poor and marginalized, most management plans for parks (when they actually address that there are people living around and in them) advocate strict protection for core 'biodiversity' zones of parks, where human use is not allowed. As an example, general national guidelines say that within national parks and reserves, it is prohibited to:

> log, exploit (excluding activities related to forest cleaning and rehabilitation), hunt animals, collect specimens under any means and forms. It is also prohib-

ited to make loud noise or do anything that causes negative impacts on living conditions and development of all plants and animals in the special-use forest … Strict protection areas within national parks and nature preservation areas should be protected strictly. Every activity that causes negative impacts to forest is not allowed. (Ministry of Agriculture 1997b)

These general guidelines make no distinction between local and extra-local resource use, or between low impact (such as collecting thatch (or making loud noises!)) and high impact (poaching with semi-automatic rifles). These kinds of generalizations in management therefore hamper attempts to find individual local solutions to conservation.[3]

This strategy is somewhat at odds with new thinking on park and protected areas management in much of the globe. More and more conservation organizations are paying attention to the fact that people live in and around most protected areas of the world, and that these local people shoulder most of the costs of conservation. For example, a May 2000 report by the World Wide Fund For Nature says that 'conservation through protected areas can at its worst exacerbate existing social inequalities; in effect putting the needs of wildlife before the needs of the poorest people' (Carey et al. 2000). Many parks in Southeast Asia have tried to combat these problems by including 'local development,' 'integrated conservation and development,' and 'buffer zones' into conservation plans. But how much is this new talk translating into how protected areas are developed and managed in Vietnam? As the following example shows, very little has changed.

Several years ago, the state of Viet Nam gave approval to the demarcation of the Pu Luong Nature Reserve, located in northeast central Viet Nam in Thanh Hoa province. Pu Luong encompasses 17,500 ha, and unlike many nature reserves in Viet Nam, the natural forest here is extensive. The province proposed this nature reserve in conjunction with support from international conservation NGOs, mainly because it supports one of only two known worldwide populations of an endemic, limited range primate species, the *vooc mong trang,* or Delacour's Langur (*Trachypithecus francoisi delacouri*). There are estimated to be 200 animals remaining here and in Cuc Phuong National Park to the northeast. This langur is one of the most critically endangered primates in Viet Nam.

Because this is a nature reserve, the forest is supposed to be protected strictly. This means no exploitation of any natural product, including for subsistence use. There are 128 households living inside the boundaries of the new nature reserve, mostly of Muong and Thai ethnicity, and 157 households in the regeneration/recovery area. Households can use bamboo to build houses and for subsistence within the recovery area, but any exploitation in the core zone is officially prohibited. This leaves the 128 families with almost no opportunity to make a living. 'It is kind of hard to tell them using the forest is illegal' a young Pu Luong park ranger told me, 'so we haven't found anybody guilty of anything yet.' The national government is currently studying how to provide investment to Pu Luong as part of its nation-wide strategy to improve nature conservation. A 'buffer zone' around the reserve has been declared in the four administrative communes nearby, and 600 households in the buffer zone are now receiving 50,000 VND/ha to protect this area for forest regeneration. But because the people actually living in the nature reserve are not supposed to be using the resources there any longer, they are not the ones receiving money for protection; at the same time, people living several kilometres away with no real dependence on the park are receiving the cash payments. So far no investment aid has been forthcoming to anyone living in the park, nor has the government even been able to pay salaries for any of the 22 people working as park staff for the past year.

The last known case of someone poaching a *vooc mong trang* was in 1990, according to the ranger, and that was an outsider who had come in to the area. He said local people were well aware of the jail sentences possible if caught with an animal. And it was clear from surveys that the langur resided deep in the limestone mountains away from human settlements in the area. Yet, the official way to manage this nature reserve was not to pay people to keep out of the langur zones and let them continuing farming on the outskirts, or to train them to spot langurs to help with the surveys and serve as local guards, or to let people continue living in their village and ensure they received money and support for reporting poachers. Rather, the plan was the same as for almost every nature reserve in Viet Nam: demarcate a border on a map, tell people they aren't supposed to be there (and thereby induce insecurity of land tenure), inform anyone living nearby that any use of this area (no matter how low-impact) was now illegal, and hope that money to pay for guards might be forthcoming in the future.

Classifying Nature, Chasing Money

Although the government and the local provinces have been involved in the designation of over one hundred national parks and reserves to date, less than half actually have management boards or budgets, as the case of Pu Luong above indicates. Many of the nature reserves also seem to contain questionable biodiversity value, in terms of unique or endangered biological resources. This is because many natural reserves are former state forest enterprises, the timber companies owned by the socialist state, and these 'nature reserves' have been stripped of much of their forest cover. These enterprises have been forced in the last ten years of market openness to find new ways to succeed, given declining budgets and a nation-wide ban on log exports from natural forests that was adopted in 1993 (Andzdec Consultants 1997). Now former timber enterprises that were about to go belly-up are sometimes transformed into 'nature reserves,' including the well-known Vu Quang Nature Reserve, famed in recent years for the discovery of several mammals previously unknown to science.

The slapdash nature of the creation of 'protected areas' has rankled international conservation organizations. A 1995 conference on Viet Nam's park system concluded that the national goal of biodiversity protection was not well served by the current disorderly system, as many reserves are too small, too isolated, or are degraded, or contain little value for biodiversity conservation (Anon. 1995). Therefore, the national Biodiversity Action Plan (funded by international donors and adopted by the government in 1995) singled out several parks for priority in funding and management for the next century, based on their perceived biological importance (see Table 17.1). This table also indicates the scope of the impact these parks may have on the large populations either inside or immediately outside their boundaries.

The new focus on biodiversity protection and national parks has yet to produce positive results. Viet Nam was recently named one of the top ten countries that is losing biodiversity the fastest.[4] Therefore, many of the newer parks (and increased protection proposed at old parks) are aimed to protect single species facing localized extinction threats. These species usually include elephant (*Elephas maximus*), gaur (*Bos gaurus*), banteng (*Bos javanicus*), tiger (*Panthera tigris*), rhinoceros (*Rhinocerus sondaicus*), several species of langurs and leaf monkeys, and the possibly-extinct kouprey (*Bos sauveli*) (see Duckworth and Hedges 1998). Mention must also be made of the three new species of mammal found in Viet Nam in recent years, the saola (*Pseudoryx nghetinhensis*), giant muntjac (*Megamuntiacus vuquangensis*) and Truong Son muntjac (*Caninmuntiacus truongsonensis*). Because not much is known about them yet, several new parks have been proposed to protect just these last three species.

The discovery of the three new mammals in the last few years has resulted not just in national attention to threatened forest habitats, but also international acclaim as well. Viet Nam's Vu Quang Nature Reserve was described in the international press as 'like opening a door into a lost and neglected place,' (Associated Press 1992) as 'one of the world's ecological hotspots... a zoologist's dream,' (Drollette 1999) and as 'a lost world that modern science has never before looked at' (Time 1992). (Not surprisingly, these reports of a 'lost world' failed to note that the lost world was previously a timber enterprise, that 20,000 people lived there, and that the heavily bombed war-time Ho Chi Minh Trail ran by it.) One scientist has said, 'Viet Nam is the place to be since all the discoveries... It's the last frontier. Every new animal makes you think, "What else is out there?"' (Drollette 1999).

It is now common to find Viet Nam – described throughout the 1980s by consultants as a dismal place for wildlife because of war and over-hunting – acclaimed as a 'global biodiversity hotspot' (Chape 1996). The press reports were particularly effective in bringing in donor money for environmental protection. 'The saola and the giant muntjac have done quite a lot to raise awareness of conservation in Viet Nam,' the head of Hanoi's WWF office has said (Drollette 1999). Less clear, however, is what effect the new awareness will have on the people who live around protected areas.[5]

Certainly one factor that is driving the interest in the establishment and management of protected areas is money. Improving the management of parks has been generating a great deal of foreign development aid for this extremely poor country. Especially because Viet Nam was shut out of many

Table 17.1 continued

Name of Protected Area (Province)	Main Reason for High Biodiversity Value	High Ethnic Minorities Population? (Name of group)	Total population in park; Total out in buffer zone
New Proposed Reserves of Biological Importance			
Song Thanh-Dakpring (Quang Nam)	Birds, Truong Son Muntjak	Yes (Katu, Gie Trieng)	11,812 in
Ngoc Linh (Kontum)	Birds	Yes (Sedang)	13,000 out
Phong Dien (Thua Thien Hue)	Edwards pheasant, other birds	No (but some Ta Oi)	17,000 out
Dakrong (Quang Tri)	Birds	Yes (Bru-Van Kieu, Ta Oi)	14,000 out

Note: The numbers of people in and out of parks is based on reporting by the park officials themselves, which is oftentimes outdated or inaccurate. Buffer zone figures may include all adjacent communes sharing an administrative border with the park, not only those communities in proximity to the park. Figures should be taken as general indicators only.

Sources: Andzdec Consultants 1996; Vo Tri Chung 2000; Ghazoul and Le Mong Chan 1994; Le Duc Giang 1995; Le Trong Trai and Richardson 1999a, 1999b; MacKinnon 1992; Ministry of Forestry 1993; Tordoff et al. 1997; United Nations Development Programme 1996; Wikramanayake and Vu Van Dung 1997; World Bank 1997.

international funding circles for the twenty years after reunification, getting international development aid is a new and important enterprise for the state. Parks have certainly proved profitable in the last few years (see Table 17.2). In fact, the money received in recent years from donors for the Vu Quang Nature Reserve is ten times the amount budgeted by the government for *all* protected areas in Viet Nam in 1991, the year before the saola was discovered.

Despite the generous donations from abroad, there are tensions between the Vietnamese state and these international funding organizations over what 'protected areas' and 'conservation' mean. The Vietnamese state has liked to put highly organized and managed landscapes in some of the national parks, like a man-made botanical garden with '100 different wood species of 52 families found in Cuc Phuong' (Ministry of Agriculture 1997b). However, a foreign consultant complained angrily that old-growth trees were cut down to make the garden, which thereby compromised the integrity of the park. In another park, the Vietnamese management board has expressed interest in capturing and taming various wild deer species for visitors to look at, while foreign consultants thought it would result in too many deaths during captures and was unnecessary for conservation. The vice-director of the Department of Environment, Science and Technology in Dac Lac province said in a meeting with me that eco-tourism was the only hope he saw in getting enough money to manage parks given poor state budgets, and he admitted that in order to attract domestic tourists, something 'interesting' had to be there. Many Vietnamese living in Buon Me Thuot city, near Yok Don National Park, often told me that to just go walking and hiking in the park was 'boring.' Much more interesting were the domesticated elephant rides offered by a nearby village, and the chance to picnic in little huts built in a clearing along the river outside the park.

Table 17.2 Funding Acquired in Recent Years for Park Management from Overseas

Name of Project	Funding Source	Term	Amount (USD):
Biodiversity Action Plan	GEF	2 years	3 million
Subregional Biodiversity Forum	UNDP	3 years	1 million
Vu Quang Conservation Project	Netherlands	5 years	2.4 million
Nam Cat Tien Natl. Park	Netherlands	5years	5.8 million
PARC Project	GEF/UNDP	5 years	6.3 million
Mom Ray/Cat Tien Natl. Parks	World Bank	5 years	32 million
Pu Mat Nature Reserve	EU	5+ years	19 million
Expanding Protected Areas Network	EU	5 years	1 million
U Minh Thuong Nature Reserve	DANIDA	n/a	2.7 million

Notes: GEF – Global Environment Facility; EU – European Union; UNDP – United Nations Development Programme; PARC – Protected Areas Management for Resource Conservation Project; DANIDA – Danish International Development Agency.
Sources: UNDP 1996; World Bank 1997; World Wildlife Fund 1996

The idea of 'developing' natural sites to make them more attractive to human visitors has a long history in Viet Nam. A number of the protected areas managed by the Ministry of Forestry harbour little biodiversity, especially the 'cultural, historical and environmental reserves'. Rather, these are places of historical importance (like Pac Bo cave in north Viet Nam, where Ho Chi Minh hid from the French in the 1940s), or else are managed landscapes of sightseeing value where one might take boat rides, see flower gardens, and so on.[6] In this sense, the state appears to recognize that human-altered areas with perhaps little in the way of biodiversity also possess value as protected landscapes. In the case of CHERs, these areas are protected precisely because they have been altered, used or valued by humans. However, a foreign consultant to Viet Nam's Tropical Forestry Action Plan review in 1990 encouraged the abolition or transference of these areas out of the protected area system, saying they were essentially valueless, and a drain on limited budgets (MacKinnon 1990).

Move Them Out, All of Them!

Officially, the core zones of national parks (and nature reserves that have management plans) are to have almost no human uses.[7] In many cases, where residence within a park or reserve is not an acceptable activity, resettlement plans exist to move local people. However, international organizations often leave out the difficult questions about resettlement from the specifics of their programmes. For example, in an EU-funded survey to delineate a new park in Quang Nam province, it was noted that 11,000 people, almost all of whom are ethnic minorities, will be living in the proposed park boundaries, but gave no indication of what is to be done with them (Wikramanayake and Vu Van Dung 1997).

The one major relocation programme that has already been adopted was in Cuc Phuong National Park in the years 1985 to 1990. The national government ordered the resettlement of all families living in the central valley of the park, involving about 550 people of Muong ethnicity. One of the interesting justifications for the move was that the Muong people were prosperous enough to be able to survive such a transition, and that they were prosperous precisely because they were 'poaching' off park resources: 'In fact the Muong people living in the park are very privileged in comparison with other peasants in the country. Besides the normal benefits from agriculture they have additional profit from illegal hunting, unlimited free fuelwood, timber for selling and unlimited pasture land for their cattle' (Szaniawski 1987). However, the local people themselves argued against resettlement in cultural, not material, terms, and forcibly resisted the resettlement: 'They affirmed that they (unlike some of the more recent settlers) had lived on the Cuc Phuong site long before the National Park was established and that it was the land of their ancestors' (Nguyen Nhu Phuong and Dembner 1994).

Another major resettlement project that has been in the works for years but not yet implemented is at Nam Cat Tien National Park, the only known habitat of the Javan rhinoceros in Viet Nam. Park project documents usually note that a major problem is the presence within the reserve of about 600 people belonging to the Xtieng minority (while ignoring at least five logging enterprises owned by the state and a state coffee farm around the edges of the park). In 1993, a plan was first proposed to move an entire Xtieng village out of the park by promising three million Vietnamese Dong (U.S. $215) per person who moved (Ministry of Forestry 1993). The plan has not yet been implemented because there is no money to pay anyone, but park staff have continued to complain that 'because of tensions, they need to resettle fifty households in this area. All of village Three and Four, as well as villages K'Lo and K'it. Mr. Khanh [of the park] says, "All of them!"' (Nhat Anh 1999).

The Dong Nai provincial authorities had previously tried to resettle some of these villagers under a different national plan to reduce shifting cultivation but almost all of the villagers have now left the old resettlement site to return to the forest around Nam Cat Tien. Villagers recently explained their motivation for opposing resettlement to a reporter (Nhat Anh 1999). One Ma man said, 'Better to be struck dead immediately and die here! If we go down [to the resettlement site] people don't know how to make a living.' The head of a Ma village scheduled for resettlement added:

> We don't know about living in the town. In making wet rice fields, compared with Vietnamese people, we will be losers. The food down there – our stomachs can't take it. Cabbages and cauliflowers stink when you eat them! Food of the Vietnamese people makes a person weak. If you eat that stuff, then you can't live in the forest like us.

And finally, a Xtieng minority woman expressed her disappointment with the government's position by saying:

> In the past, the people followed the Party, defeating the French, the Americans, and took back the homeland of our ancestors. Now life is very hard. Whatever we use, then the forest is prohibited! Our houses are shabby, and we wrote a petition asking to make new wood houses, but nobody gave us a solution. Whatever we use, we are caught and they [the national park] make a report about it. In the past, Xtieng people still lived here the way we do, so why did we have thirty-seven rhinos as well? And now if we use whatever we need, what effect does that have on the rhinos?

Resettlement is not a good solution for many reasons besides the ones expressed above by local people. In general, resettlement in Viet Nam is underfunded, as the case of Nam Cat Tien shows. Plans have been on the books for years and yet there is still no movement. Planning resettlement in this way can be even more detrimental to park management as it discourages long-term resource management by local people, who may choose to overexploit the area if they know they will be forced to leave in the future. And furthermore, removing one set of people may allow another in. This is particularly true in Viet Nam, which has very high migration rates to rural areas by farmers in search of land. Around Nam Cat Tien, the Ma and Xtieng minorities who have lived there for centuries could be forced to move, only to be replaced by Vietnamese pioneers who know little about forest management and can be expected to deforest to plant cash crops, as they have done in other parts of the province.

Problems in Paradise

The main problem affecting the management of many national parks as identified by the state, local park managers, and some international donors is encroachment and shifting agriculture (also known as swidden, or slash-and-burn agriculture) (World Bank 1997). Blame for degraded park habitats is often laid on ethnic minorities practising swidden, which is viewed by the state and park authorities as environmentally destructive: park officials often complain about the 'lazy' and 'backward' shifting agriculturalists who don't know how to make wet rice fields like Vietnamese do, and therefore waste forest lands with their itinerant agriculture. As the state agency in charge of parks complains, 'due to low living and intellectual conditions in addition with backward working methods, many of the primary and natural forests have been damaged' (Ministry of Agriculture 1997b). Yok Don park officials have added that 'Lives of people are poor, the level of intelligence is very low, lack of land leads to people swiddening. Customs and habits are backwards, languages are different, they remain deep in the area, and have difficult economies' (Ministry of Agriculture 1997a). The director of the Ben En National Park has complained that the resident groups in his park practise 'backward' farming, and that 'due to the lack of land and big population growth, they have wantonly exploited forest, practiced burn-out clearing, and fished by explosives' (Le Duc Giang 1995).

However, it is usually unclear what 'shifting agriculture' really means and who practises it around parks (McElwee 1999). Over-generalizing that all swidden is bad, and that all swidden is practised by 'backward' ethnic minorities, has made it difficult to understand the situation and take steps to address it. In fact, in many areas of Viet Nam, swidden agriculture is not the number one cause of deforestation at all. It appears that it is the uncontrolled migration of lowland Vietnamese (*Kinh*) people. In a recent report about forest loss in two provinces, a Canadian research team found that swidden contributes far less to national deforestation than other causes such as agricultural expansion by lowland *Kinh*, logging (both legal and illegal), and fuelwood collecting. The report was unequivocal:

> We were able to isolate several processes in time and space and to represent them cartographically, allowing us to point to *Kinh* agricultural expansion as the one major instrumental cause of deforestation... Even if some evidence suggests that a few members of ethnic minorities are involved in forms of agriculture that may lead to deforestation, the consequences are in no way comparable to those of the agricultural practices of the *Kinh* colonists. In broad terms, it can be estimated that for each hectare of forest destroyed by the agricultural practices of minorities, at least twenty hectares is destroyed by those of the *Kinh* pioneers (DeKoninck 1998).

The problem of spontaneous migration and resource degradation is enormously complicated. The World Wildlife Fund must be congratulated for recognizing the impact that migration has on conservation when they funded a study on the impacts of migration around Yok Don National Park (Huynh Thu Ba 1998). In this study, an elder of the Ede minority said that,

> Before the migration in the early 1990s, our tribe used to practice shifting cultivation with a fallow cycle of between three to seven years. These lands were more fertile and we never used fertiliser. In 1992 we stopped shifting cultivation. There is nowhere to move. There are too many people and the forest is too far away from us.

The study went on to conclude that Yok Don would continue to suffer encroachments as long as the provincial authorities turned a blind eye to migration.

There are other problems affecting protected areas that are equally complicated. These include wildlife poaching and illegal timber cutting – often by authorities. A new word has appeared with increasing frequency in the Vietnamese press – '*lam tac*.' A *lam tac* is a person who poaches and deforests with impunity, usually because he has connections. One particularly egregious example was the Tanh Linh forest enterprise case prosecuted in 1999. The thirty-six defendants in national court were accused of 'violating forest protection rules, irresponsibility, corruption and illegally stockpiling military weapons' (Long Huong 1999). The gang reportedly cut down 53,429m^3 of trees – with a value of 1.6 million U.S. dollars – in various wildlife sanctuaries and protected forests with the tacit co-operation of local officials. Twenty-nine district and provincial officials were eventually under indictment, including the former deputy director of the provincial Binh Thuan Department of Agriculture and Forestry. If twenty-nine out of thirty-six defendants in this case are government officials, it becomes easy to see why the illegal timber problem is not just a case of small-scale cutting of firewood by local people. Rather, it is high level, often organized criminal activity, involving many branches of the government.[8]

Many of these reports on the true nature of conflicts over resources in Viet Nam are coming from an unlikely source: the Hanoi press. These hard-hitting articles have exposed a variety of reasons for poor protected area and forest management, including expansion of coffee planting and fruit tree planting by state enterprises into protected areas; high level corruption in provincial forest protection departments; complicit involvement in the wildlife trade by customs, border guards, and the army; and confusing policy directions from the central government that have allowed loopholes in the wildlife trade by officials (e.g. a policy No. 433/KL named 'permission to exploit common wild animals and plants in the scope of management') (Cao Hung 1999; Huy Ha and Anh Tuan 1999; Le Huan 1999; Nguyen Tuan 1999; Xuan Quang 1999). However, because these reports have been entirely in Vietnamese, and not reprinted in the single English language newspaper, they have not been publicized outside Viet Nam. And there has been little discussion of these problems inside conservation circles or by park managers.

Conclusion

What is really needed to improve both protected areas and human lives in Viet Nam? I am mainly arguing for more diverse solutions which recognize that people must be included in conservation plans, as they are an essential part of the landscape (and have been for many years). 'Buffer zones' added as an afterthought to strictly protected parks are not enough. The problems facing Viet Nam's parks are many-fold and need to be addressed at each specific park. Instead, generalizations and misunderstandings about the role of humans in landscapes are hampering efforts. It has been easier to blame shifting agriculturalists than admit that sometimes the management boards of parks are corrupt, or that central government policies are encouraging migration, which directly leads to forest destruction.

There are some causes for hope, where local parks are acknowledging that people can be included in conservation. Medicinal plant collecting is being investigated in Ba Vi National Park, where the Dao people have a long history of traditional medicine (Tran Cong Khanh 1995). In Hoang Lien Son Nature Reserve, people have been allowed to plant cardamom under the forest canopy, as it is a cash crop that needs shade to survive, and thus provides an incentive for forest protection (Sobey 1997). In another case, where firewood is the problem facing a park, other areas can follow the lead of Tam Dao National Park and allocate park budgets to plant fuelwood species specifically for the needs of local people (Xuan Quang and Quang Thien 1999). In other areas, park officials have allocated protected forest land to people for protection in '*kiem lam nhan dan*' (people's forest protection committees) and pay villagers well for doing so. In other words, the solutions must be local, not international, and must recognize the specific and very real needs of Viet Nam's diverse peoples.

Notes

1. There is mention that the last Governeur General of French Indochina, Leopold Cadière, proposed in 1941 to set up five nature reserves in Viet Nam – two areas in Son La province of Tonkin, two in the coastal forests around Danang (Tourane) and one area in the former capital city Hue (Cao Van Sung 1995). The Bach Ma park outside Hue was specifically to be set up to protect the Edwards Pheasant *(Lophura edwardsi)*, the first example of a species-driven park in Viet Nam (Soer 1997). However, the Japanese invasion of Indochina wiped out these plans, and they were never to be revived by the French.

2. The former Republic of Viet Nam also made a number of war-time park openings. In 1965, ten protected areas were proposed and demarcated by Saigon authorities. According to the IUCN in 1974, South Viet Nam had over 750,000 ha of land in protected area status. However, after reunification, many of these areas were not renewed as protected areas. In fact, out of the original ten areas, only two are now part of the national park system in present-day Viet Nam.

3. The most egregious case of overgeneralization is to be found in the management plans of Yok Don National Park and Vu Quang National Park, which were both made by the same consultant, and which are in some passages verbatim copies of one another when assessing the impact of the park on local people. 'However the density of people is not high and the area of

other production zones is large enough to meet real needs. There does not seem to be a real dependence on the reserve itself by local people. Their needs could be met in other ways' is found verbatim in both documents, despite the fact that the parks are quite different. Vu Quang has no minorities living around it; Yok Don has many. Yok Don is a riparian and dry deciduous park, while Vu Quang is mountainous and the site of some of the highest rainfall in Viet Nam. Claiming that these two areas both have exactly the same human problems indicates a neglect to look into the details of the socio-economic situation around parks (see MacKinnon et al. 1989; MacKinnon 1992).

4. The other countries in the top ten are Indonesia, South Africa, Ethiopia, Myanmar, Madagascar, Cameroon, Malaysia, Ivory Coast, and the Philippines (Nguyen Tuan and Dang Ba Tien 1999).

5. All of the rhetoric about the new species already has a downside. Due to the international attention paid to the new species, to the exclusion of attention to older species seemingly worthy of protection like gaur and banteng, some biodiversity inventories in Viet Nam have begun emphasizing that the saola and giant muntjac might be found in areas far from their original sightings. There is no evidence that the saola has ever lived in the dry dipterocarp forests of Yok Don National Park, hundreds of kilometres from the montane wet forests where it was first sighted, yet it was mentioned as a possible mammal found in Yok Don in a 1995 inventory. The problem has become so troubling that international scientists doing a compilation of all large mammal sightings in Indochina question whether any biodiversity inventories coming out of Viet Nam in recent years are in fact scientifically accurate, as many of them are riddled with exaggerations and plagiarism (Duckworth and Hedges 1998).

6. In fact, the preservation of 'cultural legacies' (*di san van hoa*) is mandated in the 1992 Constitution, Article 34: 'The government and society will protect and develop the cultural relics of people: take care of the work of protection, putting in museums, repairing, and conserving the effects of legacies of history, the revolution, cultural relics, artistic works, and famous landscapes.'

7. As one example, a management plan for Yok Don National Park admits that human use is currently low and may be sustainable, but that it is too difficult for park staffers to monitor it: 'Modest levels of hunting or collecting from the reserve on a sustained and controlled basis would be theoretically possible without destroying the habitat of the wildlife or threatening wild populations. However, the practical problems of controlling such harvesting and the permanent danger of fires make these activities incompatible with conservation objectives' (MacKinnon et al. 1989). And while the report said that the levels of harvesting in the park were so small as to 'not be worth the measures that would be necessary to check that everyone was keeping to the proper permitted levels,' the fact that this small harvesting might be allowed without strict checks by the authorities was not considered.

8. For another example of government involvement, the army has a large stake in the timber industry. When the former Prime Minister ordered the end to raw log exports from Viet Nam in 1993, he also forced three army-owned sawmills in Qui Nhon to close. In response, the army daily newspaper, Quan Doi Nhan Dan, lashed back that it was 'necessary to export timber products to earn money for reafforestation,' a particularly interesting justification.

References

Andzdec Consultants. 1996. *Viet Nam Biodiversity Conservation and Rural Development Project*. Hanoi.
—— 1997. *Restructuring of State Forest Enterprises*. Hanoi.
Anon. (ed.). 1995. *Proceedings of the National Conference on National Parks and Protected Areas of Viet Nam*. Hanoi.
Associated Press. 1992. 'Lost World of Unknown Creatures'. *Sydney Morning Herald,* 28 July.

Cao Hung. 1999. 'Gan 300ha rung bi tan pha, ai chiu trach nghiem? [Nearly 300ha of Forest Are Deforested, Who Has Responsibility?]', *Lao Dong (Labour)*. Hanoi, 10 August.

Cao Van Sung. 1995. *Environment and Bioresources of Viet Nam: Present Situation and Solutions*. Hanoi.

Carey, C., Dudley, N. and Stolton, S. 2000. *Squandering Paradise? The Importance and Vulnerability of the World's Protected Areas*. Gland, Switzerland.

Chape, S. 1996. *Biodiversity Conservation, Protected Areas and the Development Imperative in Lao PDR: Forging the Links*. Vientiane.

DeKoninck, R. 1998. *Deforestation in Viet Nam*. Ottawa.

Drollette, D. 1999. 'The Last Frontier: In Viet Nam, zoologists discover animals long hidden by rugged terrain and political strife'. *Newsday*, 27 April.

Duckworth, J.W. and Hedges, S. 1998. *A Review of the Status of Tiger, Asian Elephant, Gaur and Banteng in Viet Nam, Lao, Cambodia and Yunnan Province (China) with Recommendations for Further Conservation Action*. Hanoi.

Ghazoul, J., and Le Mong Chan. 1994. *Site Description and Conservation Evaluation: Nui Hoang Lien Nature Reserve, Lao Cai Province, Viet Nam*. Hanoi.

Huy Ha and Anh Tuan. 1999. '4,000 ha rung phong ho Dau Tieng dang bien thanh 'vuon phong ho' [4000 Hectares of Protected Forest in Dau Tieng are Becoming 'Protected Gardens']', *Lao Dong (Labour)*. Hanoi, 23 August.

Huynh Thu Ba. 1998. *Human Migration and Resource Utilization*. Hanoi.

Le Duc Giang. 1995. 'Ben En National Park – Natural resources, potential of scientific research and tourism development'. In *Proceedings of the National Conference on National Parks and Protected Areas of Viet Nam*. Hanoi.

Le Huan. 1999. 'Vi sao rung Tay Nguyen bi pha [Why Are the Central Highlands Being Deforested?]', *Lao Dong (Labour)*. Hanoi, 14 May.

Le Trong Trai and Richardson, W.J. 1999a. *A Feasibility Study for the Establishment of Phong Dien (Thua Thien Hue Province) and Dakrong (Quang Tri Province) Nature Reserves, Viet Nam*. Hanoi.

—— 1999b. *An Investment Plan for Ngoc Linh Nature Reserve, Kon Tum Province, Viet Nam*. Hanoi.

Long Huong. 1999. 'Ngay thu 3 xet xu vu an pha rung Tanh Ling [Day Three of the Trial of the Case of Deforestation in Tanh Linh]', *Lao Dong (Labour)*. Hanoi, 5 April.

MacKinnon, J. 1990. *Review of the Natural Conservation Systems, National Parks and Protected Areas: Forestry Sector Review Tropical Forestry Action Plan*. Hanoi.

—— 1992. *Draft Management Plan for Vu Quang Nature Reserve, Huong Khe district, Ha Tinh Province, Viet Nam*. Hanoi.

MacKinnon, J., Laurie, A. et al. 1989. *Draft Management Plan for Yok Don Nature Reserve, Ea Sup District, Dak Lak Province, Viet Nam*. Hong Kong.

McElwee, P. 1999. 'Policies of Prejudice: Ethnicity and shifting cultivation in Viet Nam'. *Watershed*, 5(2): 33–8.

Ministry of Agriculture and Rural Development. 1997a. *Bao Cao Tom Tat: Noi dung Hoat dong cua Vuon Quoc Gia Yok Don trong nhung nam qua [Summary Report: Contents of Activities of Yok Don National Park in the Past Years]*. Dak Lak, Viet Nam.

—— 1997b. *Review Report on Planning, Organization and Management of Special Use Forest*. Cuc Phuong, Viet Nam.

Ministry of Forestry. 1993. *Management Plan: Cat Tien National Park*. Hanoi.

Mishra, H. 1994. 'South and Southeast Asia'. In J. A. McNeely, J. Harrison and P. Dingwall (eds). *Protecting Nature: Regional Reviews of Protected Areas*. Gland, Switzerland.

Neumann, R.P. 1998. *Imposing Wilderness: Struggles over Livelihood and Nature Preservation in Africa*. Berkeley.

Nguyen Duc Khang and Phung Tien Huy. 1995. 'Restoration of Forest Ecological System in Ba Vi National Park'. In *Proceedings of the National Conference on National Parks and Protected Areas of Viet Nam*. Hanoi.

Nguyen Nhu Phuong and S. Dembner. 1994. 'Improving the Lifestyles of People Living in or Near Protected Areas in Viet Nam'. *Unasylva*, 176.

Nguyen Tuan. 1999. 'Vu Quang, khu bao ton khong yen tinh [Vu Quang, a Nature Reserve That Is Not Calm]', *Lao Dong (Labour)*. Hanoi, 21 May.

Nguyen Tuan and Dang Ba Tien. 1999. 'Bao gio Viet Nam bao ve duoc dong vat hoang da [When Will Viet Nam Protect Wild Animals?]', *Lao Dong (Labour)*. Hanoi, 5 June.

Nhat Anh. 1999.'Song cung te giac [Living Together with the Rhinoceros]', *Lao Dong (Labour)*. Hanoi, 7 July.

Sobey, R.T. (ed.). 1997. *Biodiversity Value of Hoang Lien Mountains and Strategies for Conservation: Proceedings of a Seminar and Workshop, 7–9 Dec. 1997*. Sapa, Viet Nam.

Socialist Republic of Viet Nam. 1995. *Biodiversity Action Plan For Viet Nam*. Hanoi.

Soer, A. 1997. *Results of a Buffer Zone Workshop in Bach Ma, August 1997*. Hanoi.

Szaniawski, A. 1987. *Assistance to Strengthen Cuc Phuong National Park – Field document No. 4*. Hanoi.

Time Magazine. 1992. 'Journey into Viet Nam's Lost World'. 10 August.

Tordoff, A., Siurua, H. et al. 1997. *Ben En National Park*, London.

Tran Cong Khanh. 1995. 'Studies on Potential Medicinal Value of Plants of the National Parks and Protected Areas in Viet Nam Towards Their Sustainable Utilization and Conservation'. In *Proceedings of the National Conference on National Parks and Protected Areas of Viet Nam*. Hanoi.

Tran Lien Phong. 1995. 'The Role of Protected Areas in the Biodiversity Action Plan'. In *Proceedings of the National Conference on National Parks and Protected Areas of Viet Nam*. Hanoi.

United Nations Development Programme. 1996. *Creating Protected Areas for Resource Conservation Using Landscape Ecology*. Hanoi.

Viet Nam News Agency. 1998. 'Scientists Suggest Ways to Protect Natural Resources'. *Viet Nam News*. Hanoi, 3 April.

Vo Tri Chung. 2000. *Combining Local, National, and International Measures of Biodiversity Management in the Central Highlands: Implications for Delineation of Protected Areas in Viet Nam*. Hanoi.

Wege, D.C. et al. 1999. *Expanding the Protected Areas Network in Viet Nam for the 21st Century*. Hanoi.

Wikramanayake, E. and Vu Van Dung. 1997. *A Biological and Socio-Economic Survey of West Quang Nam Province, with Recommendations for a Nature Reserve*. Hanoi.

World Bank. 1997. *Staff Appraisal Report: Forest Protection and Rural Development Project*. Hanoi.

World Wildlife Fund for Nature. 1996. *Vietnam Nature Reserve Reveals Another 'New' Species*. Hanoi.

Xuan Quang. 1999. 'Rung Mai Chau 'chay mau' [The Forest of Mai Chau Are Bleeding]', *Lao Dong (Labour)*. Hanoi, 26 June.

Xuan Quang and Quang Thien. 1999. 'Kinh nghiem 'cho cui ve rung' [Experiences from Carrying Firewood to the Forest]', *Lao Dong (Labour)*. Hanoi, 12 August.

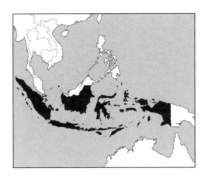

18

The History of Displacement and Forced Settlement in West Kalimantan, Indonesia

IMPLICATIONS FOR CO-MANAGING DANAU SENTARUM
WILDLIFE RESERVE[1]

Reed L. Wadley

Introduction

Co-management or community-based management is seen by many conservationists as a humane and practical alternative to eviction and punitive sanctions
of people who inhabit and use lands designated for conservation (Wells et al.
1992; Kemf 1993; Pimbert and Pretty 1995). In its ideal form, co-management
involves 'the active participation in management of a resource by the community of all individuals and groups having some connection with, or interest
in, that resource' (Claridge 1997b: 19), often involving some sort of economic
development component. Such 'integrated conservation and development projects' are not always easy solutions (Western and Wright 1994; Vandergeest
1996), but are today regarded as the standard approach to conservation.

Ever in demand, however, are new concepts, values, methods, and practices that enhance the abilities of area managers to work cooperatively with
local people and build on existing systems (Pimbert and Pretty 1995: 44). In
conjunction with studies aimed at understanding present local resource use,
attention needs to be given to settlement history in conservation areas as an
important component of the local condition. Such history, including settlement prior to conservation activities, is of great value in efforts to co-manage

protected resources. It shows how local people came to be where they are now, and how they view the local landscape and use their histories to lay claim to resources. It also indicates how area resources have been shaped by the people who have relied on them for centuries; and it provides insight into present and future conditions, such as how resource competition between area residents is structured and perceived (Moore 1998).

This chapter focuses on the history of population change in and around Danau Sentarum Wildlife Reserve (DSWR) in West Kalimantan, Indonesia, and its implications for co-management. After introducing the DSWR area and its peoples, I describe pre-conservation demographic history using a combination of oral and colonial accounts. This history shows frequent population displacement and forced resettlement under conditions of endemic, indigenous warfare and colonial efforts to suppress it. I also consider the effect of these conditions on the state of the DSWR environment, and then discuss local perceptions and constructions of local history. The intention is to highlight to DSWR managers overlooked aspects of local history and perceptions, and to provide an example of what may prove useful in other conservation areas.[2]

Danau Sentarum Wildlife Reserve

The DSWR core is a network of lakes and seasonally flooded forests, located about 700 km up the Kapuas River from the provincial capital, Pontianak (Figure 18.1). The lakes area (ca 80,000 ha) was gazetted as a wildlife reserve in the early 1980s because of its unique flora and fauna and its comparatively good condition (Giesen 1987).[3] Formal management began in 1992 with a conservation project funded jointly by the British and Indonesian governments. Its primary purpose was to develop a management plan for the reserve using co-management principles. Initial project efforts involved studying patterns of local resource use and collaborative work with local people (e.g. Colfer et al. 1996, 1999; Wickham, 1997).

Located within and around the reserve core are numerous communities from two different ethnic groups – Iban swidden farmers and Melayu fishers who rely to various degrees on its resources. The Muslim Melayu inhabit the reserve core while the nominally Christian Iban live largely in the surrounding hills.[4] Those who live within the reserve boundaries have been in a potentially problematic status because Indonesian conservation law does not allow for human habitation of wildlife reserves. If the law had been enforced in this case, a large number of people would have been evicted and alienated from their homes. In February 1999, however, the reserve was designated a national park allowing residents to remain within the conservation area. (Because the field research described here was conducted prior to this new designation, I will continue to refer to DSWR rather than DSNP.)

The DSWR Melayu are legal inhabitants of and typically have close relations with the larger towns along the upper Kapuas. These towns are now

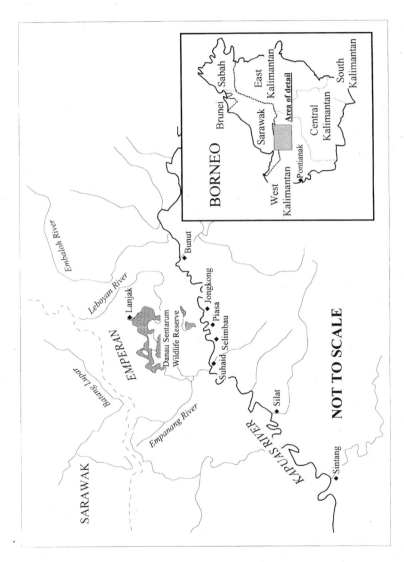

Figure 18.1 The Upper Kapuas River and Danau Sentarum Wildlife Reserve, West Kalimantan

district centres, but were once minor sultanates, generally under indirect Dutch colonial control from the mid-1800s until their abolition in 1916. People from these towns have traditionally come into the lakes area in large numbers during the dry season when fish are most accessible. An estimate of the (primarily Melayu) population in 1995 showed a permanent population within DSWR of 6575, with the population rising to 8055 individuals in the dry season. Permanent population growth within the Reserve, during the previous ten years, was about 41 per cent; population density was around 6.4/sq. km (Aglionby and Whiteman 1996). However, those Melayu who live within the reserve are strongly connected to the local landscape. They have developed complex systems of rights and responsibilities to the land and its resources within the lakes area (Harwell 1997), and they engage in local management practices on which the conservation project has tried to build (Claridge 1997a).

The Iban have resided in the surrounding hills for centuries, and have long-established systems of tenure and use rights, some that extend seasonally into the lakes area. They practise swidden cultivation in the lakes catchment area incorporating forest and other resource management. In contrast to the Melayu orientation toward the Kapuas, the Iban look to nearby Sarawak, Malaysia, where they have deep cultural, social and economic ties. Iban men regularly go to Sarawak and Brunei for better wage labour opportunities (Wadley 1997), and social interactions are routine between the interrelated communities along the international border. In the two border districts to the north of DSWR, the Iban population was around 4600 in 1995 (Wadley and Kuyah n.d.) with a density of around 2.5–3.0/sq. km.

The Settlement History of DSWR and Vicinity[5]

Any reference to settlement history in and around DSWR must deal with native orientations to the local landscape. For the Iban, there are two important areas. The first is the Batang Lupar, a principal river in Sarawak, whose upper course runs roughly parallel to the international border. The second is the stretch of low-lying hills and flats south of the international border and north and east of the lakes. The Iban call this the Emperan (literally, flat land). For the Melayu, the most important feature, besides the lakes themselves, is the Kapuas River which has served as both natural resource and means of communication for centuries.

Before the mid-nineteenth century, DSWR settlement history comes exclusively from local oral histories. As such there is much detail of movement that has been lost over the generations. However, what is clear from the abundant work done on Iban pre-history is that people ancestral to today's Iban (and Kantuk) originated in the middle Kapuas basin several centuries ago and migrated across the low-lying hills that today form the international border between Malaysia and Indonesia. Equally important but overlooked is that the geography of low hills along the border presents no physical barriers to travel (Pringle 1970: 210). What is more, the Iban have occupied the Batang Lupar river system in Sarawak for a very long time (Pringle 1970: 213–14). It is thus likely that they have lived in or at least made frequent use of the area north of the lakes for centuries. Even prior to actual settlement, Iban from the Batang Lupar would have been well aware of the Emperan.

Local accounts of pre-Iban settlement in the lakes area are contradictory, probably reflecting the movement of various peoples back and forth through the easily travelled region.[6] According to one account, when the Iban first migrated into the Emperan from the Batang Lupar,[7] they displaced peoples said to be related to the Embaloh. Other Iban claim that before the Iban moved into the area there were no people living in the region west of Lanjak. Still others say there were non-Iban (sometimes glossed as Embaloh) living west of Lanjak. Embaloh themselves generally agree that they occupied portions of the area prior to Iban entry.

By and large, the local Melayu originate from indigenous Dayaks[8] along the upper Kapuas who in the course of the last few centuries took on Islam as both a religion and an identity. They are most strongly associated with the small kingdoms that arose along the upper Kapuas even prior to the entry of Islam.[9] For these native kingdoms, the surrounding peoples fell into one of several statuses. There was one or another form of subject population with differing tax and tribute obligations, and there were the free peoples, like the Iban, who the rulers considered either enemies, allies, or potential subjects. Most Melayu communities in the DSWR area are derived from the population of subjects under most direct influence of these kingdoms.

Raiding and Trading

Although the entry of the Iban into the Kapuas drainage was often peaceful, warfare was the common theme in their relations with other groups (Pringle 1970: 253), and such relations with the Melayu sultanates along the upper Kapuas are recounted in early stories (Sandin 1967: 86). Oral accounts show a pattern of Iban attacking Kapuas settlements, Melayu relocating, and counter-attacking the Iban who would then flee the area (von Kessel 1850: 183; Enthoven 1903: 136, 159–62; Bouman 1952: 55–6, 72–4; Brooke 1990, II: 208). This pattern of fight-and-flight seems to have characterized life in the area. In 1861 Charles Brooke reported people in the upper Batang Lupar fleeing at a report of raiders. He wrote, 'The inhabitants of this river are much scattered, and there is little unity among them. They ... fly like a flock of sheep on hearing of an enemy. We passed many remains of villages' (1990, II: 191–92).

However, an emphasis on warfare would capture only part of the picture, as the Iban made regular alliances with Melayu rulers along the Kapuas and along the north coast (Bouman 1952: 52, 73–4; Pringle 1970: 61, 149). In addition, Iban genealogies are replete with honorific titles that Melayu rulers most likely bestowed on their Iban allies. Trade was likewise an important link between the Iban and the Melayu, especially since the

Emperan provides fairly easy access from the upper Kapuas to the north coast. Iban traded regularly with the Kapuas states for goods such as salt, tobacco, and cooking pots (von Kessel 1850: 198; Pringle 1970: 55). And Iban leaders settled in key places along the trade route between the Kapuas and the north coast (Niclou, 1887: 38; Sandin 1967: 7).

Early Colonial Contacts

This broad sweep of local history provides some impression of the population flux as raiding and revenge raiding pushed people out of the area, and as the desire for trade and farmland pulled people back. Early European accounts of the area give some further indications of this flux.

The first European to traverse the area in 1823 observed active fishing huts in the lakes and double-palisaded longhouses along the Kapuas. This Dutch official made no mention of any Iban inhabitants in the Emperan, but he did observe that the areas upriver from Sintang were better inhabited than below it.[10] This might have been the result of the upper Kapuas kingdoms pushing the Iban back to the Batang Lupar river in the 1820s (Enthoven 1903: 161), opening the lakes to fishing and the banks of the upper Kapuas to resettlement.

By 1852, another European described Iban settlements in the Emperan as lying 'only at rare intervals' from each other (Pfeiffer 1856: 73). A more telling sign of the impact Iban raiding might have had on human habitation along the Kapuas was the 'fewer inhabited places' along the upper Kapuas than along the Batang Lupar (Pfeiffer 1856: 75). In areas below Sintang and not subject to Iban raiding, '[t]he banks of the river were more or less inhabited' with 'many little villages' (1856: 82). Likewise, an island in the western lakes was seen in the early 1860s as being uninhabited (Hunnius 1863: 176; Kater 1883: 6), its people probably having fled for fear of Iban attack and cholera epidemics. Yet in 1867 Melayu were living there again, while Iban lived nearby (Hunnius, 1863: 176; Beccari 1904: 183, 186) where the Dutch had encouraged them to settle in the 1850s (Niclou 1887: 41).

According to reports from the 1850s, Iban were established in the Emperan when the Dutch first made contact with them. The Dutch, however, regarded them as newcomers, thinking they had only migrated to the area since the 1830s (von Kessel 1850: 198). While some Iban migrants to the Emperan were probably newcomers, it is likely that others were actually resettling land they or their close ancestors had farmed in the past. Local geography and the flux of warfare make this likely. One other important factor to the north was the establishment of the Brooke kingdom of Sarawak in the 1840s.[11] The first 'White Rajah' sought to establish a British presence on the north coast of Borneo and suppress local piracy (which consisted in part of joint Iban–Melayu coastal raiding). His efforts to do this using military force may have sparked some of the Iban migrations during this time (Pringle 1970: 78), although how this affected Iban movement into the Emperan is unknown.

Suppressing Headhunting

During their early presence in the 1850–60s, the Dutch called the Iban the terror of the Kapuas (Kater 1883: 4). And in the early 1870s, Dutch reports frequently summarized the situation in the area by saying that headhunting was the order of the day. Rumours continually circulated of planned Iban attacks on Kapuas settlements.[12] The rulers of the Kapuas kingdoms also planned pre-emptive attacks on the Iban which the Dutch forbade.[13] For both small and large raids, Iban warriors made frequent use of the lakes to attack into the Kapuas by boat,[14] and Kapuas Melayu obstructed passages with tree trunks to prevent Iban raiders from gaining access to the Kapuas (Teysmann 1875: 302).

During the 1870s and 1880s, continued Iban raiding and Dutch measures to stop it appear to have created a situation that promoted settlement insecurity. The Dutch began to take increasingly aggressive actions to stop the raiding. In troubles with border-dwelling Iban, Dutch policy coalesced to move the 'rebels' to more accessible areas near the lakes, and some Iban actually moved to the lakes under this plan. In Sarawak, Rajah Charles Brooke planned to move 'his' Iban away from the border as a solution to the problem of raiding into Dutch-held territory.[15] Military patrols began to attack rebellious Iban, confiscating freshly severed heads, burning long-houses and destroying rice fields. This led, on numerous occasions, to Emperan Iban hiding their valuables, abandoning and even torching their own longhouses, for fear of Dutch reprisals against headhunting.[16] The Dutch forces also cut down fruit trees,[17] which was intended to remove inherited and inheritable claims to the area.

Within the lakes, settlement insecurity appears to have continued. The Dutch built a small fort in the western lakes where their patrols were based. It was also designed to protect the Melayu settlement there, but during a large Iban attack on the Kantuk, the Melayu abandoned the area, despite Dutch patrol presence.[18] In neighbouring areas, displacement of long-established populations occurred as well. To the east, Embaloh abandoned their dwelling areas in the upper Leboyan after sustained attacks from Sarawak Iban until 1884 (Enthoven 1903: 58).[19] Reflecting the repeated raiding they endured, their clustered longhouses along the middle Leboyan were fortified with palisades (Enthoven 1903: 59). To the west, Kantuk began an exodus from the Empanang that accelerated after 1881.[20] Iban raids were partly responsible, but another reason was that the Dutch awarded the Empanang territory to Selimbau in 1881, and many Kantuk refused to be subjects of that Melayu kingdom (Enthoven 1903: 70–1).[21]

The constant raiding and headhunting by the upper Batang Lupar and Emperan Iban along the border peaked in 1886. A Sarawak government punitive expedition received Dutch permission to cross the border and deal with both groups. In the aftermath of what is known as the Kedang Expedition, some eighty longhouses on both sides of the border lay in ashes and about twenty people had been killed. On Dutch-controlled territory, there were forty-one destroyed longhouses and sixteen people killed.[22] The Dutch took advantage of the situation to resettle the Emperan Iban away from the border to designated places close to the lakes. They gave them the choice to move or have their longhouses burned down and rice fields slashed. Related Iban in Sarawak were also resettled away from the border.[23]

On both sides, Iban were forbidden from farming close to the border. Repeated attempts to farm there resulted in the slashing of fields, but by 1890–95, they were tacitly allowed back depending on the inclination of local Dutch officials (Enthoven 1903: 227). This ban might have directed some Iban migration from the Lanjak area into the lower Leboyan and eastern lakes (Molengraaff 1900: 120). By the late 1890s, the situation had calmed down enough for the Dutch to permanently withdraw their military forces from the area. But the Iban's headhunting reputation continued, causing panic in trading settlements whenever they showed up to sell forest products (Enthoven 1903: 233).

From the late 1890s until 1909, there was another period of unrest along the border involving Iban rebellions in Sarawak. The Emperan and the Kapuas were not directly affected, except for the increased insecurity, although regular Dutch patrols ensured safety. In an effort to prevent further problems, in 1917 the Dutch military resettled troublesome Iban in the upper Leboyan by force to the middle and lower Leboyan and to the lakes (Bouman 1952: 83).[24] Iban continued to encroach on Embaloh lands along the Leboyan and Embaloh rivers (Bouman 1952: 54; King 1976). In the 1930s, there was some movement back into the upper Leboyan by Iban previously resettled downriver.[25] By the 1930s, it appears that Melayu were settled permanently on the lower Leboyan (de Mol 1933–4: 85) and at other places, a process that accelerated after the 1960s.

Environmental Effects

Given the anecdotal evidence, it is hard to know exactly the environmental effects from this movement of people in and out of the lakes area. The effects were probably periodic as people fled the area in times of unrest and returned when it was safe. Thus, for example, the impact on the fisheries was probably beneficial as there would be some dry seasons when very few people from either the Kapuas or the Emperan fished. The already sparse human population would have been made periodically even more sparse, so reducing the impact on the local forest and aquatic environments.

In a few accounts, Dutch officials mention the local economy, but during the late nineteenth century, the troubles with Iban raiding overshadowed more mundane concerns. Officials did observe that the lakes were a source of fish, honey and wax for sale among the Melayu, and birds' eggs for home consumption among the Iban (Hunnius 1863: 176; Gerlach 1881: 294; Enthoven 1903: 150–51). Later on in the early 1900s, they mentioned the importance of lakes for crocodile hunting and fishing, mainly by Melayu from the Kapuas (Bouman 1952: 62–3). There is also mention of the entire

Melayu population in the area fishing in the lakes during the dry season, with the Dayaks being less involved.[26]

Regarding the surrounding forests, the Dutch noted areas of denuded forest in the lakes and attributed them to a combination of Melayu and Iban dry-season burning and prolonged drought (Gerlach 1881: 293; Enthoven 1903: 148–9; Molengraaff 1900: 88). In the 1890s, little old growth remained in the Emperan; most vegetation, except that on the steep slopes of mountains, consisted of sword-grass, shrubs, and secondary forest (Molengraaff 1900: 90, 101; Enthoven 1903: 227–8). Officials called Iban farming *roofbouw* (literally, plunder farming) because of its rapid depletion of forests and soil fertility (Bouman 1952: 58). They mentioned frequent Iban movement in search of fresh old forest to farm.[27] Besides early attempts to control farming along the border and subsequent (failed) efforts to encourage irrigated rice cultivation (Bouman, 1952: 59–60),[28] the Dutch also attempted to regulate Sarawak Iban migration into Dutch territory.[29]

If we consider that colonial officials may have been fairly accurate in describing the situation, their observations must still be placed in context. The Dutch were seeing Iban practices under conditions of intense instability brought on by Iban raiding itself, revenge attacks from the Kapuas, and colonial punitive expeditions from both sides of the border.[30] Furthermore, local population density is likely to have increased from colonial resettlement efforts. This may also partly explain the lack of old growth in the Emperan observed during the 1890s. From another angle, a Dutch official in West Borneo in the 1850s observed that because of swiddening, there appeared to be no *true* primary forest (von Gaffron 1858–9: 224).[31] Thus, while the Dutch decried forest destruction, it may well have been that much of the forest the Iban had been farming in the Emperan was in fact very old secondary growth, farmed and abandoned in the past.

However, there may be something of an institutional bias against swidden cultivation that coloured colonial observations. For example, not until the 1930s is there any mention of swamp rice cultivation,[32] even though the Iban had long made swamp swiddens and had been farming swampy lands in the eastern lakes areas since the 1890s. In addition, a geological survey in the 1890s mentioned the lack of older growth forest in Iban areas, but also provided photographs showing swiddens cut from at least older secondary forest (Molengraaff 1900: 88, 120). This suggests the officials missed a good deal of the local system's complexity. Most Emperan Iban today practise settled dry swidden agriculture in a long-fallow cycle with supplementary swamp rice swiddens on a short fallow. Recent GIS research shows something close to sustainability for the current system (Dennis et al. n.d.).

As testament to the reduced threat of raiding and the decreased need for continual Dutch attention to military matters, we begin to see more and more references to economic pursuits from the 1920s. For example, the Dutch built a salt storehouse in 1918 in the lakes because of the abundant fish harvest that dry season.[33] They set up beacons for navigating the lakes in the late 1920s and renovated them in 1934.[34] They began to set regulations controlling *tuba* fishing[35] in the lakes and rivers by Melayu and Dayaks in the late 1920s, and to look into improved fishing methods in the 1930s.[36] As another sign of stability, lumbering of *Fagraea* and *Shorea* trees in the lakes appears to have increased during this time.[37] Dutch officials also mention various other productive activities such as hunting for snake and monitor lizard in the lakes for their skins, hunting crocodile for its gall bladder by Melayu and Dayaks, bee cultivation for honey and wax by Melayu and some Dayaks (de Mol 1933–4: 85), and gold mining by Melayu in 1932.[38]

Local Perceptions of Local History

Equally important as the historical details of people's movement and resource use are local perceptions and constructions of that history. These reveal the ways people see themselves, their places in the local landscape, and their claims to resources there. Local oral history may be considered a particular form of indigenous knowledge, but one that is routinely manipulated (in the broad sense of the word) for contemporary interests and purposes. Outside views, such as that of the Dutch and Indonesian governments and conservationists, are also important to consider and have the most direct implications for attempts to co-manage local resources. Here the focus will be on the local level.

Looking at the Melayu and Iban, as principal stakeholders in DSWR and surrounding lands, we see that relations between these two groups are overtly peaceful now. Even in the past, individual relations were quite cordial and close ties of kinship have existed between some communities. However, the history I have outlined precludes smooth interaction on a more general level and contributes to a certain amount of distrust.[39] In large measure, it is that history of raiding and flight that continues to influence how people view each other. For example, some Melayu see the Iban as newcomers from the north much as the Dutch did. As the Iban expanded south into the lakes and east to the Leboyan and Embaloh rivers, the Melayu and the Embaloh came to consider them to be usurpers of their lands. For their part, the Iban view the Melayu as newcomers themselves who are increasingly encroaching on land where the Iban claim traditional rights of access.

Each group views itself as having a long-established presence and claims in the area, and both are correct. The Melayu's long connection to the fisheries in the lakes probably goes back for centuries, but they appear to have not been able to settle there permanently until the Dutch succeeded in stopping Iban raiding. The Iban's long habitation of the border hills and their seasonal use of the lakes is equally ancient, and they most likely viewed the sparse and seasonal settlement in the lakes as an invitation to settle there themselves, eventually coming into conflict with Melayu claims (Colfer et al. 1997). Differing notions of resource claims between the two groups exacerbate this (Harwell 1997: 10; 2000).

The underlying distrust is seen in contemporary explanations for historical phenomena. For example, in the 1920s when the Iban first began cultivating para rubber (*Hevea brasiliensis*), someone had a dream that rubber trees supernaturally ate rice from their swiddens; news of the dream spread widely, and in response, people began to fell their rubber trees (Freeman 1970: 268). A contemporary, local Iban explanation for the event is that the Melayu tricked the Iban smallholders, telling them that their rubber would eat the more valuable rice. This was done, some say, because the Melayu did not want to see the Iban get ahead of them economically.[40]

Despite the emphasis on past headhunting and raiding, there have been many important, peaceful interactions including the trade and alliances mentioned above, and knowledge of these can be used by area managers in building alliances and coalitions today. Other important Iban-Melayu connections may provide avenues to cooperation. For example, in local accounts of the Iban sacking of Selimbau (prior to Dutch entry into the area), the boy-heir to the throne is said to have been captured, and, as was the custom among the Iban with child war-captives, an Iban family adopted him. His Iban name was Minsut, and when he was an adult, the Melayu from Selimbau asked that he return to take the throne. They paid a ransom of two large ceramic jars filled with gold, and Minsut took the throne to become Pangeran Suma Raden Dra Abang Berita. Several decades ago the gold was reportedly stolen, and Iban and Melayu joined together to look for it.[41] The gold was never found, but an attentive manager can use such knowledge in bringing people together to help preserve and manage the 'gold' of DSWR.

Another connection is that of marriage, and there is fairly regular intermarriage between Iban and Melayu communities that lie near each other. For example, an Iban longhouse on the lower Leboyan River has close social and kin ties with an adjacent Melayu community, and the two communities share some use areas. Likewise, other Melayu can cite ancestors (often women) who were originally Iban. This is not to say that marriage is always free of tension. Indeed, many Iban regard the occasional marriage of Iban women to Melayu men as a threat to their ethnic viability. Marriage to the Muslim Melayu requires Iban to become Muslim themselves, a procedure that means a change in ethnic identity. Yet awareness of these connections and divisions can help the area manager in negotiating the variously cooperative and divisive local interests.

Conclusion

Based on the above overview of settlement history and the local perception of that history, several general points may be drawn:

- Both Iban and Melayu have a long presence in the vicinity of DSWR, and equally long use of its natural resources.
- Because of periodic abandonment of the lakes and Emperan in the past as raiding and counter-raiding occurred, the human impact on the local

environment (especially the fisheries and forests) has probably not been great until the colonial suppression of warfare and the onset of permanent habitation in the lakes.

- Both Iban and Melayu perceive the other to be the newcomer, intruding on long-established rights of use and access.
- Points of mutual interest exist within oral history, as do important social connections.

For conservation managers, these points can serve as beacons in navigating the currents of the local situation. First is the realization from both oral and documentary history that both groups are equally indigenous to the area, and that the local and colonial perception of one or the other group being new-comers is probably based on the conditions created by indigenous warfare (involving *both* Iban and Melayu as aggressors and victims). Second, the fairly good condition of the lakes and surrounding forests may be in part the result of that warfare, which until its suppression provided for periods of abandonment and regeneration. Third, the general attitude of distrust and suspicion between the two groups can be balanced with some attention to points of mutual interest, whether in oral histories or in genealogies.

Thus, in addition to and in conjunction with efforts to understand present local resource use, attention should be given to settlement history, and local perception and construction of that history, as a critical component of the local condition. Conservation managers will be able to identify events that shape local perception, whether negatively or positively, keeping in mind that contemporary concerns also shape the way history is perceived and the way it is used by local people (Brosius et al. 1998; Moore 1998). Knowing this history and the ways it is used and presented can help managers in understanding the local view of things. The manager who makes an effort to understand local history from a variety of perspectives will be in a better position to use that history in dealing with resource conflict by providing a tool in building co-management. In addition, local people will better respect and want to cooperate with a manager who has made some effort to understand their history and who can draw on that knowledge in trying to co-manage local resources.

It is important to underscore, however, that co-management is rife with politics, broadly defined. Given a true option to choose, local people might not elect to become involved in co-management systems that come from the outside. From their perspective, co-management may be another attempt by outsiders, with government support, to wrest control of local resources from local hands. Their choice to join co-management projects may be conditioned by the realization that such involvement may be the only practical way to secure long-term rights and claims. The conservation project may also be viewed as another resource to be used for local ends (Tsing 1999). Thus, within this context, the construction and use of history by locals will be geared toward certain practical goals, which may not coincide with conservation aims. The manager must deal with those potentially conflicting constructions of local history within a co-management framework for conservation.

Notes

1. This article is based in part on field research (1992–94) funded by the National Science Foundation (Grant BNS-9114652), Wenner-Gren Foundation for Anthropological Research, Sigma Xi, and Arizona State University, sponsored in Indonesia by the Balai Kajian Sejarah dan Nilai Tradisional, Departemen Pendidikan dan Kebudayaan. Additional field study (June 1996) was done with the Center for International Forestry Research (CIFOR), Wetlands International, and the Indonesian Directorate of Forest Protection and Nature Conservation. Archival research in the Netherlands was supported by the International Institute for Asian Studies in collaboration with CIFOR's 'Adaptive Co-Management of Forests' project. Thanks to Carol Colfer for her comments. The above agencies and individuals do not necessarily share the conclusions and opinions drawn here, for which the author alone is responsible.

2. Space does not allow consideration of the impact of DSWR itself on the indigenous peoples (see e.g. Harwell 2000).

3. Recently an expansion of the boundaries to the north and east of the reserve core has been sought. This would put DSWR at 197,000 ha.

4. To the west and east of them live the Kantuk and Embaloh, respectively, whose subsistence economies resemble that of the Iban (King 1976; Dove 1981).

5. For a more detailed account, see Wadley 2000.

6. On Iban expansion, see Sandin 1967; Morgan 1968; Pringle 1970.

7. Some oral accounts have ancestral Iban settling in the Emperan prior to migration to the Batang Lupar (Morgan 1968: 159).

8. Dayak is the general term for the non-Muslim, indigenous peoples of Borneo.

9. These upper Kapuas kingdoms were Sintang, Silat, Suhaid, Selimbau, Piasa, Jongkong, and Bunut (Enthoven 1903; Bouman 1924).

10. Register der Handelingen en Verrigtingen, C. Hartmann, 1823–1825. Department of Historical Documentation, Koninklijk Instituut voor Taal-, Land- en Volkenkunde, Leiden.

11. The British adventurer, James Brooke, began the dynasty that was continued by his successors, Charles and Vyner, into the 1940s.

12. Algemeen Rijksarchief (The Hague), Ministrie van Koloniën, Mailrapport 1872 No. 499, 1873 No. 50 and 483, 1874 No. 196 (hereafter, MR date/number).

13. MR 1873/553.

14. MR 1873/405.

15. MR 1871/649, 1873/206.

16. MR 1874/196 and 420, 1875/835, 1876/1000, 1877/43 and 65, 1878/360, 1879/285, 757, 771, and 788.

17. MR 1880/196.

18. MR 1877/43, 1878/219. Subsequent patrol posts were set up in the late 1870s throughout the Emperan to control raiding from Sarawak; in the early 1900s, various Iban longhouses were designated as patrol posts during border unrest.

19. MR 1886/689.

20. MR 1879/771.

21. The Dutch government recognized Selimbau claims over a great portion of the lakes area in 1880 (Enthoven 1903: 140, 145).

22. MR 1886/293 and 342; see Niclou 1887; Pringle 1970.

23. MR 1886/293, 342, and 364.

24. Algemeen Rijksarchief (The Hague), Memorie van Overgave by Burgemeestre, 1934, KIT No. 999, Bijlage: 'Onze Verhouding tot Sarawak en de Batang Loepar-Bevolking', p. 29 (hereafter, MvO, KIT/No.).

25. MvO, KIT/999, p. 2; MvO, KIT/999, Bijlage: 'Onze Verhouding', p. 29.

26. MvO by Scheuer 1932, KIT/997, p. 4, 25–26; MvO by Werkman 1930, KIT/995, p. 18.

27. MvO, KIT/995, p. 5.

28. MvO, KIT/997, p. 22.

29. MvO, KIT/995, p. 6. Efforts in Sarawak to control internal Iban migration included the creation in the early 1930s of the Lanjak-Entimau Protected Forest outlawing Iban from moving into the Batang Lupar headwaters. Because of this, the reserve, now called the Lanjak-Entimau Wildlife Sanctuary, was dubbed a 'political forest' (Pringle 1970: 280).

30. Brooke (1990 II: 160) quotes a local observer as saying, 'After a man has had his house plundered and burnt, and finds resistance hopeless, it is some time before he again recovers himself sufficiently to build another and better one.'

31. Von Gaffron claimed that the forests he saw were 130–50 years old, which would certainly be classified as old growth forest today. His point was, however, that given the widespread practice of swidden cultivation, they were not likely to be virgin forest, but rather very old succession forest.

32. MvO by Oberman 1938, MMK/265, p. 36.

33. MvO by James 1921, MMK/262, p. 12.

34. MvO, KIT/995, p. 2; MvO, KIT/999, p. 3.

35. This refers to catching fish using poisons from such plants as *Derris* spp.

36. MvO, KIT/995, p. 19; MvO, MMK/265, p. 64.

37. MvO, KIT/997, Bijlage C; MvO, KIT/995, p. 7.

38. MvO, KIT/995, p. 17 and 25; MvO, KIT/999, p. 15 and 17.

39. An incident involving fish poisoning by Iban in 1994 reinforced these inter-ethnic suspicions (Aglionby 1995b).

40. Dove (1996) suggests that the dream warned against an over-involvement in the market economy at the expense of the traditional, mixed subsistence economy.

41. Thanks to Emily Harwell for the details of Minsut's story.

References

Aglionby, J.C. 1995. 'The Issue of Poisoning in DSWR: A Stakeholder Analysis'. In J. C. Aglionby (ed.). *The Economics and Management of Natural Resources in Danau Sentarum Wildlife Reserve, West Kalimantan, Indonesia*, pp. 108–15. Bogor: Wetlands International/PHPA/ODA.

Aglionby, J.C. and Whiteman, A. 1996. *The Utilisation of Economic Data for Conservation Management Planning: A Case Study from Danau Sentarum Wildlife Reserve*. Pontianak: Conservation Project, Indonesia-UK Tropical Forest Management Programme.

Beccari, O. 1904. *Wanderings in the Great Forests of Borneo*. London: Archibald Constable.

Bouman, M.A. 1924. 'Ethnografische Aanteekeningen omtrent de Gouvernementslanden in de Boven-Kapoeas, Westerafdeeling van Borneo'. *Tijdschrift voor Indische Taal-, Land- en Volkenkunde*, 64: 173–95.

—— 1952. 'Gegevens uit Smitau en Boven-Kapoeas (1922)', *Adatrechtbundels*, 44: 47–86.

Brooke, C. 1990. *Ten Years in Sarawak*, 2 vols, reprint of 1866 edition. Singapore: Oxford University Press.

Brosius, J.P., Tsing, A.L. and Zerner, C. 1998. 'Representing Communities: Histories and Politics of Community-based Natural Resource Management'. *Society and Natural Resources*, 11: 157–68.

Claridge, G. 1997a. *Interim Report on DSWR Participatory Management Experience*. Bogor: Wetlands International.

—— 1997b. 'What is successful co-management?' In G. Claridge and B. O'Callaghan (eds). *Community Involvement in Wetland Management: Lessons from the Field*, pp. 19–24. Kuala Lumpur: Wetlands International.

Colfer, C.J.P., Wadley, R.L. and Widjanarti, E. 1996. 'Using Indigenous Organizations from West Kalimantan'. In P. Blunt and D. M. Warren (eds). *Indigenous*

Organizations and Development, pp. 228–38. London: Intermediate Technology Publications.

Colfer, C.J.P., Wadley, R.L., Harwell, E. and Prabhu, R. 1997. *Inter-generational Access to Resources: Developing Criteria and Indicators.* Working Paper No. 18 (August). Bogor: Center for International Forestry Research.

Colfer, C.J.P., Wadley, R.L. and Venkateswarlu, P. 1999. 'Understanding Local People's Use of Time: A Pre-condition for Good Co-management'. *Environmental Conservation,* 26: 41–52.

Dennis, R.A., Colfer, C.J.P. and Puntodewo, A. n.d. 'Forest Cover Change Analysis as a Proxy Sustainability Assessment Using Remote Sensing and GIS in Danau Sentarum Wildlife Reserve, West Kalimantan'. Unpublished ms.

Dove, M.R. 1981. 'Subsistence Strategies in Rain Forest Swidden Agriculture, Vol. I and II'. Ph.D. dissertation, Stanford University.

—— 1996. 'Rice-eating Rubber and People-eating Governments: Peasant versus State Critiques of Rubber Development in Colonial Indonesia'. *Ethnohistory,* 43: 33–63.

Enthoven, J.J.K. 1903. *Bijdragen tot de Geographie van Borneo's Wester-afdeeling.* Vol. I. Leiden: E. J. Brill.

Freeman, J.D. 1970. *Report on the Iban.* London: Athlone Press.

Gaffron, H. von, 1858–59. 'Nota omtrent den Getah-pertja-boom', *Natuurkundig Tijdschrift voor Nederlandsch Indië,* 16: 224–8.

Gerlach, L.W.C. 1881. 'Reis naar het Meergebied van den Kapoeas in Borneo's Westerafdeeling'. *Bijdragen tot de Taal-, Land- en Volkenkunde van Nederlandsch-Indië* 29: 285–327.

Giesen, W. 1987. *Danau Sentarum Wildlife Reserve: Inventory, Ecology and Management Guidelines.* Bogor: World Wildlife Fund and PHPA.

Harwell, E. 1997. *Law and Culture in Resource Management: An Analysis of Local Systems for Resource Management in the Danau Sentarum Wildlife Reserve, West Kalimantan, Indonesia.* Bogor: Wetlands International – Indonesia Programme.

—— 2000. 'The Un-Natural History of Culture: Ethnicity, Tradition and Territorial Conflicts in West Kalimantan, Indonesia, 1800–1997'. PhD dissertation, Yale University.

Hunnius, C.F.W. 1863. 'Beschrijving eener Reis naar de Boven-Kapoeas van den 24sten April tot den 4 Mei 1862'. *Geneeskundig Tijdschrift voor Nederlandsch Indië,* 10 (n.s. 5): 174–82.

Kater, C. 1883. 'Iets over de Batang Loepar Dajakhs in de "Westerafdeeling van Borneo"'. *De Indische Gids,* 5: 1–14.

Kemf, E. 1993. *The Law of the Mother: Protecting Indigenous Peoples in Protected Areas.* San Francisco: Sierra Club Books.

Kessel, O. von. 1850. 'Statistieke Aanteekeningen omtrent het Stroomgebied der Rivier Kapoeas (Westerafdeeling van Borneo)'. *Indisch Archief.* 1: 165–204.

King, V.T. 1976. 'Some Aspects of Iban-Maloh Contact in West Kalimantan'. *Indonesia* 21: 85–114.

de Mol, G. A. 1933–4. 'Inzameling van Was en Honig in het Merengebied van de Westerafdeeling van Borneo'. *Landbouw,* 9: 80–86.

Molengraaff, G. A. F. 1900. *Borneo-Expeditie: Geologische Verkenningstochten in Centraal-Borneo (1893–94).* Leiden: E. J. Brill.

Moore, D.S. 1998. 'Clear Waters and Muddied Histories: Environmental History and the Politics of Community in Zimbabwe's Eastern Highlands'. *Journal of Southern African Studies,* 24: 377–403.

Morgan, S. 1968. 'Iban Aggressive Expansion: Some Background Factors'. *Sarawak Museum Journal,* 16: 141–85.

Niclou, H.A.A. 1887. 'Batang-Loepars. – Verdelgings-oorlog. Europeesch-Dajakshe Sneltocht'. *Tijdschrift voor Nederlandsch Indië,* 1: 29–67.

Pfeiffer, I. 1856. *A Lady's Second Journey round the World.* New York: Harper and Brothers.

Pimbert, M.P. and Pretty, J.N. 1995. *Parks, People and Professionals: Putting 'participation' into Protected Area Management.* Gland: WWF, IIED, and UNRISD.

Pringle, R. 1970. *Rajahs and Rebels: The Ibans of Sarawak under Brooke Rule, 1841–1941.* London: Macmillan.

Sandin, B. 1967. *The Sea Dayaks of Borneo before White Rajah Rule.* London: Macmillan.

Teysmann, J.E. 1875. 'Verslag eener Botanische Reis naar de Westkust van Borneo'. *Natuurkundig Tijdschrift voor Nederlandsch Indie,* 35: 271–586.

Tsing, A.L. 1999. 'Becoming a Tribal Elder, and Other Green Development Fantasies'. In T.M. Li (ed.). *Transforming the Indonesian Uplands,* pp. 159–202. Amsterdam: Harwood Academic.

Vandergeest, P. 1996. 'Property Rights in Protected Areas: Obstacles to Community Involvement as a Solution in Thailand'. *Environmental Conservation,* 23: 259–68.

Wadley, R.L. 1997. 'Circular Labor Migration and Subsistence Agriculture: A Case of the Iban in West Kalimantan, Indonesia'. PhD dissertation, Arizona State University.

—— 2000. 'Warfare, Pacification, and Environment: Population Dynamics in the West Borneo Borderlands, 1823–1934'. *Moussons,* Vol. 1: 41–66.

Wadley, R.L. and Kuyah, F. n.d. 'Iban Communities in West Kalimantan'. In V. H. Sutlive (ed.) *An Encyclopedia of Iban Studies.* Kuching: The Tun Jugah Foundation, forthcoming.

Wells, M. and Brandon, K. with L. Hannah. 1992. *People and Parks: Linking Protected Area Management with Local Communities.* Washington DC: World Bank, WWF, and USAID.

Western, D. and Wright, R.M. (eds). 1994. *Natural Connections: Perspectives in Community-based Conservation.* Washington DC: Island Press.

Wickham T. 1997. 'Case Study No. 5: Community-based Participation in Wetland Conservation: Activities and Challenges of the Danau Sentarum Wildlife Reserve, West Kalimantan, Indonesia'. In G. Claridge and B. O'Callaghan (eds). *Community Involvement in Wetland Management: Lessons from the Field,* pp. 157–78. Kuala Lumpur: Wetlands International.

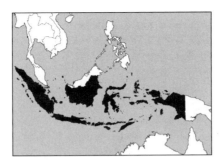

19

Planning for Community-based Management of Conservation Areas

INDIGENOUS FOREST MANAGEMENT AND CONSERVATION
OF BIODIVERSITY IN THE KAYAN MENTARANG NATIONAL
PARK, EAST KALIMANTAN, INDONESIA

Cristina Eghenter

Indigenous People and Conservation Areas

Biodiversity conservation, sustainable exploitation of natural resources, and rights of indigenous people are dominating NGOs and government agendas in several countries of Southeast Asia. The issues are not simply juxtaposed. A 'natural connection' (Western 1994) is assumed to exist between the interests of biodiversity conservation and those of indigenous people. Moreover, in the rhetoric of conservation organizations, it is often implied that sustainable use and conservation of biodiversity can be achieved insofar as the management of forests and conservation areas is granted to local people (van den Top and Persoon 1998).

Some critics, however, have noted that the current trend to entrust 'traditional' communities with the management of forest areas is still based on some 'unproven' assumptions about forest people (Wells 1994–5; van den Top and Persoon 1998; Eghenter and Sellato 1999). These pertain to the following aspects: the supposed privileged relationship of local people with nature; the static view of local people's future as an unconditional projection of their past; and the belief that local institutions (and local leaders) are endowed with a 'natural' capacity to manage a protected area in the interest of biodiversity conservation. Unless recognized and dealt with, these assumptions can ultimately undermine the efforts to effectively protect the rights of local communities and preserve biodiversity in conservation areas.

This chapter describes the experience of the Kayan Mentarang project of WWF Indonesia with regard to planning for community-based management in the Kayan Mentarang National Park, East Kalimantan, Indonesia. It shows the strategies adopted by the project to open a dialogue with local people and establish a common ground for collaborative management of the national park. Initial 'unproven' assumptions are highlighted as well as the challenges that exist when trying to negotiate between the priorities of nature conservation and the economic aspirations of local people.

This case study illustrates how difficult it is to mediate between local and external priorities, and between different ideas of conservation. It urges conservationists to ask local people first about their vision and hopes for their forest, and what they would like to conserve for the future, instead of imposing external views. Finally, this case study is presented as an example of how planning for community-based management, if conducted as an open and participatory process, may ultimately lead to results that may not be those anticipated by conservationists, but nevertheless must be accommodated and respected (cf. Headland 1998).

Local People as Partners in Management

Since the early 1980s, in conservation and protected areas management circles, there has been an increasing emphasis on the participation of local people in the management of conservation areas and the need to balance conservation priorities with the development needs of the communities living in and around the conservation area. Evidence from anthropological, human ecological, and archaeological studies had shown that most wilderness areas had been modified or managed by humans at one point in time, and that human induced disturbances had been part of the natural landscape as we know it (Sponsel et al. 1996; Headland 1997).

The recognition that anthropogenic influence was not necessarily incompatible with the conservation of natural resources had important implications for the theory and practice of protected areas management. It convinced conservation specialists of the value of the participation of resident communities in the protection and management of the environment. The new view drew strength from findings that showed that traditional activities of local communities in certain cases had played an important role in the conservation of specific environments, preventing soil erosion and the loss of biodiversity (Pimbert and Pretty 1994).

While it was becoming apparent that people could have an important role in the management of conservation areas, it was also clear that conservation priorities and communities' needs were not always compatible, and could give rise to conflicts in designated protected areas (Wells 1994–5). Local people were, for the most part, economically dependent on the natural resources of the conservation area. The regulation and limitation of their use for conservation purposes would require that alternative and compensatory means of livelihood be provided outside the protected areas. Moreover, recognition of the rightfulness of local peoples' claims to the land based on a long history of settlement in the protected area would discredit initiatives of strict enforcement of protection measures such as forced resettlement or prohibition to exploit natural resources.

In response to these challenges, conservation specialists started to plan new initiatives designed to involve local communities and link biodiversity conservation with the creation of economic incentives to promote sustainable use of natural resources. Integrated Conservation and Development Projects (ICDP) and community-based management exemplify the new approach. It is assumed that local communities have a greater interest and greater accountability in the sustainable management of resources over time than does the state or other distant stakeholders. It is believed that local people, precisely because of their long-term residence in the area, possess a wealth of knowledge about the natural environment and ecological processes, and that they are more able to effectively manage resources through local management strategies and traditional forms of tenure (Brosius et al. 1998). Participation of local people and adoption of local management practices are seen as essential for the achievement of a conservation programme that is biologically, socially, and economically sustainable.

In the Kayan Mentarang conservation area, WWF started a project in collaboration with the Directorate of Nature Protection and Conservation (PKA) of the Indonesian Ministry of Forestry and LIPI, the National Institute of Research, in 1991 (Eghenter and Sellato 1999). The project has been working to secure the communities' support and participation in the sustainable management of forest resources inside the conservation area. To achieve this goal, the project employed various strategies like research, community mapping, biological surveys, and participatory planning in order to set the conditions for collaboration and openness in the establishment of a joint management. The experience, however, shows that the agendas of local communities and conservation managers might, at times, diverge. The legitimacy of local claims to natural resources and the existence of active traditional institutions in the communities certainly provide a strong basis for the creation of a sustainable community-based management. The same factors, however, might also, depending on the social and economic circumstances, encourage local people to challenge the existence of a national park and the importance of biodiversity conservation.

The Kayan Mentarang National Park

Stretched along the mountainous interior of East Kalimantan, Indonesian Borneo, the Kayan Mentarang National Park lies at the border with Sarawak to the west and Sabah to the north. With 1.4 million hectares, it is the largest protected area of rainforest in Borneo and one of the largest in Southeast Asia. A strict nature reserve since 1980, the area was declared a National Park in October 1996 (see Figure 19.1).

Mixed dipterocarp forests of mainly *Shorea* spp. dominate about half the reserve which lies below 1000m, while at higher altitude we find predominantly plants of the fagaceae family. Moss forests rich in ferns grow on upper mountain slopes and the highest tops of sandstone hills that reach the elevation of 2500m. The forests of the interior are also prime habitat for many protected animal species, among others: the sun bear (*Helarctos malaynus*); five species of langurs or leaf monkeys (*Presbytis* spp.); pangolin (*Manis javanica*) and clouded leopard (*Neofelis nebulosa*), the largest feline in Borneo. The rhino (*Dicerorhinus sumatrensis*) was thought to be extinct in East Kalimantan by the 1950s as a result of hunting, but local people claim to have seen traces indicating its presence in remote parts of the park as recently as 1996.

Extensive archaeological remains in the form of stone burials occur in the reserve and date back from about 300 years ago. They are evidence that people have co-habited with the environment in this part of Borneo for centuries. As a result, forests have evolved under human and natural pressures. In some cases, permanent vegetation changes have taken place. For example, vast expanses of grasslands interrupt the endless forest coverage at the upper reaches of the Bahau River. According to a local tradition, people intentionally managed the grasslands and burned them to divert the attention of large mammals from the rice fields. At the same time, they also created hunting grounds for wild cattle (*Bos javanicus*), a sort of local game reserve.

About 15,000 Dayak people now live inside or in close proximity to the Kayan Mentarang National Park. Roughly half of these people, mostly Kenyah, a small number of Kayan, Saben and Punan, are primarily shifting cultivators. The rest, mostly Lun Dayeh and Lengilu in the north, are mainly wet-rice farmers. The inhabitants of the park and surrounding areas depend on hunting, fishing, and collecting wild plants for their subsistence needs. They also exploit forest products such as rattan, *gaharu* or aloeswood (*Aquilaria* spp), and bezoar stones for commercial trade.

The communities living in and around the park are still 'traditional communities,' largely regulated by *adat* or customary law in the conduct of their daily affairs and the management of natural resources. The role of traditional institutions, presently reflected in semi-formal institutions of the Indonesian state like the *lembaga adat,* or customary council, and the *kepala adat* or customary chief, is key to understanding the communities' views of rights and the way they deliberate on issues of forest management as well as social responsibilities.

Building a Common Understanding of Forest Uses: Research and Mapping

The long-term goal of the Kayan Mentarang project was the establishment of conservation management and sustainable economic development in the Kayan Mentarang National Park and surrounding areas. Not much information about local biodiversity and uses of the forest was available at the time the area was declared a protected area. The project staff conducted biological

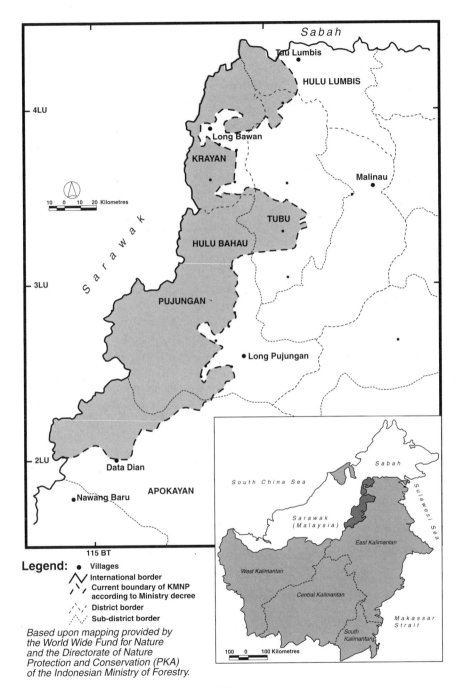

Figure 19.1 Kayan Mentarang National Park

surveys, traditional social science research, and community mapping to document human–environment interactions and collect data on traditional forms of forest management. The data could support both legal and ecological arguments for local people and institutions to take a leading role in forest management and the economic development of the area (Eghenter and Sellato 1999).

The Research Programme of 'Culture and Conservation'

Most of the initial field activities of the WWF Kayan Mentarang project were conducted under 'Culture and Conservation', an inter-disciplinary research programme supported by the Ford Foundation (1991–7). The programme had four main goals: to investigate local communities' knowledge of, and attitudes towards, their natural environment; to identify past and present interactions between people and the forest; to demonstrate the existence of practices of land tenure based on traditional legal systems; and to train Indonesian, particularly Dayak, researchers to carry out field studies. Research questions were framed in ways that allowed for empirical testing. What do local people know about forest resources? What has been the rate and frequency of exploitation of forest resources? Which area(s) did they exploit and under what circumstances? What rules do and did exist with regard to the exploitation of particular forest resources? Questions of this kind have the advantage of making concepts like 'traditional conservation' or 'traditional practices' intelligible by situating them in specific social, historical, and environmental context(s). They allow the collection of a wide range of data on the origin, determinants, and reinforcers of certain practices, and trace the causal influences of their changes and continuities. Moreover, the data thus obtained facilitate the design of a better management system which conforms to real local practices and needs, as opposed to those assumed by conservation managers and WWF staff.

The field data showed the strong co-incidence and co-dependence of people and forests in this part of the interior of Kalimantan. This required that conservation efforts be based on the recognition of the importance of the natural as well as the human components of the environment (Dove and Nugroho 1994). Accordingly, the project used the results of research activities as evidence to recommend a change of status for the conservation area from a *cagar alam* or strict nature reserve, as it had been gazetted in 1980, to a *taman nasional* or national park. According to the current regulations of the Indonesian government, in a nature reserve all human settlements are in principle excluded and human activities are by definition illegal. In a national park, instead, local people are allowed to reside and use natural resources in 'traditional use' zones where 'traditional' practices by local people and sustainable patterns of exploitation of forest resources are permitted.

An evaluation team sent by the Ministry of Forestry endorsed the WWF recommendation in 1994. The conservation area was finally designated as a national park with a decree of the Ministry of Forestry two years later, in October 1996.

Community Mapping and Local Resource Use

In the initial phase of 'Culture and Conservation', methods other than traditional research were used to explore aspects of local land tenure and resource management. A pilot project in community mapping took place in late 1992, focusing on 'the location and nature of forest-tenure boundaries ... and indigenous ways to organize and use space and how these might conflict with or support forest protection' (Sirait 1994: 411).

In Indonesia, community mapping has emerged as a 'counter-mapping' strategy to the authoritative mapping by government agencies in recent years (Peluso 1995). Official maps had been drawn on 'empty' charts with little or no consideration of already existing claims to the area, in particular those by local communities that had been living off that land for several generations. Maps of such imprecision often showed forested territories as wild and uninhabited, officially non-settled and therefore ready for exploitation. The boundaries of conservation areas as well as those of timber concessions were established by the Ministry of Forestry prior to any consultation with local communities living in or near the area.

Community maps produce information on how local people view and manage their territory and the resources within. Maps help communicate this information to outsiders and, by doing this, become a powerful medium of negotiation with government agencies over community forest access and use rights (Eghenter 1999).

After the pilot project, the community mapping programme of the Kayan Mentarang project grew into an independent programme intending to document local land use and management of natural resources throughout the entire area of the park. It included a strong training component targeting local people and local government. The transfer of expertise from the project staff to community representatives was expected to encourage the participation of local people in the management of the conservation area. Moreover, the mastery of mapping techniques could increase the ability of local people to control, manage, and monitor the information contained in the maps to prevent its misuse for the economic benefit of a few individuals and/or outside companies (Eghenter 1999).

In 1998, the WWF team completed training and mapping in all the communities of the national park area. Maps were initially compiled at village level. Subsequently, representatives of villages belonging to one *wilayah adat*, the customary land under the leadership of a customary chief, met and combined the village maps into customary land maps which are now used in various negotiations. It was expected that these maps would provide crucial evidence for the official inventory of customary lands, an initiative which was recently launched by the government and will de facto recognize the role of local, 'traditional' people in planning the development of the nation (see Figure 19.2).

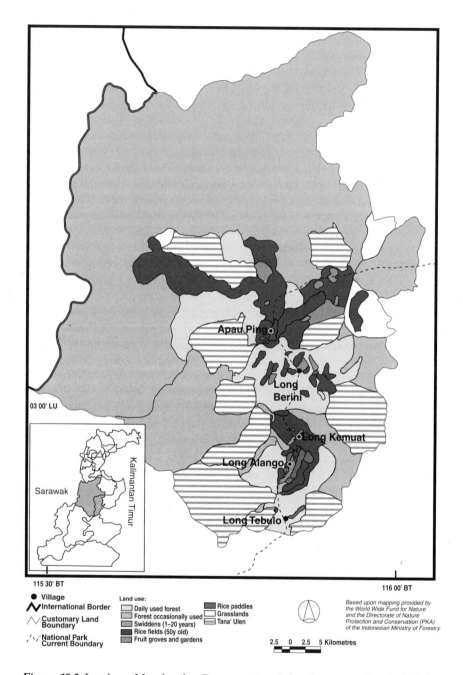

Figure 19.2 Land-use Map by the Communities of the Customary Land of Hulu Bahau

Is Traditional Forest Management the same as Biodiversity Conservation?

One assumption often made is that local people have traditionally managed the forest in the interest of biodiversity conservation and that they are, by definition, potential 'natural' managers of conservation areas. Conservation organizations like WWF have also fostered the idea that traditional management of resources is inherently conservative and local people are the custodians of well-adapted management systems (e.g., Sorensen and Morris 1997).

During the implementation of the 'Culture and Conservation' programme, several researchers had documented the existence of a form of forest tenure known as *tana ulen* among Kenyah communities along the Bahau and Pujungan rivers (Blajan 1999). WWF staff engaged in mapping exercises in the same communities highlighted the role of this tenure arrangement as common property regulated by customary law to the advantage of the entire community and the environment (Tirusel ST. Padan 1992). But had WWF discovered a model of traditional conservation of biodiversity or had they found what they wanted to prove?

Based on ethnographic and historical information, we know that *tana ulen* is *tana* or land which is *m/ulen* or restricted, prohibited because it is claimed by someone, individual, family, or community. As such, the land is off limits to outsiders and may not be accessed or exploited without permission of those who claim it (Eghenter 2000).

In the past, the area declared *tana ulen* was an expanse of primary forest at the upper reaches or along the entire course of a river, often a smaller tributary of a major river. It was a forest rich in natural resources, in particular forest products such as rattan (*Calamus* spp), *sang* leaves (*Licuala* sp.), hardwood for construction (e.g., *Dipterocarpus* spp, *Shorea* spp, *Quercus* sp.), fish and game, all of which had high use value for the local community. The area was strategically located near the village where management and control were easier. As a general rule, no forest could be cleared to open rice fields in *tana ulen*. The territory varied in size from 3000 hectares to up to 12,000 hectares, the size being largely dependent on the natural boundaries of the watershed of the *ulen* area (Eghenter 2000).

Tana ulen functioned as forest reserves managed by the aristocratic families of the community. Unlike other ethnic groups of central Borneo, Kenyah people are stratified in social classes that include: aristocrats (among whom the longhouse chief and customary chief were always chosen), lower aristocrats, commoners, and descendants of slaves. *Tana ulen* was 'opened' for exploitation only on certain occasions like celebrations, the arrival of guests, or ritual events of the lifecycle. These could be either communal celebrations or more private affairs. The decision on when to open a *tana ulen* rested with the aristocrats who held the privilege to manage the forest reserve. They also got undeniable personal benefits from it in the form of a percentage of all forest products collected (Simon Devung and Rudy 1998).

In addition to game, fish and minor forest products, valuable timber was also an important resource in *tana ulen*. The imposition of restricted access to areas with abundant resources was one way to ensure constant availability. Although we can only speculate on the reasons behind the establishment of *tana ulen*, it can be argued that the main conservation factor in such practices was the desire on the part of the aristocratic families to regulate the use of the forest to prevent depletion of resources needed in collective rituals and other family affairs.

As a result of the incorporation into the Indonesian state after 1945 and the conversion to Christianity in the 1950s and 1960s, the aristocratic class started to lose its exclusive grip on power in the communities of the interior. The change in status of *tana ulen* areas is one example. Responsibilities for the management of the forest reserves were transferred to the customary councils that now oversaw *tana ulen* forests on behalf of the entire community and according to customary law (Blajan 1999).

Nowadays, *tana ulen* is, in principle, off limits to all outsiders, including people from nearby villages. Collection of specific products may be done only on a collective basis. Collection takes place when there is need to obtain produce or money for village celebrations at Christmas and New Year, and special projects like the building of a community centre. Also, the market value of certain forest products affects decisions regarding collection. Collection of forest products is restricted by customary regulations with regard to times of collection (only every 2–3 years, for example), tools and methods employed (no chemical poison may be used to catch fish in the streams, for example), and quantity and kind of products harvested (no wild cattle may be hunted and only five wild pigs may be killed on one hunting expedition).

With little regard for the social and historical complexity of *tana ulen* tenure arrangements and their circumstances, the staff of the WWF Kayan Mentarang project interpreted *tana ulen* as a cogent example of conservation ethic among local communities and good evidence for justifying the adoption of indigenous practices in the management of the national park. In the language and rhetoric of WWF reports, *tana ulen* became forest communally protected and a refuge for wildlife and plant species (Sorensen and Morris 1997). *Tana ulen* was the old way communities had been continuously caring for nature and managing natural resources sustainably (Tirusel ST. Padan 1992). The WWF staff did not hesitate to hail the discovery of *tana ulen* as the model for community-based management of the conservation area. The historical antecedents and variations of current *tana ulen* practices were neglected and the role of aristocrats downplayed. Cross-comparative efforts to explore forest management in communities other than Kenyah ones in the National Park area were also limited.

Dove and Nugroho (1994) argue that whether *tana ulen* conserve for the benefit of the communities or conserve for the privilege of the ruling class of their former managers might not have any relevance with regard to the conservation goals of the WWF project. In other words, the social and historical aspects of *tana ulen* would not matter much as long as the main objective of protection of biodiversity pursued by the national park was secured. However, precisely the aspect of the environmental consequences of *tana ulen* practices had been strangely neglected by the WWF Kayan Mentarang project. Such was the enthusiasm raised by the belief that there was a perfect connection between traditional practices and biodiversity conservation, that an important issue was overlooked. This was the need to assess whether resource management practices, however constructed and/or interpreted by the various actors, do or do not 'conserve' natural resources, or, under what circumstances these practices become factors in promoting sustainable use of natural resources (Eghenter 2000).

Was the forest better protected under *tana ulen*? Under what conditions could regulations of *tana ulen* guarantee a sustainable exploitation of forest resources by local communities over a long term? For example, low population density (0.24/sq. km in the Pujungan sub-district of the interior of East Kalimantan), limited technology, and a discontinuous demand for forest products could have contributed to low levels of exploitation (hence sustainable) more than the strict enforcement of *tana ulen* regulations. Key issues of this kind were not addressed as the project prepared to design the management plan for the park. The mere existence of *tana ulen*, however sketchily documented, was deemed sufficient evidence that a local arrangement for the conservation of the forest was already in place, and that it was sustainable.

In addition to the neglect of environmental and ecological aspects, other important considerations on social aspects of *tana ulen* and their possible impact on local communities were ignored. Was the communal tenure arrangement of *tana ulen* socially sustainable? What would be the social implications of using this management model where the role and influence of the chief of aristocratic descent appear to have been so prominent and decisive? Didn't its application risk the legitimation of the power status quo in a community and the maintenance of social discrimination? Moreover, if traditional leaders were once the stewards of the forests, their power is often now wielded on behalf of modern developmental goals and personal economic and political interests. Some members of customary councils may prove to be principal catalysts for accelerated resource exploitation.

Since 1996, a new approach has been slowly emerging at the WWF project. It is recognized that *tana ulen* might have not been at any time a tenure arrangement for protecting biodiversity nor a perfect model of community-based management of forest resources. Amid contradictions and the legacy of old, unproven assumptions, the project agenda has been modified to reflect a new emphasis on working with the communities to evaluate and revise current customary regulations on the use of natural resources, and elaborate new ones if needed. While acknowledging the strong influence of traditional leaders in local affairs, the project works with them and other community representatives to devise flexible structures for the management of 'traditional use' zones in the national park. Biological surveys and participatory inventories of forest products are now being conducted in several parts of the conservation area to help support, from a biological point of view, the claim that forests managed and regulated by local communities can be a viable alternative for the national park.

On the one hand, the initial misinterpretation of *tana ulen* is the result of the 'a-historicization' and 'genericization' of local management practices (Brosius et al. 1998: 159). As the authors convincingly argue, it is precisely the lack of historical context and local perspective that have contributed to the failure of many local experiments of community-based management. On the other hand, the *tana ulen* case made clear the political relevance of this indigenous form of forest tenure for drawing the attention of the Ministry of Forestry and other government agencies on issues of customary rights (Eghenter 2000).

Subsequently, the project made specific efforts to secure the official recognition of *tana ulen* and grant exclusive rights over natural resources to local people in the conservation area. The efforts were motivated by the realization that local communities should not bear all the costs of conservation. It was a way to impose a form of benefit sharing between the park and its residents by which communities would enjoy permanent usufruct rights in return for better stewardship and sustainable use of forest resources in and around the conservation area.

Establishing the Bases for Community-based Management

The community maps gave a clear indication of the extent of the area inside the park which was still used by local people to harvest forest products or for agricultural purposes. This overlap of claims could trigger a potential conflict over the rights of local people to use the forest in the conservation area the way they had been doing for centuries. In order to ease management decisions, the project used community maps to run participatory zonation planning meetings with the communities at the end of 1998. The main purpose was to collect recommendations for the revision of the external boundaries of the park and the allocation of 'traditional use' zones inside the park.

The meetings took place in all the customary lands of the park. Influential individuals, traditional leaders, and other community representatives, women and men, took part. The participants debated their willingness to support the park and which part of their land they would agree to have inside the conservation area. They also examined management options with regard to the role of customary law in the management of the conservation area.

In most areas, communities recommended that fallow land, new rice fields, gardens, and forest near the settlement used on a daily basis be excluded from the national park. They also agreed that old secondary and primary forest be left inside the park provided that the management of the park be based on their *adat* regulations and the collection of economically valuable forest products be allowed. In other areas, the people suggested that the park area be extended to include all of their customary territory. In this, they were motivated by the concern to protect their forest from the illegal logging of nearby timber concessions (see Figure 19.3).

Figure 19.3 Recommendations for new Boundary of the Park by the Communities

During the meetings, the communities expressed their preoccupation that the government would not support community-based management of the park once WWF leaves the area. Their negative experience and repeated disappointment made it difficult for them to trust promises of management arrangement that had to do with the government. The name 'national park' did not make it any easier. It only heightened the suspicion that the forest, by being included in the park and defined as 'national', was being alienated from their control and appropriated by the central government based in the far away capital of Jakarta. All community leaders requested that *adat* or customary regulations be made the law within the park and the communities be put in charge of the management of the 'traditional use' zones. They also stressed that park rangers be local people.

Concerns and requests of this kind are particularly significant in the Indonesian context where participatory planning of conservation areas and involvement of local people still does not have a strong mandate. The rights of access and use of resources in the park area are still contingent upon the establishment of strict regulations that allow only 'traditional' collection of forest products for local consumption and require that residents be granted special permission from the park authorities. The communities regard these as unacceptable constraints to their economic activities that are for the most part dependent on the exploitation and sale of forest products like *gaharu* and bezoar stones. They also feel that local customary law and local leadership are not given due recognition and respect.

Combining Traditional Regulations with Principles of Conservation Management

The idea that the management of the conservation area could be built using local models of resource management practices remained an important consideration, but it needed a lot of work. The experience with *tana ulen* challenged the validity of quick schemes that intend to transpose indigenous forest management models in unqualified ways.

The results of the zonation meetings and the feedback from the communities helped plan the following phase in the design of collaborative management. The project staff analysed local customary regulations to gain a sense of the local view on conservation, and assess the extent to which regulations governing management of natural resources could be adequate measures with regard to conservation management. For example, a closer look reveals some interesting aspects. Regulations include both old restrictions as well as more recent ones that were adopted by the communities following changes in environmental conditions and the availability of certain resources. The principle of sustainability, for example, is expressed in the importance of not wasting animals or forest products by collecting more than needed or harvesting them in ways that would hamper their future reproduction or growth. Moreover, it is recognized that certain modes of exploitation that employ chemicals or sophisticated technology may be damaging in the long term. In recent deliberations, some customary councils have imposed temporary bans on the hunting of animal species that are perceived as threatened or vulnerable. Regulations also commonly state that trees at the headwaters of rivers may not be cut, suggesting basic criteria of watershed protection. These and other elements reveal local concerns over renewable supplies and the need to guarantee future availability of natural resources. While the regulations highlight a utilitarian sense of conservation (Western 1994), they are still clearly in agreement with conservation management principles. As such, they can be effectively integrated to form a strong and legitimate basis for community-based management of the national park.

The WWF staff subsequently compiled preliminary directives for the management of the 'traditional use' zones of the park. Their conceptual and practical justification drew upon the following considerations: recognize the exclusive rights of communities and guarantee their permanent usufruct rights; adopt customary regulations where these stress conservation and sustainable use; allow the imposition of customary fines for most infractions; share management responsibilities; take into consideration the local economic situation and levels of exploitation; introduce and refine tools such as quotas and seasonal harvesting, and other measures of animal population management, only when conditions require. The draft is now being discussed in local meetings with the communities. The feedback from the first few rounds of discussions was positive. The communities felt that their concerns were being addressed and community-based management of the park was becoming more meaningful and real to them.

When Plans for the Forest Might Diverge: Community Rights vs Nature Interests

Many challenges to the realization of a collaborative management of the Kayan Mentarang National Park remain. In large part, they reflect the emergence of differing priorities between some conservation managers and local communities. In particular, managers who still assume that the economic and social circumstances of local people are a projection of the past, and future development needs an extension of present ones, are likely to be surprised at the eroding support for conservation areas on the part of local people (cf. Persoon and Van Est 1998).

The Case of 'Return' Migrations

Participatory methods like community mapping and participatory planning, which were used by the Kayan Mentarang project, can produce an outcome that may be used for purposes other than conservation management. In principle, this is an indication of a successful implementation of the activities and transfer of skills that are now being used on behalf of the specific interests of the participants (Eghenter 1999). But what if these needs may in the long term conflict with the objectives of the conservation area? In Kayan Mentarang, the notion of an ancestral land that the community maps helped 'prove' poses new challenges to the integrity of the conservation area and a powerful dilemma to its managers. Two different groups recently asked that all of their customary lands be excluded from the national park. They intend to claim back the land from where their relatives left for Malaysia decades ago, and might intend to return in the future. Similarly to the case of the Pataxo Indians of Brazil who demand the restitution of their ancestral land which is now a national park (*The New York Times*, 1 December 1999), the leadership of the two groups claim the control of the forest in the Kayan Mentarang National Park for their economic and social development goals. Some individuals have plans to clear the forest to establish oil palm plantations and build roads that would connect their villages to Malaysia.

The extent and degree of these 'return' migrations are difficult to determine, but the current period of social and political reform in Indonesia further encourages such initiatives prompted by the legitimate demand for the acknowledgment of indigenous land claims. Management solutions like the creation of enclaves are under examination by the Directorate General of Nature Protection and Conservation (PKA) in order to mediate between nature interests and the needs of the people. On the one hand, it is true that the conservation area would not survive plans for large plantation schemes or logging operations inside the enclaves. On the other hand, a veto to return would be equivalent to a virtual displacement of those communities.

The prospect of 'return' migrations to areas of the park is complicated by the fact that many of these areas are claimed by several 'local' groups. This, in part, is the result of voluntary migrations that have taken place over the centuries. As such, the legitimacy of the claims of those who want to return might be challenged by some local customary chiefs who would argue that people can no longer lay claims over the area since they abandoned it voluntarily. The 'forced' relocation by the government, as part of its plan to resettle and 'regroup' scattered and isolated villages in the most distant territories in the 1970s and early 1980s, has further resulted in blurred and disputed boundaries of ancestral domains. Moreover, it is difficult to define who is local or how close to the national park a community should be considered local (Wild and Mutebi 1996) and therefore enjoy the right to exploit resources in the park. The situation is likely to further increase divisiveness and trigger land conflicts in the communities.

Local Managers of the Forest

The recent strides towards democratization and political openness in Indonesia will make it possible finally for people to obtain long overdue recognition of their customary rights and be in control of their lands. In the area of the national park, community mapping and the close collaboration between WWF and local customary councils over the last several years have raised the social and political profile of the councils and increased the visibility of some of their most active leaders. This has brought about a renewed confidence in the ability of local institutions to manage the forest on their own. In some cases, communities do not fight the idea of a conservation area and the importance of protecting forest resources for future generations, but challenge the need for an external, government institution like a national park to do that. They claim they can do it on their own as they have been doing for centuries before. After all, in their view, the forest is still there as irrefutable evidence of good local management.

Concluding Remarks

The case of the WWF Indonesia Kayan Mentarang project is an illustration of the challenges encountered when planning for community-based management of conservation areas. The experience confirms that attention to history and local perspectives can help make the right policy decisions. This might imply harder work than just 'discovering' something indigenous and using it, but the outcome might be a more sustainable, and long-term, management arrangement for the conservation area (cf. Ingerson 1997).

It is important to avoid making generic assumptions with regard to traditional management practices and conservation ethic. While we cannot assume that local people are conservationists, we cannot, equally, make the opposite assumption that local people are not interested in conservation. Open, collaborative processes like participatory planning can provide important opportunities for discussing and deciding on the present and future options for local forests.

Maybe, in the end, the Kayan Mentarang National Park will be smaller in size and the managers will be local people. Such an outcome would actually indicate that a real connection has been established between local and external conservation interests (cf. Western 1994).

References

Blajan, K. 1999. 'Jaringan Pemasaran Gaharu, Pengelolaan Hutan, dan Dampak Sosiologis, Ekonomis, dan Ekologisnya di Kawasan Sungai Bahau'. In C. Eghenter and B. Sellato (eds). *Kebudayaan dan Pelestarian Alam. Penelitian Interdisipliner di Pedalaman Kalimantan,* pp. 181–200. Ford Foundation and WWF Indonesia.

Brosius, P., Lowenhaupt Tsing, A.L. and Zerner, C. 1998. 'Representing Communities: Histories and politics of community-based natural resource management'. *Society and Natural Resources,* 11: 157–68.

Dove, M.R. and Nugroho, T. 1994. 'Review of *Culture and Conservation* 1991–1994: A Sub-project funded by the Ford Foundation, World Wide Fund for Nature, Kayan Mentarang Nature Reserve Project in Kalimantan, Indonesia'. WWF Kayan Mentarang, MS.

Eghenter, C. 1999. *Mapping Peoples' Forests: The role of community mapping in planning community-based management of conservation areas. A review of three projects of WWF Indonesia,* Biodiversity Support Program, Washington DC.

—— 2000. 'What Is Tana Ulen Good For? Considerations on Indigenous Forest Management, Conservation, and Research in the Interior of Indonesian Borneo'. *Human Ecology,* 28(3): 331–57.

Eghenter, C. and Sellato, B. (eds). 1999. *Kebudayaan dan Pelestarian Alam. Penelitian Interdisipliner di Pedalaman Kalimantan.* Ford Foundation and WWF Indonesia.

Headland, T. 1997. 'Revisionism in Ecological Anthropology'. *Current Anthropology*, 38(4): 605–30.

—— 1998. 'Managing the Natural Resources of the Sierra Madre: What is the role of the Agta?' IIAS/NIAS workshop on Co-management of Natural Resources in Asia: A Comparative Perspective. Philippines, 16–18 September.

Ingerson, A.E. 1997. 'Comment on T. Headland's "Revisionism in Ecological Anthropology"'. *Current Anthropology*, 38(4): 615–16.

Persoon, G. and Van Est, D. 1998. 'Co-Management of Natural Resources: The concept and aspects of implementation'. IIAS/NIAS workshop on Co-management of Natural Resources in Asia: A Comparative Perspective. Philippines, 16–18 September.

Pimbert, M. and Pretty, J. 1994. 'Participation, People and the Management of National Parks and Protected Areas: Past failures and future promise'. United Nations Research Institute, Social Development and International Institute for Environment and Development, and World Wide Fund for Nature, MS.

Peluso, N. 1995. 'Whose Woods Are These? Counter-mapping forest territories in Kalimantan, Indonesia'. *Antipode*, 27(4): 383–406.

Simon Devung, G. and Rudy, A.K. 1998. *Sistem Pemilikan Tanah Tradisional Pada Masyarakat Adat. Di Kawasan Taman Nasional Kayan Mentarang*. Pusat Kebudayaan dan Alam Kalimantan dan WWF Kayan Mentarang, Samarinda.

Sirait, M. 1994. 'Mapping Customary Land in East Kalimantan, Indonesia: A tool for forest management'. *Ambio*, 23(7): 411–17.

Sorensen, W.K. and Morris, B. (eds). 1997. *The Peoples and Plants of Kayan Mentarang*. Jakarta: Unesco.

Sponsel, L., Headland, T. and Bailey, R. 1996. *Tropical Deforestation: The Human Dimension*. New York: Columbia University Press.

The New York Times, 1 December 1999, 'Indian Tribe Wants Brazil's Plymouth Rock Back'.

Tirusel ST. Padan, 1992. 'Tanah Ulen (Tanah Adat) pada masyarakat Dayak di desa Long Uli, Kecamatan Pujungan, Kabubaten Bulungan, Kalimantan Timur'. WWF Kayan Mentarang, MS.

van den Top, G. and Persoon, G. 1998. 'Dissolving State Responsibilities for Forests: The cooperation between international and local actors in Northeast Luzon'. IIAS/NIAS workshop on Co-management of Natural Resources in Asia: A Comparative Perspective. Philippines, 16–18 September.

Wells, M. 1994–5. 'Biodiversity Conservation and Local Peoples' Development Aspirations: New priorities for the 1990s'. *Rural Development Forestry Network Paper* 18a: 1–24.

Western, D. 1994. 'Linking Conservation and Community Aspirations'. In D. Western and R.M. Wright (eds). *Natural Connections*, pp. 499–511. Island Press.

Wild, R.G. and Mutebi, J. 1996. 'Conservation through Community Use of Plant Resources: Establishing collaborative management in Bwindi Impenetrable and Mgahinga Gorilla National Park, Uganda'. *People and Plants Working Paper* no 5: 1–45.

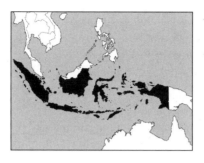

20

Resettlement and Natural Resources in Halmahera, Indonesia[1]

Christopher R. Duncan

Introduction

Generally, when people write their donation cheques to conservation organizations they think about helping endangered species or saving the tropical forests depicted in nature documentaries. The effect of conservation upon the people who live in these forests and depend on them for their survival is rarely mentioned in fundraising circles. The glossy brochures sent out by conservation NGOs do not contain images of people being moved to resettlement sites against their will; yet, this is often the result. Although these conservation groups may have no specific agenda for using force to protect biodiversity,[2] their support of governments that either lack the capacity to properly manage resources or that intend to control 'national' resources at any price, contributes to the disenfranchisement of indigenous people with resource claims (Contreras 1992; Peluso 1993). Some conservationists have even begun to openly advocate the removal of people from parks (MacKinnon 1994; Kramer and van Schaik 1997), stepping away from past attempts to incorporate indigenous people (see Brown and Wyckoff-Baird 1992; Kemf 1993).

In Southeast Asia, governments have often appropriated the ideology of conservation and biodiversity interests to advance their policies for the sedentarization, resettlement and incorporation of ethnic minorities (McElwee and Duncan 1999). By creating parks and resettling local people outside of these areas into state-established villages, many countries in Southeast Asia are continuing a long history of efforts to control people through resettlement and sedentarization (Duncan n.d.b). In this way, parks are not only about biological resources, but also about creating citizenries. Environmentalism is merely giving these programmes a rosier face and new justifications. Throughout Southeast Asia, as successive governments adopt new conservation strategies and map out new protected areas, local people (often peripheral minorities) are displaced.

In Indonesia, the government has long tried to resettle forest-dwelling populations throughout the archipelago. In the 1960s and 1970s, threats to national security justified these programmes; a fear of 'communist' sympathizers in the hinterlands served as the rationale for removing these groups from their land and placing them in resettlement sites. At the new millennium, the government's desire to resettle peripheral minorities remains, but the justification for it has changed to one of environmental protection; now indigenous populations need to be removed from forests because of the threat they pose to the environment and biodiversity. The politics of these programmes are further complicated by the fact that logging and plantation crops constitute a large percentage of the country's GNP, especially on the Outer Islands.[3] In this chapter, I examine how government resettlement programmes, extraction of natural resources, and wildlife conservation interact on the island of Halmahera in the province of North Maluku in eastern Indonesia. In particular I look at how they affect the Forest Tobelo who live in the island's interior.

The Forest Tobelo, often referred to with the derogatory term 'Tugutil', are the Tobelo-speaking forest-dwellers of Halmahera. Reliable figures concerning their population are not available and their exact distribution has not been mapped; however, an estimate of 3000 can be put forward with some credibility (Duncan 1998: 48–54). In the 1990s, many of the Forest Tobelo, particularly those living on the northeastern peninsula of the island (the focus of this chapter), still followed a shifting residence pattern in the island's interior, spread out across numerous river valleys. They subsisted largely on hunting pig and deer, processing sago, and foraging for other foods. They occasionally planted small swiddens with fruit trees, palms, bananas and cassava, but these were of secondary importance to their hunting and sago production. By the 1990s many Forest Tobelo communities had settled on or near the coast as a result of missionization and planted coconut and cacao groves; however, there remain numerous groups living in the interior.

A History of Forest Tobelo Resettlement

In the last thirty-five years, the Forest Tobelo have been the focus of numerous resettlement efforts by various government and non-government agencies with only limited success. Many of these programmes are instituted with the interests of other constituents in mind, and often the idea is to remove the Forest Tobelo in order to open up areas of Halmahera for natural resource exploitation. Success has been limited because the very nature of the project – to turn Forest Tobelo into villagers, to assimilate them into coastal communities

– runs counter to their wishes. The Forest Tobelo define themselves in opposition to coastal villagers and have no desire to move into their communities. None of this is taken into account in planning the resettlement sites; in fact, government bureaucrats have little knowledge about, or interest in, Forest Tobelo realities. But government planners do have an interest in removing them from the forest so it can be exploited, logged or mined more easily.

In Central Halmahera, the government at all levels has a long and well-documented history of attempting to sedentarize these mobile populations.[4] In 1952, the village head of Lolobata, whose wife was of Forest Tobelo descent, convinced eighty Forest Tobelo to create a settlement called Para-Para next to the village of Dodaga, and seventy-five individuals to settle next to Lolobata (Martodirdjo et al. 1985: 214). Residents stayed on until two months after the general election of 1955, when they all returned to the forest as a result of conflicts with villagers (Martodirdjo et al. 1985: 214). The Department of Social Affairs made the next attempt in 1964 in conjunction with the Indonesian takeover of West Papua. Central Halmahera was considered part of the Province of the Struggle for the Return of West Irian with a capital in Tidore off the coast of Halmahera, and as the Indonesian government turned its attention to developing its new citizens in Papua, this also included a focus on the Forest Tobelo. A Department of Social Affairs field officer sent to the Wasile district managed to persuade a number of Forest Tobelo to settle on the coast and a few families even converted to Christianity (Martodirdjo et al. 1985: 215). However, when Jayapura became the capital of Irian Jaya attention to Central Halmahera waned.[5] By the beginning of the 1970s all but a handful of Forest Tobelo had returned to the forest to live.

In 1978, the Department of Social Affairs implemented the Programme for the Development of the Social Prosperity of Isolated Communities *(Pembinaan Kesejahteraan Sosial Masyarakat Terasing,* hereafter referred to as PKSMT) in Wasile District near the village of Dodaga (Direktorat Bina Masyarakat Terasing 1993: 40). This programme sought to raise the level of 'civilization' and 'social welfare' of those groups deemed 'backwards' and to incorporate them into the larger nation state. The term *'masyarakat terasing'* is frequently translated as 'isolated society', or 'isolated community'. The government defined them as 'societal groups that live, or move nomadically, in geographically remote and isolated places, and who are socio-culturally isolated and still backwards compared with Indonesian society in general' (Departemen Sosial RI 1994).[6] Once groups were chosen to be the object of this programme they became subject to (often forced) resettlement in more manageable locations. In the process the government tried to intervene into virtually every aspect of their lives, changing settlement patterns, agricultural techniques, religion, and even trying to change eating habits (see Duncan n.d.a).

In Wasile, the original plan for the PKSMT settlement located it three kilometres from the village of Dodaga, but it was eventually built right next to the village on land already owned by villagers.[7] A number of coconut groves were cleared to make room for the new houses, a factor that would later plague the success of the settlement. The government offered those villagers who suffered financial losses an opportunity to have one of the new houses, but they all refused. The project went well for the first three months and fifty Forest Tobelo families moved into the new houses. Although they took advantage of the help offered to them and exploited the resources that they received, a large number of them continued to return to the forest for periods of time. At the end of three months a number of people became ill and died, including the local healer and his wife. The Forest Tobelo thought these deaths were the result of sorcery aimed at them by the villagers who had not been compensated for their lost land. In response, virtually all of the people in the project returned to the forest. In the beginning of 1981, the village head of Dodaga started allowing villagers to move into the vacant homes. The settlement suffered a further setback in 1982 when the government began transmigration projects in the region. One of the transmigration sites was placed between the resettlement compound and the Forest Tobelo who still lived in the forest, making it more difficult for the two groups to maintain contact. When I visited Dodaga in 1994, only a few Forest Tobelo families still lived in the PKSMT houses. Most of them were empty or had been occupied by other members of the community. Other attempts to resettle the Forest Tobelo in Wasile as part of the PKSMT programme have also taken place since 1983 with varying levels of success. All of these projects have been turned over to the regional government for administration.[8]

In 1993 the Tunggal Agathis Wood Industries Company (a subsidiary of Barito Pacific) built a resettlement site called Tanjung Lili for the Forest Tobelo of the Lili, Waisango, and Afu rivers in Maba District. Barito built this settlement as part of the Department of Forestry's Forest Concession Rights Holders' Forest Village Development Programme (*Hak Pengusahaan Hutan Bina Desa*, hereafter referred to as HBD). The Department of Forestry runs this programme to foster the development of those groups located within, or next to, a logging concession. As part of their legal agreement with the government, the timber company takes responsibility for development efforts and decides what programmes to implement. The government essentially sub-contracts governance to private timber companies. If the concessionaires fail to fulfil their HBD obligations they risk losing their concession; however, these penalties are rarely enforced. In essence the programme justifies removing people from forests so they can be cut down.

The HBD programme has two main stated goals: 1) to increase the 'welfare' of communities in and around forest concessions that timber companies are exploiting, and 2) to ensure the sustainability of forest resources (Planning Bureau 1993: 15). One aim is to improve the farming methods of the target population in order to increase productivity and to ensure sustainability; in other words, to turn swiddeners and foragers into sedentary farmers who will not compete with the timber concessionaire. According to the government, groups are chosen for these development projects because:

> Generally people in and around the forest live in dispersed, isolated locations that are difficult to reach. The fertility of their land is relatively low and continues to decrease. Their education, knowledge and skills are very low and can

not support their way of life. They do not easily accept change and their existence is extremely dependent on nature. They are easily satisfied and [as a
result] their activities are less than productive. In general they rely on swidden
agriculture and they clear forest, but they are not ready for agricultural sugges-
tions and acquire them with difficulty. (Direktorat Jenderal Reboisasi dan Rehabilitasi Lahan 1993: 6)

The HBD programme also seeks to change the settlement patterns of the
people involved because 'the existing settlements of the intended target beneficiaries are mostly developed following the local natural features and hence
not well arranged' (Planning Bureau 1993: 29).

In the case of Tanjung Lili, the HBD resettlement project was a relative success; there are currently sixty-two households of Forest Tobelo living in Tanjung Lili, with coconut plantations nearby and their children attending school
on a regular basis. The local timber company takes credit for this success, but
the presence and success of Western missionaries from the New Tribes Mission, who arrived in 1982, is the main reason the settlement has succeeded,
not the efforts of the timber company (see Duncan 1997; 1998: 218–61).

Although the Forest Tobelo of Tanjung Lili were the beneficiaries of an
HBD programme when the Barito Pacific concession was centred at Tanjung
Lili, other Forest Tobelo populations on the island have not received equal
attention, if any at all. When Barito Pacific had its base on the coast near
Miaf, the company did not provide any development assistance or compensation to the Forest Tobelo who remained in the interior, many of whom suffered the loss of fruit trees and swiddens as a result of the timber company's
activities; the same is true for other concessions. There are several possible
explanations for the difference in Barito Pacific's approaches. One is that the
presence of Western missionaries and pressure from them led the timber
company to create the settlement at Tanjung Lili. However, the most likely
explanation is that the government had demarcated the land behind Miaf
and around the mouth of the Akelamo River as transmigration areas for landless farmers from Java. Any move to create HBD settlements in these areas
would have conflicted with transmigration plans.

Exploiting Halmahera's Natural Resources

For most of the twentieth century, government and industry planners largely
ignored Halmahera; however, this changed in the 1980s when timber companies began to move in, and eventually the entire island was divided up
between several major firms. The Department of Forestry did not consider the
presence of the Forest Tobelo in the interior when concessions were handed
out and their land use patterns were ignored. The only concern that timber
companies had with the Forest Tobelo was the threat they were believed to
pose to the forest company employees, as the Forest Tobelo were considered
violent savages (Duncan 2001). When the first surveyors from a logging company arrived near the Waisango and Lili rivers, they did so with an armed

escort of thirty police officers to protect them. They quickly realized that the Forest Tobelo did not pose a threat and proceeded to cut down the forest at a phenomenal rate, far exceeding the boundaries and quotas granted them by their concessions, and completely ignoring local property claims.[9]

In Halmahera, after timber companies had exhausted the potential of a concession, government-sponsored transmigration schemes often followed. Before the programme was cancelled in 2000, transmigration represented the most ominous threat that the Indonesian state posed to the forest and the Forest Tobelo on Halmahera. The Indonesian transmigration programme moved landless peasants and other impoverished rural and urban populations from the overcrowded islands of Java and Bali to the more sparsely populated parts of the archipelago. The government claimed that the programme served to relieve the demographic stress on Java and Bali, and help incorporate the Outer Islands into the archipelago. In Maluku, transmigration projects began in the mid-1950s, but increased considerably after 1980, focusing on the larger islands (Goss 1992). I interviewed one official at the provincial office of the Department of Transmigration who expressed his hope that Maluku would eventually look like Lampung in southern Sumatra where numerous highways cris-cross the province, connecting the many transmigration sites to each other and to cities.

There are eighteen transmigration sites on Halmahera with a total of 28,616 transmigrants, and there were plans for increasing the transmigrant population to 68,616, or almost 25 per cent of the island's population (KWPM 1995: Lampiran 2). On the northeastern peninsula of Halmahera there are currently three transmigration sites. According to a member of a Department of Transmigration survey team who I interviewed, transmigration was to extend from the tip of the peninsula all the way down to the district capital of Buli, a solid line of new villages connected by the planned Trans-Halmahera Highway with a projected population of 13,500 people (KWPM 1995: Lampiran 4). However, the transmigration programme was one casualty of the fall of former President Suharto's government in 1998 and the new government has announced that it will cease the large-scale shipping of transmigrants throughout Indonesia. Furthermore, the outbreak of religious and ethnic violence on Halmahera in 1999–2000 resulted in the removal of thousands of transmigrants against their will.

In addition to the former threat of transmigration, the establishment of a timber plantation (*Hutan Tanaman Industri*, hereafter referred to as HTI) behind the village of Miaf is of particular interest to a discussion of resettlement and conservation in Central Halmahera.[10] This plantation was opened in 1982 and has expanded to include 23,500 hectares (Bubandt 1998a: 212). Whereas regular timber logging operations remove selected trees and then move on, timber plantations require the clear-cutting of vast areas to be replanted with a single species of tree (a process that many officials refer to as reforestation). The destructive tendencies of timber plantations are well known to the Forest Tobelo, and many people at the HBD of Tanjung Lili expressed dismay that the regular logging company was moving on (the same logging company that they blamed for the destruction of the forest and of some of their gardens). They feared the arrival of an HTI would place a freeze on all garden expansion, ensuring that many of them would become landless labourers working for the plantation. Ironically, government officials pursued the establishment of a timber plantation in the region to provide a source of income for the Forest Tobelo and to sedentarize them (Bubandt 1998a: 222). However, this plan to 'improve' the living conditions of the Forest Tobelo and to 'civilize' them by using them as labour on plantations would provide them with less security and less income than the continued expansion of their coconut and cacao groves, but it would provide the plantations with an inexpensive source of labour.

Having weathered these threats from the industrial arm of the government, the Forest Tobelo now face a new one, which while sounding benign, could have a similar affect on their ability to utilize the forest. This new threat is the proposed Lalobata-Aketayawe National Park that covers a significant portion of the land where many still live and forage.

Conservation Comes to Halmahera

In 1996 the Indonesian government began considering a proposal to demarcate approximately 350,000 hectares in the interior of Halmahera as the Lalobata-Aketayawe National Park (Suherdie et al. 1995: 4).[11] The main organization behind the establishment of this reserve has been Birdlife International with funding from the World Bank. Birdlife's interest in the region stems from their larger project to document and protect the Endemic Bird Areas of the world. In this project Birdlife International uses birds as indicators of the overall biodiversity for a region. As a result they have named 221 Endemic Bird Areas in the world as 'biodiversity hot spots' (ICBP 1992). An Endemic Bird Area (EBA) is defined as an 'area which encompasses the overlapping breeding ranges of restricted range species such that the complete range of two or more restricted range species are entirely included within the boundary of the EBA' (Stattersfield et al. 1998: 20). A restricted range bird species is 'a land bird which is judged to have had a breeding range of less than 50,000 km² throughout historical times (since 1800)' (Stattersfield et al. 1998: 20).

The North Maluku region is home to forty-three restricted range bird species, thirty-eight of which live on the island of Halmahera (Stattersfield et al. 1998: 535). Furthermore, Halmahera is home to four endemic species (the Halmahera Cuckoo-Shrike *Coracina parvula*, the flightless Drummer Rail *Habroptila wallacii*, the Sombre Kingfisher *Todirhamphus funebris* and the Dusky-brown Oriole *Oriolus phaeochromus*). Three of the restricted range species found on Halmahera are threatened: the White Cockatoo *Cacatua alba* and the Chattering Lory *Lorius garrulus* are threatened by the bird trade, and the Moluccan Scrubfowl *Megapodius wallacei* is threatened by hunting, egg-collecting and forest destruction. Halmahera also represents the only place outside of New Guinea (and its associated islands) where birds-of-paradise can be found. As a result the area is considered an important one in terms of bird conservation and the park has been proposed primarily to protect endemic bird species in the region. In preparation for the establishment of the park, the Directorate of Forest Protection and Nature Conservation (PHPA) has opened up new offices in the region, including one in Buli in Central Halmahera. PHPA staff, with support from Birdlife International, have also started monitoring the trade in lories and cockatoos (Birdlife International Indonesia Programme 1997).

The proposed Lalobata-Aketayawe National Park is one component of the larger World Bank Maluku Conservation and Natural Resources project that includes Manusela National Park on Seram, and two proposed marine parks, Aru and Banda National Parks, as well as several other areas (World Bank 1998). Proposals for the protected area on Halmahera suggest the area be demarcated in various ways. Two proposals (one from the National Conservation Plan for Indonesia (UNDP/FAO 1981) and one from PHPA) have the park divided into two separate areas; the northern part of the park would be the Lolobata Game Reserve (*Suaka Margasatwa Lolobata*), and the southern half would be the Aketayawe Nature Reserve (*Cagar Alam Aketajawe*) (Suherdie et al. 1995). The Birdlife International proposal connects the two parks into one protected area but does not specify under what IUCN category the park would fall (Suherdie et al. 1995).

The IUCN category of the protected area will affect the Forest Tobelo in various ways. The Indonesian government defines a 'nature reserve' as an IUCN Category I area 'in which no management or human interference with the environment is permitted' (Sumardja et al. 1984: 214). A 'game reserve' is defined as an IUCN Category IV area 'in which the natural balance of the environment must not be disturbed but low levels of management, visitor use, and utilization are permitted' (Sumardja et al. 1984: 214). Under both of these designations the growing of food crops, the growing of tree crops, hunting and fishing are prohibited (Sumardja et al. 1984: 221). If a park were established along either of these lines, the Forest Tobelo would no longer be allowed to utilize their land or forest resources.

In 1996, I visited the PHPA office in Ternate to inquire about the impact of the park on the Forest Tobelo who currently live within its proposed boundaries. The PHPA staff informed me that all people would have to move out of the park and would be prohibited from hunting within its borders due to the need to 'protect biodiversity.' I asked if they would still be allowed to hunt pig and deer (both introduced species). They explained that all wildlife was of equal importance in this new protected area and no hunting would be permitted.

In Ambon, the former provincial capital, I checked this information with a member of the NGO that was working on the park and he said that the PHPA staff were incorrect in their statements. However, he did add that the NGO 'had been advised' not to include the Forest Tobelo in any of their management plans. The government did not want people living inside the proposed park, just as they did not want people living in the forest prior to the discovery of its 'importance for biodiversity'. Although government officials may want to protect the forest as a natural environment, they do not want to preserve it as a place of refuge for the Forest Tobelo. As a result the NGO had, as of 1996, made no concessions for them in the management plan, for they feared it might interfere with the government's approval of the park.

The above statements are ironic when one considers that the World Bank is putting forth the idea of the park as an Integrated Conservation and Development Project (ICDP). As an ICDP, the park would 'involve local stakeholders in the design and management of protected areas' and 'promote local stakeholders' involvement in natural resource conservation while reducing poverty in remote rural areas' (World Bank 1998). The exclusion of the Forest Tobelo from the park, should this occur, would be contradictory to these goals and result in increasing poverty in rural remote communities who will lose access to their traditional lands. However, whether or not the Forest Tobelo will be recognized as 'stakeholders' by the government organizations involved remains to be seen.

As the park is being set up primarily for the protection of birds, it is worth examining how the Forest Tobelo utilize birds in their diet. They consider every bird species, with the exception of various owls, as a viable food source. In the past it was taboo to eat the Common Scrubfowl *Megapodius freycinet* or its eggs. However, as several Forest Tobelo groups have converted to Christianity they have abandoned that taboo, and the bird has become a major food source. The Moluccan Scrubfowl *Megapodius wallacei* is also hunted for its flesh and its eggs are gathered. This species usually buries its eggs on the beach and they are easily spotted in passing. These two species are also caught through the use of snare traps. Other birds that move about on the forest floor, such as the Nicobar Pigeon *Caloenas nicobarica*, the Blue-breasted Pitta *Pitta erythrogaster* and the endemic Ivory-breasted Pitta *Pitta maxima* are occasionally caught with these same snares. High-flying birds, such as Blyth's Hornbill *Rhyticeros plicatus ruficollis*, and various pigeon species, are highly sought after but are rarely caught due to the limited range of the Forest Tobelo's bird hunting spear guns. Only a few individuals have acquired air rifles that make it easier to shoot these birds. In addition, male children spend a large amount of time developing their hunting skills by hunting birds with small spear guns and slingshots. These efforts are usually aimed at catching small songbirds (which are eaten as snacks), primarily the Yellow-bellied Sunbird *Nectarina jugularis frenata*, the Black Sunbird *Nectarina sericea auriceps*, the Dusky Myzomela *Myzomela obscura*, the Metallic Starling *Aplonis metallica* and the Moluccan Starling *Aplonis mysolensis*. Despite this plethora of strategies for catching birds, the majority of protein in Forest Tobelo diets is from wild pig and deer and from river and marine fish.

The capture and sale of parrots for the caged bird trade has become a major source of income for Forest Tobelo who still live in the interior and have not planted extensive coconut groves. This trade focuses on the White Cockatoo and the Chattering Lory, both endemic and threatened species (Lambert 1993). The most popular species for trade is the Chattering Lory, and these birds have largely disappeared from settled areas. In 1996, these birds were captured and sold to villagers who then sold them in Tobelo, or to gold miners who transported them back to their villages elsewhere in Indonesia to sell for a profit. A small trade is also underway with Philippine fishing boats that frequent the area and which land to trade fish and liquor for food and birds. The White Cockatoo and the Red-cheeked Parrot *Geoffroyus geoffroius* are also sold in smaller numbers. Moluccan Red Lories *Eos squamata* are captured and sold as well, often as 'female' Chattering Lories to unsuspecting buyers, as they do not command a high price otherwise. An additional source of cash income is available through the sale of swift nests; however, Forest Tobelo rarely search for these and during my period of fieldwork only one party went looking for them (with no success).

The involvement of the Forest Tobelo in the bird trade increased only after logging began in the region. Prior to the arrival of timber companies their primary source of income was the sale of canari nuts. Individuals in the forest had access to a number of trees over an area and could either collect the nuts and deliver them to a village for sale, or merely sell the rights of harvest to an interested villager. Unfortunately, the timber industry also prizes these trees for making plywood and is reluctant to leave any standing. They proceeded to cut down most of the canari trees despite agreeing to honour Forest Tobelo property claims.[12] The loss of canari nuts as a major source of income has led many Forest Tobelo in Central Halmahera to shift to the trapping and selling of birds.

If the proposed park is established, and more importantly, if the regulations are enforced by the understaffed and under-funded PHPA, the effects on the Forest Tobelo communities still living in the area of the park could be significant.[13] Most lowland areas suitable for agriculture have already been claimed by coastal villagers or been consumed by vast transmigration projects (many of which expropriated their land from the Forest Tobelo). Currently the Forest Tobelo remaining in the interior have been unable to hold on to their land in the face of encroachment by the government or neighbouring villagers. Those who have resettled on the coast and tried to register their land with the government have been told that in order to get deeds for their property they need to have a government-issued identity card, but to get that they need a birth certificate. However, since most of the Forest Tobelo were born prior to having significant contact with the local government, none of them have one of these, and to obtain these they have been told they need an identity card. Thus they are caught in a Catch-22 seemingly designed to prevent them from registering their land. Assuming they are evicted from within the park boundaries, and their swiddens located in the park are destroyed, or planted over, they will be forced to become agricultural labourers in the gardens of coastal villagers and transmigrants, or labourers on timber plantations.

PHPA officials hope that the establishment of the park will lead to an influx of eco-tourists flocking to Halmahera to see the endemic bird species, particularly the Wallace's Standard-wing *Semioptera wallacei* (one of the birds-of-paradise). At least one US-based birding tour operator had plans to visit the reserve, and there was a steady trickle of hard-core birdwatchers into Halmahera prior to the outbreak of violence in 1999. Already at the village of Labi-labi some minor developments have been made and a small house has been built to accommodate birdwatchers. All of this development must be seen in light of the eco-tourism that briefly began in the Sidangoli region of Halmahera. Birdwatchers went there to see the Wallace's Standard-wing and other endemics, but the area has now been heavily logged and many of these species can no longer be found there. The role the Forest Tobelo would play in this eco-tourism remains uncertain. It seems most likely that, with their limited grasp of the Indonesian language, and their lack of contacts in Ternate and other urban centres where these tours would most likely be arranged, they would miss out on any benefits to be gained.

Conclusion

The Indonesian government's new concern with conserving the forest and biodiversity of Halmahera must be examined against a backdrop of twenty years of heavy logging, expansion of transmigration, mining, and plantation development in the region. The demarcation of a new 'biodiversity' park has come after all of the flat lowland areas have been utilized for transmigration, after the timber concessions have exploited the valuable timber from the region, and after mining surveys have been undertaken and mining projects begun in several locales. It would seem that only now that the former options are all exhausted, has the Indonesian government turned to conservation (and the hoped for benefits of eco-tourism and an influx of conservation funds) as the future for Halmahera. Furthermore, plans for establishing an IUCN category I nature reserve further legitimates the government's ever-present desire to remove forest-dwelling populations from the interior of the island, a programme it has been slowly pursuing over the last half-century.

The idea that the few Forest Tobelo still living in the interior of the north-eastern peninsula of Halmahera pose a threat to the island's remaining forest seems rather ludicrous when compared to the government's previous plans for the island of logging and transmigration. The changing rhetoric for resettling the Forest Tobelo merely exemplifies the point that, regardless of the reasons, the Indonesian government wants ethnic minorities settled where it can control them, and where they do not pose a threat to Indonesian notions of stability and order. This, not conservation, remains the paramount problem for the government. The problem of controlling peripheral citizens may also expand in the future, given recent elections and increasing cries for separatism and independence in other parts of the archipelago. It remains to be seen what effect these will have on the Forest Tobelo of Halmahera.

Epilogue

Since this paper was originally presented in the fall of 1999, Birdlife International's plans for the park have been discarded (Paul Jepson, pers. comm.).

The proposed park plans were superseded by other regional agendas for mining, timber plantations, and transmigration. Currently discussions are underway to attempt to demarcate a much smaller park. However, these discussions must cope with new legislation on regional autonomy that make districts responsible for a large portion of their own budget. These new laws encourage district level governments to generate income through the exploitation of natural resources, as they will receive a certain percentage of the revenues. An additional obstacle to establishing the park has been the religious and ethnic violence that has plagued the region since August of 1999 and has continued through the beginning of 2001. Tens of thousands of refugees have fled the island for neighbouring Sulawesi, Ternate and Tidore, and thousands more have been internally displaced within Halmahera. A number of Forest Tobelo communities that had resettled on the coast have returned to the interior to wait until the violence subsides. This violence has brought all government projects to a virtual standstill.

Notes

1. The fieldwork upon which this chapter is based was undertaken in 1995–6 and was funded with a grant from the Yale Council for International and Areas Studies under the auspices of The Indonesian Institute of Sciences (LIPI) and the Universitas Pattimura in Ambon. I also wish to thank Pamela McElwee, Larry Lohman and the other conference participants for their comments on the original version of this paper.

2. Some conservationists have openly advocated the use of military force to protect nature reserves, see Kramer and van Schaik 1997: 224 and Terborgh 1999: 201.

3. In 1996 income derived from forest-related industries constituted approximately 10 per cent of Indonesia's GDP (Barber and Schweithelm 2000: 2).

4. The Indonesian government is not the first to try and resettle the Forest Tobelo; the Dutch colonial government also made such efforts (Miete 1933).

5. The province of Maluku was divided into the two provinces of Maluku and North Maluku in 1999.

6. In 1999 the Department of Social Affairs was dissolved and the Body for Coordinating National Social Prosperity (*Badan Kesejahteraan Sosial Nasional*) replaced it at the national level. Furthermore, the programme for the Development of Social Prosperity of Isolated Tribes was changed to the programme for the Development of Geographically Isolated Adat Communities (*Pembinaan Kesejahteraan Sosial Komunitas Adat Terpencil*). For more on this programme and recent changes to it see Duncan n.d.a.

7. This section is primarily based on information taken from Martodirdjo 1984, see also Huliselan 1980.

8. Forest Tobelo populations elsewhere on Halmahera, primarily North Halmahera, have been subjected to the PKSMT programme as well (see Duncan 1998: 250–55; Safwan 1995).

9. In Central Halmahera, logging companies remove 200,000 cubic metres of wood annually (KSKHT 1992: 82 cited in Bubandt 1998a: 220). However, this total represents the 'official' statistics reported to the government, and the timber companies often exceed their government quotas. The Forest Tobelo at Tanjung Lili had seen numerous instances where timber company employees resorted to destroying logs in advance of 'surprise' inspections by government officials (which were almost always announced beforehand).

10. For an in-depth discussion of the political economy of timber plantations in Halmahera see Bubandt 1998a and 1998b.

11. The name Lalobata is a misspelling of 'Lolobata', a village located on the west coast of the peninsula. This typographical error first appeared in the National Conservation Plan (UNDP/FAO 1981) and has persisted (MacKinnon et al. 1995).

12. Ironically, in 1996 the timber company offered to provide the Forest Tobelo of Tanjung Lili with canari seedlings in an effort to increase their income. The Forest Tobelo were promised one hundred seedlings each, but eventually only received five per household.

13. The PHPA has numerous problems in enforcing its mandate, as one review states:

> Although PHPA is the primary agency responsible for managing Indonesia's protected areas, it faces a daunting series of constraints and limitations in carrying out its mandate, including general lack of stature within its own Ministry, lack of support from and cooperation with other government agencies and ministries, inadequate capacity and ability in monitoring and evaluations of protected areas, insufficient funding, an undermotivated staff which is also insufficient in numbers and in training ... (British Council 1996 as cited in Wells et al. 1999: 57).

References

Barber, C.V. and Schweithelm, J. 2000. *Forest Fires and Forest Policy in Indonesia's Era of Crisis and Reform*. Washington DC: World Resources Institute.

Birdlife International Indonesia Programme. 1997. Halmahera Report. http://www.kt.rim.or.jp/~birdinfo/indonesia/halma.html.

British Council. 1996. *Indonesia: ADB Institutional Strengthening for Biodiversity Conservation Study*. Manila: Asian Development Bank.

Brown, M. and Wyckoff-Baird, B. 1992. *Designing Integrated Development and Conservation Projects*. Washington DC: Biodiversity Support Program.

Bubandt, N. O. 1998a. 'The Race for Resources: Timber estates, transmigration and the political economy of natural resources in Halmahera'. In S. Pannell and F. von Benda-Beckmann (eds). *Old World Places, New World Problems: Exploring Resource Management Issues in Eastern Indonesia*, pp. 205–42. Canberra: Centre for Resource and Environmental Studies, The Australian National University.

—— 1998b. 'Profile of a Modern Utopia: Timber plantations as exemplary centers of Indonesian modernity'. *Canberra Anthropology*. 21: 28–59.

Contreras, Antonio P. 1992. 'The Political Economy of State Environmentalism: The hidden agenda and its implications on transnational development in the Philippines'. *Capital/Nature/Society*, 2: 66–85.

Departemen Sosial RI. 1994. Keputusan Menteri Sosial Republik Indonesia, Nomor 5/HUK/1994. Jakarta: Departemen Sosial RI.

Direktorat Bina Masyarakat Terasing. 1993. Data dan Informasi Pembinaan Masyarakat Terasing: Tahun 1993. Jakarta: Direktorat Bina Masyarakat Terasing, Direktorat Jenderal Kesejahteraan Sosial, Departemen Sosial.

Direktorat Jenderal Reboisasi dan Rehabilitasi Lahan, Departemen Kehutanan. 1993. Program HPH Bina Desa Hutan untuk Meningkatkan Kesejahteraan Masyarakat Sekitar Hutan. *Kehutanan Indonesia*, 3: 6–10.

Duncan, C.R. 1997. 'Social Change and the Reformulation of Identity among the Forest Tobelo of Halmahera Tengah'. *Cakalele*, 8: 79–90.

—— 1998. 'Ethnic Identity, Christian Conversion and Resettlement among the Forest Tobelo of Northeastern Halmahera, Indonesia'. PhD dissertation, Yale University.

—— 2001. 'Savage Imagery: (Mis)representations of the Forest Tobelo of Indonesia'. *The Asia Pacific Journal of Anthropology* 2(1): 45–62.

—— n.d.a 'Implementing Development Among Isolated Societies: Indonesian Government policies for incorporating peripheral minorities'. In C. Duncan (ed.). *Legislating Modernity: Southeast Asian Government Programs for Developing Minority Ethnic Groups.*

Duncan, C.R. (ed.). n.d.b (forthcoming) *Legislating Modernity: Southeast Asian Government Programs for Developing Minority Ethnic Groups.*

Goss, J. 1992. 'Transmigration in Maluku: Notes on Present Conditions and Future Prospects'. *Cakalele*, 3: 87–98.

Huliselan, M. 1980. 'Masalah Pemukiman Kembali Suku Bangsa Tugutil de Kecamatan Wasilei, Halmahera Tengah'. In E.K.M. Masinambow (ed.). *Halmahera dan Raja Ampat: Konsep dan Strategi Penelitian,* pp. 169–86. Jakarta: LEKNAS-LIPI.

ICBP (International Council for Bird Preservation). 1992. *Putting Biodiversity on the Map: Priority Areas for Global Conservation.* Cambridge: International Council for Bird Preservation.

Kemf, E. (ed.). 1993. *The Law of the Mother: Protecting Indigenous People in Protected Areas.* San Francisco: Sierra Club Books.

Kompas Online. 2000. Program Transmigrasi Jadi Korban Politik. http://www.kompas.com/kompas-cetak/0012/07/UTAMA/prog01.htm.

Kramer, R.A. and van Schaik, C.P. (eds). 1997. *The Last Stand: Protected Areas and the Defense of Tropical Biodiversity.* New York: Oxford University Press.

KSKHT (Kantor Statisk Kabupaten Halmahera Tengah). 1992. *Halmahera Tengah Dalam Angka 1992.* Soa Sio: Kantor Statisk Kabupaten Halmahera Tengah, Soa Sio.

KWPM (Departemen Transmigrasi dan Pemukiman Perambah Hutan, Kantor Wilayah Propinsi Maluku). 1995. Penyelenggaraan Transmigrasi dan Pemukiman Perambah Hutan di Propinsi Maluku: Sebegai Laporan dalam Rangka Kunjugan Kerja Meneteri Transmigrasi dan Pemukiman Perambah Hutan di Propinsi Maluku. Ambon: Departemen Transmigrasi dan Pemukiman Perambah Hutan, Kantor Wilayah Propinsi Maluku.

Lambert, F.R. 1993. *The Status of and Trade in North Moluccan Parrots with Particular Emphasis on Cacatua alba, Lorius garrulus, and Eos squamata.* Gland, Switzerland: IUCN/SSC Trade Specialist Group.

MacKinnon, J. 1994. 'Analytical Status Report of Biodiversity Conservation in the Asia-Pacific Region'. In *Biodiversity Conservation in the Asia Pacific Region.* Gland: Asian Development Bank and World Conservation Union.

MacKinnon, J.L., Fuller, R., Harper, M.E., Hugh Jones, T., Knowles-Leak, R., Rahman, D., Robb, D. and Vermeulen, J. 1995. Halmahera '94: A University of Bristol Expedition (Final Report).

Martodirdjo, H.S. 1984. *Gejala Perubahan Sosial dalam Kehidupan Orang Tugutil di Halmahera Tengah.* Jakarta: LEKNAS-LIPI.

Martodirdjo, H.S., Jayaputra, A. and Agung, I. 1985. *Persentuhan Antar Sistim dan Perubahan Sosial (Studie Kasus Orang Tugutil di PMT Wasile dan Orang Woda di Pulau Woda, Halmahera Tengah).* Jakarta: LEKNAS-LIPI.

McElwee, P.D. and Duncan, C.R. 1999. 'Conservation at What Cost?: The Logic of Sedentarization, Resettlement and Displacement in Southeast Asia'. Paper presented at the conference on Displacement, Forced Settlement and Conservation held by the Refugee Studies Programme, St. Anne's College, University of Oxford, Oxford.

Miete, A. 1933. 'De To-goetils'. *Koloniaal Tijdschrift,* 25: 336–7.

Peluso, N.L. 1993. 'Coercing Conservation? The Politics of State Resource Control'. *Global Environmental Change,* 3: 199–217.

Planning Bureau, Secretariat General, Ministry of Forestry. 1993. 'Study of Enhancing the Roles of Local Communities in the Development of Forestry Resources. The Natural Resources Management Project: Final Report'. Jakarta: P.T. Ardes Perdana.

Safwan, SH. 1995. 'Kebijaksanaan dan Program Pembinaan Kesejahteraan Sosial Masyarakat Terasing di Propinsi Maluku'. Paper presented at the Seminar dan Lokakarnya Sosial Budaya Masyarakat Terasing di Lokasi Dusun Iloa Kecamatan Tobelo, Kabupaten Maluku Utara, Kantor Wilayah Departemen Sosial Propinsi Maluku, Ambon, Indonesia.

Stattersfield, A.J., Crosby, M.J., Long, A.J. and Wege, D.C. 1998. *Endemic Bird Areas of the World: Priorities for Biodiversity Conservation.* Cambridge: BirdLife International.

Suherdie, Ir. H.E., Basuki, Ir. M., Cahyadin, Y. and Poulsen, M. 1995. Preliminary Evaluation of Boundaries for a Protected Area in Halmahera, North Maluku, PHPA/Birdlife International-Indonesia Programme – Memorandum Teknis, no. 8. Jakarta: PHPA/Birdlife International-Indonesia.

Sumardja, Effendy A., Harsono and MacKinnon, J. 1984. 'Indonesia's Network of Protected Areas'. In J. A. McNeely and K. R. Miller (eds). *National Parks, Conservation and Development: The Role of Protected Areas in Sustaining Society,* pp. 214–23. Washington DC: Smithsonian Institution Press.

Terborgh, J. 1999. *Requiem for Nature.* Washington DC: Island Press.

UNDP/FAO. 1981. *National Conservation Plan for Indonesia: Vol. 7: Maluku.* Bogor: UNDP/FAO.

Wells, M., Guggenheim, S., Khan, A., Wardojo, W. and Jepson, P. 1999. *Investing in Biodiversity: A Review of Indonesia's Integrated Conservation and Development Projects.* Washington DC: The World Bank.

World Bank. 1998. Indonesia-Maluku Conservation and Natural Resources (MACONAR) Project Information Document. URL: http://www.worldbank.org/pics/pid/id59991.txt.

21

Welcome to Aboriginal Land[1]

Aɴaɴɢu Ownership and Management of Uluɽu-Kata Tjuꞵa National Park[2]

Graham Griffin

Uluɽu sits fairly in the centre of the Australian continent. Its landscape is remote, wild and harsh. Few people live anywhere within hundreds of kilometres. Yet, Uluɽu is a place of great cultural and symbolic significance to Australians. Uluɽu-Kata Tjuꞵa National Park, with the great monolith Uluɽu as its centrepiece, is owned by Aboriginal traditional owners and has a resident community of over 300 Pitjantjatjara and Yankunytjatjara Aboriginal (Aɴangu) people. Its tenure history reflects the changing fortunes of conservation and Aboriginal land rights movements in Australia. It has become an environmental and cultural symbol both nationally and internationally. It is a place of substantial economic importance, and a place where Territory and Federal governments contest control over the land.

The Original Inhabitants

Uluɽu-Kata Tjuꞵa National Park lies within a larger region occupied by Aboriginal people belonging to the 'western desert' language group (Figure 21.1), of which Pitjantjatjara is a dialect spoken across large areas of the north east (Berndt 1959; Meggitt 1962; Tindale 1974; Peterson 1976). Aboriginal people have lived in central Australia for at least 30,000 years, but probably not

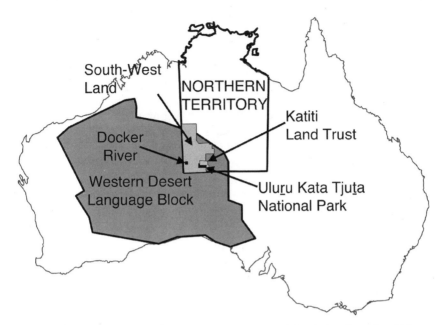

Figure 21.1 Location of Uluṟu Kata Tjuṯa National Park (black), the Katiti Land Trust Area (light grey), the South-West Land including the Petermann, Haasts Bluff and Lake Mackay Land Trust Areas (medium grey within the Northern Territory), and the extent of Western Desert language block (dark grey).

continuously (Gould 1977; Smith 1989; Thorley 1998). It seems likely that contemporary Aṉangu culture was established no more than 5000 years ago (Smith 1989; Layton 1993). Aṉangu retained almost exclusive use of the area until the 1930s. Aṉangu have strong physical and spiritual connections with the land, evident in traditional forms of land ownership, and individual responsibility for maintenance of knowledge and the environment (Harney 1957; Mountford 1965; Tindale 1974; Layton 1986). Numerous sacred sites, rock engravings and cave paintings provide evidence of the long history of occupation (Mountford 1965; Layton 1993). Aṉangu persisted historically as hunters and gatherers, collecting a wide variety of vegetable foods (Gould 1969; Goddard and Kalotis 1988) and hunting the diverse reptile, mammal and bird fauna, of which they had an extensive knowledge (Finlayson 1952; Lockwood 1964; Burbidge et al. 1988). They imbued the landscape with profound cultural and physical significance. Aṉangu use of fire (Latz and Griffin 1978), amidst the inevitable lightning-initiated fires (Griffin et al. 1984), extensively altered the vegetation and affected animal populations (Burbidge 1985; Griffin and Friedel 1985).

The Environment

The park includes two main ecosystems: the extensive desert ecosystems which cover much of inland Australia include a complex of desert dune fields, sand plains and salt lakes (Mabbutt 1984; Williams and Calaby 1985); the Cainozoic red earth plains occur adjacent to low mountain ranges of crystalline and sedimentary rocks (Perry et al. 1962). The vegetation of the region is dominated by spinifex (mainly *Triodia* spp.) grasslands on the sand sheets and dune fields, *Acacia* shrub lands on the red earth plains and low *Acacia* shrub lands and spinifex grasslands on the mountains (Perry et al. 1962). The unusually diverse fauna in the deserts and uplands includes a remarkably rich reptile fauna (Pianka 1969), a modest array of mammals, and a variety of bird species (Williams and Calaby 1985).

The hot, arid climate (Austin and Nix 1978) has an average annual rainfall of about 250mm (ANPWS 1991). Most falls in summer, with extremely high inter-annual variation (Fleming 1978). Mean monthly temperatures range from a maximum of 39°C in January to a minimum of 4°C in July.

The Image

Uluru lies in wild, remote, dry desert country, in the very centre of the continent, and has come to represent many facets of Australia and modern Australians' image of themselves. For both Australians and others, it is now the most readily identified image of Australia (Fiske et al. 1987; Breeden 1994; Hill 1994). Images of Uluru have been used extensively in product and tourism advertising (Brereton 1990; Haines 1992; Rowse 1992), music (for example Warumpi Band 1988; Porter and McCormack 1997; Williamson and Williams 1999), and through the sensational reporting of local events (for example Thornhill 1984; Bryson 1985; English 1986; Reynolds 1989). Such is its symbolic value that a national (albeit unofficial) popular vote on the design of a flag for the proposed republic of Australia chose a design incorporating the Southern Cross over a stylized image of Uluru (*The Australian*, 17 December 1993; http://www.ausflag.com.au/new/93–1.html).

European Discovery and Possession

Europeans first explored the park area in 1872 and 1873 (Gosse 1874; Giles 1875). The first major scientific expedition to central Australia studied the biology and geology of the area (Spencer 1896). As part of the process of colonizing and owning new lands, it was promptly renamed 'Ayers Rock' (Hartley 1988). Few visitors came to the area over the coming decades. In 1920, the Petermann Reserve had been declared, to protect tribal Aboriginal people from the effects of contact with Europeans (Layton 1986). By 1926, there was a well-established horse and camel trail to Uluru, mainly used for mineral prospecting (Terry 1930; Gill 1968; Cartwright 1996). Early pastoral interest in the area led to massive land speculation around the turn of the century (Bauer 1964; Hartwig 1965), most of which faded quickly. Motorized vehicles first accessed the area in 1930 (Terry 1930) and these tracks were extended and expanded in the next few years by prospectors for the legendary Lasseter's gold reef (Idriese 1932; Clune 1957; Cartwright 1996). The popular and widely distributed *Walkabout* magazine began to promote tourism, with articles on Uluṟu in 1941, 1949 and 1950 (Rowse 1987). Bowing to pressure from local tour operators and government officials to open the area up for tourism, the Federal government excised the Ayers Rock–Mount Olga National Park from the Petermann Reserve in 1958, under section 103 of the *Northern Territory Crown Lands Ordinance 1931–1957*. The then Federal Minister for Aboriginal Affairs justified the decision to excise it from the reserve saying, 'Ayers Rock and Mount Olga now have little ceremonial significance for the natives. Water supplies in the area are not permanent, and hunting resources are sufficient for only a few scattered bands.'

Bill Harney, the first resident ranger at Uluṟu, was employed by the Native Welfare Branch in 1957. The Branch administered the park until it was taken over by the Northern Territory Reserves Board (later to become the Conservation Commission of the Northern Territory (CCNT)) in 1958. At that time, the Aṉangu presence was accepted in the park. Harney (1963) records that Aṉangu often travelled through the park and lived there for months at a time. When the first infrastructure and substantial residences were constructed there in 1959, Aṉangu were discouraged from visiting the park. In addition, protracted drought, punitive police patrols and government policies of settling people in reserves, all contributed to a depopulation of the area (Toohey 1980). However, in 1962, as many as sixty Aṉangu still lived in the park (Rowse 1987). By 1967, when camping facilities and tourist accommodation were built, Aṉangu were almost entirely absent. Although Aṉangu were not forcibly displaced from the park, the Native Welfare Branch of the Federal Department of the Interior became uncomfortable with the image of ragged Aboriginals selling artefacts and begging around the Uluṟu area, and developed the Docker River settlement, about 200 km to the west, in part to entice them away (Rowse 1987; Cartwright 1996). In the mid 1970s, Aṉangu began to move back to the park from communities on surrounding reserve land (Dunlop and Anson 1991), even though the CCNT attempted to resist this influx (Hodges 1977; CCNT 1981). Resident Aṉangu formed themselves into an incorporated body under Northern Territory legislation. This body, Uluṟu Community, was later to become the Muṯitjulu Community. The Native Welfare Branch helped secure a lease within the park to develop a store and fuel outlet (Ininti Store and Garage), an Aboriginal-owned enterprise. It began operation in 1972 (Cartwright 1996). Aṉangu continued to return to the park and by 1981 about 30 lived there more or less permanently, selling arts and crafts to tourists. By 1986, following Aboriginal control of the park, a small community of about 140 Aṉangu were living permanently near the store and camping grounds (Altman 1987; Dunlop and Anson 1991). In 1989, there were an estimated 129 Aṉangu residents (Ditton 1990) and by 1998, the community had grown to over 300 (UKTBM and PA 1999). Conservative projections of population growth by Taylor (2000) predict the Muṯitjulu population will be about 850 by 2021.

process only applied to unalienated crown land, unlike the pre-existing tenure regime of the national park (Toohey 1980).

At that time, conservation interests had substantial influence with the Federal government (Toyne 1994). This reflected a broader commitment in the Australian community to increasing conservation areas across the continent (Mosley 1978), particularly in the arid regions (Messer and Mosley 1983). The park was designated a Biosphere Reserve in 1977 (Australia National Commission for UNESCO 1983), and was listed on the Register of the National Estate under the *Australian Heritage Commission Act 1975*. Land rights for Aboriginal people were strongly debated and contested at all political and social levels, following the conversion of large areas of Aboriginal reserve land and unalienated crown land elsewhere in the Northern Territory to Aboriginal freehold title (Keon-Cohen 1980; Peterson and Langton 1983; Nutting 1994; Whittaker 1994).

Anangu persisted with their claim to Uluru (Toohey 1980; Lester 1993), prompting the Northern Territory government, in 1982, to offer a perpetual lease or Territory freehold. The offer included a condition that there would be no further claims to Aboriginal pastoral leases, stock routes, reserves or public purpose land in the Northern Territory. Anangu declined this offer, recognizing it to be an inferior form of title to that available under the *Aboriginal Land Rights (Northern Territory) Act* and that acceptance would disadvantage other Aboriginal people seeking to claim lands. In 1983, the Federal Government promised Aboriginal freehold title to the park but not until 1985 was title to the park granted. It would be leased back to the Federal Parks Service for 99 years to be managed as a national park in accordance with the wishes of the traditional owners. The Federal Government was able to grant title as the park was still held by the Commonwealth and was excluded from the land grants to the Northern Territory under the self-government legislation (*Northern Territory Self-Government Act 1978*). The negotiation for the grant of Uluru occurred outside the land claim process established under the *Aboriginal Land Rights (Northern Territory) Act*. The Northern Territory government saw it as a direct challenge to its role, as the body elected to control and manage lands within its own borders. In addition to the transfer of ownership to Aboriginals, the new regime excluded the Territory parks service, which was replaced by its Federal counterpart. The then Federal Labor Government amended the *Aboriginal Land Rights (Northern Territory) Act* (*Aboriginal Land Rights (Northern Territory) Amendment Act 1976*) in 1985 to establish a lease arrangement between the traditional owners (through a Land Trust) and the director of the Federal Parks Service, leading to formal 'handback' in the same year. For the next three years, the Northern Territory government, controlled by the conservative Country Liberal Party (CLP), railed against the Federal Government's right to manipulate land rights and control land (Gibson 1994). In 1983 and 1984, the CLP exploited divided public opinion, and campaigned on the Uluru handback issue and Territory versus Commonwealth rights (see for example the full page campaign advertisements in *The Northern Territory News*, November 1983, November 1984, and all daily national newspapers in October 1985, cited in Gibson 1994). It achieved substantial victories in the Territory election and in the Territory seat in the Federal elections.

Concurrent with the political arguments that raged during this period, opponents of handback prophesized that Aboriginal ownership would dramatically diminish the park's conservation value and effectively stop tourism to it (English 1986; Hollingsworth and Downing 1987). In terms of visitor numbers, Uluṟu was the Northern Territory's premier tourist attraction, and tourism was a major income-generator for the Territory at that time (Northern Territory Tourist Commission Annual Reports 1973/74–1976/77, 1980/81–1982/83, 1984/85–1986/87). The CLP government bolstered its election campaign arguments by claiming that this predicted loss of tourism would seriously affect the Territory's economy (Gibson 1994).

Handback

The 1985 handback was conditional upon the land being leased to the Commonwealth for operation as a national park. The Commonwealth *National Parks and Wildlife Act* was amended to establish the park under joint management with the traditional owners, with a board of management (*National Parks and Wildlife Amendment Act 1975*). This board had a majority of six nominees of the traditional owners, along with the director of the parks service and three other appointees of Federal ministers, representing interests in the environment, conservation and tourism. While it was never specified, it was assumed that the traditional owner members of the Board would represent the interests of the Muṯitjulu community. Aṉangu and the Muṯitjulu community are concerned that this is not explicit, nor has it always proved practical (Ditton 1990). The park is legally held by the Uluṟu Kata Tjuṯa Land Trust, on behalf of traditional owners. While many of them live in the community, there are sometimes tensions between the interests of resident and non-resident traditional owners. Moreover, both in Uluṟu and elsewhere in the Territory, the focus on rights stemming from traditional ties to country has sometimes placed other Aboriginal residents in an ambiguous position. The Muṯitjulu community is central to the existence of the park yet its physical existence on land in the park is neither legally defined nor recognized (UKTBM and PA 1999), and it has no formal representation in the management of the park.

The Director and Aṉangu jointly control the use and management of their land under the terms of the lease, which also provides the means for generating income both for the Director and for traditional landowners. It provides for an annual rental (currently AUS$150K) and for 25 per cent of the park entrance fees to be paid to the traditional owners. In 1996 the park entrance fees generated AUS$4.4m. In addition, the park service provides most community services and employment for Muṯitjulu residents, contributing in 1996 another AUS$1m to the economy of the Muṯitjulu community. The lease terms (except for the 99-year term) are renegotiable every five years and the conditions can be altered by agreement between the Land Trust and the Director of National Parks.

Joint Management

The park has operated successfully since handback. Both Anangu and non-Anangu have argued that the joint management arrangements, put in place under the *National Parks and Wildlife Act* are working well (Tjamiwa 1992; Willis 1992). The services and resources available to visitors have improved significantly over the last fourteen years (UKTBM and PA 1999). The majority of visitors to the park express high levels of satisfaction with their visit, although in recent years the proportion of visitors expressing concern over crowding and site damage has risen (Environmental Science and Services 1991; Centre for Leisure Research 1992). The park has gained an international reputation as a place of great environmental and cultural significance. An ever increasing number of visitors come from all corners of the globe to see Uluru, to experience the remote desert environments, and, most importantly, to experience Aboriginal culture (Department of Environment, Sport and Territories 1994). The park has been twice declared a World Heritage site, first in 1987 for its natural values, and again in 1994 for its cultural values (Department of Environment, Sport and Territories 1995). The park is still the centrepiece for tourism advertising by the Northern Territory Government (see for example: http://www.nttc.com.au/) and the Australian Tourism Commission (see for example: http://www.aus-in-shanghai.com/tourist/part1.html). Indeed, promotional material plays on Aboriginal ownership and culture as major attractions.

Young et al. (1991), Craig (1992), DeLacey (1994) and Nutting (1994) have pointed to Uluru as a model for joint management, to be emulated by other Australian States and Territories. Legislation for joint management in other states has been loosely based on the Uluru model (*New South Wales National Parks and Wildlife Amended Act 1990; Queensland Nature Conservation Act 1992*). While these forms of title and management are not as encompassing as the Uluru model, they do give Aboriginal people some rights and access to park lands (Woenne-Green 1994). Different models in other parts of the world are developing, and provide contrasts and alternatives to Uluru (see, for example, Durbin 1992; Wells et al. 1992; Notzke 1993).

As in other parts of arid Australia, conservation issues include management of feral animals and plants, loss of indigenous biota and their re-introduction, and the management of fire-prone landscapes. The dramatic influx of visitors to the park and the development of an infrastructure to service their access have placed substantial pressure on the natural resources (UKTBM and PA 1999). In addition, some issues result directly from the existence of a large human population within the park. Uluru has acted as an oasis for Anangu in the region, drawing many people in to live there over recent years. Having had limited contact with the world beyond their lands, most Anangu retain strong ties to their lands and land use traditions. They continue with hunting and food gathering activities that have important social, cultural and economic benefits. However, the number of people using the limited resources now places unprecedented pressure on the biota. This is a common experience in other areas surrounding Aboriginal communities (Altman and Allen 1992; Baker and the Mutitjulu Community 1992; Head 1994; Bomford and Caughley 1996). The Mutitjulu community attempts, through the park board of management, to affect the level of resource use although means of regulating the now large and socially diverse community are limited. For example, despite the obvious ecological problems entailed, Anangu prefer to use wood for fuel. The large community requires substantial volumes of firewood to meet its needs, and vehicle access to harvest wood inflicts damage on the soils (UKTBM and PA 1999).

The Mutitjulu site is not formally recognized by the Northern Territory government as a town and does not fully receive many of the normal essential services. The Northern Territory government argues that it cannot provide services and infrastructure on land and towns outside its jurisdiction. To compensate for this, the park service contributes almost a fifth of its operational budget towards providing essential community services and employment. Again, the crux is Territory versus Commonwealth rights.

A number of issues have arisen that stem directly from the cultural significance of the park area to Anangu, Anangu ownership and management of the park, and the existence of a resident community at Uluru. Visitors repeatedly express a strong interest in Anangu art and ritual life, including the desire to photograph it. In addition, images of Anangu art and ceremony have a high commercial value. Many aspects of Anangu culture are not open for public access (Harney 1963; Mountford 1965; Layton 1986). In response to Anangu concerns about limiting access to and use of images of themselves and their important sacred sites, the board of management in consultation with traditional owners are developing formal guidelines to intellectual and cultural property, including a prohibition on unauthorized commercial photography. The latter in particular has drawn strong criticism from commercial photography interests (Gower 1991; Thompson 1991). The *Northern Territory Aboriginal Sacred Sites Act 1989* makes unauthorized entry of a sacred site anywhere in the Northern Territory a criminal offence. It provides stiff penalties for entering or interfering with a sacred site, and allows for site fencing if requested by Aboriginal traditional owners. The regulations of the *National Parks and Wildlife Act* also contain protective provisions. However, the high visitor numbers, and the Anangu decision to retain the visual integrity of the area by only constructing nominal barriers round certain key sites, make policing difficult, and occasional intrusions have led to site damage (Gale and Jacobs 1987; Marcus 1988). More insidious and difficult to control is the ever-increasing appropriation of traditional environmental knowledge, particularly bioprospecting for commercial interests (Baker et al. 1992; Fourmile 1998).

Visitors express a strong desire to experience Anangu culture directly, including personal interactions with Anangu. Aspects of Anangu culture could, potentially, be made publicly accessible. However, few Anangu possess the skills, or desire, to deal with the enormous visitor population. If only half the annual visitors to the park wished to meet Anangu (a conservative estimate), this would mean that every man, woman and child in the Mutitjulu community would have to meet almost 600 visitors a year. In the face of these overwhelming numbers, the Mutitjulu community have chosen to close their community to visitors, to maintain their own privacy, and to allow them to follow their own lives, separate from the demands of tourism.

Uluru has been used frequently as a spectacular symbol, an easily identified image of Australia. Despite retaining control over new commercial photography, however, Anangu have little control over the use of existing images in advertising products or causes. Effective regulation of these public uses of imagery, although difficult to implement, is crucial if Anangu are to retain control over their culture and sacred places (UKTBM and PA 1999).

Anangu are frustrated that the park, their land, and their culture are the attractions for a major tourism industry that generates many millions of dollars in revenue for external operators. Anangu have limited resources to engage with and benefit from the tourism industry, and many are in any case ambivalent about personal involvement with it. They are confronted by the vexing problem of how to tap the plentiful tourism dollars, to help bring the Mutitjulu community out of its third world living conditions.

The Uluru model of ownership and joint management has succeeded in many areas of shared resource use, knowledge, and income generation. The park is managed to the highest world standards and recognized as the most important symbol of Australia. The traditional Aboriginal owners of the park have shared in some of the benefits, but in doing so have exposed their culture to high pressures and interests. They have profited little from the massive tourism industry that has besieged them and their homelands. Anangu are engaged in a difficult task, protecting traditional concerns within the context of an inevitably dynamic culture, while under intense scrutiny as custodians of the national icon, and as exemplars of Aboriginal ownership, control, and joint management.

Notes

1. Sign at the entrance to Uluru National Park. From Pititjantjarra: '*Pukulpa pitjima anangu ku ngurakutu*'.

2. I am grateful to Sarah Dunlop for substantial input to repairing my grammar and expression, and for alerting me to numerous sources of information about Uluru and its history. Graham Lightbody of the Central Land Council contributed his extensive knowledge of the park history, especially around the time after handback.

References

Altman, J.C. 1987. 'The Economic Impact of Tourism on the Mutitjulu Community, Uluru (Ayers Rock–Mount Olga) National Park'. Working paper No. 7. Department of Political and Social Change, Research School of Pacific Studies, The Australian National University, Canberra.

Altman, J.C. and Allen, L.M. 1992. *Living off the Land in National Parks, issues for Aboriginal Australians.* Canberra: Aboriginal Studies Press.

ANPWS, Australian Natonal Parks and Wildlife Service. 1991. *Uluru Ayers Rock–Mount Olga National Park Plan of Management.* Canberra: ANPWS.

Austin, M.P. and Nix, H.A. 1978. 'Regional Classification of Climate and its Relation to Australian Rangeland'. In K. M. W. Howes (ed.). *Studies of the Australian Arid Zone. III. Water in Rangelands*, pp. 9–17. Melbourne: CSIRO.

Australia National Commission for UNESCO. 1983. *Australia's Biosphere Reserves: Conserving Ecological Diversity.* Canberra: Australian Government Publishing Service.

Baker, L.M. and the Mutitjulu Community. 1992. 'Comparing Two Views of the Landscape: Aboriginal traditional ecological knowledge and modern scientific knowledge'. *Rangeland Journal,* 14: 174–89.

Baker, L.M., Woenne-Green, S. and Mutitjulu Community. 1992. 'The Role of Aboriginal Ecological Knowledge in Ecosystem Management'. In J. Birckhead, T. DeLacey and L. Smith (eds). *Aboriginal Involvement in Parks and Protected Areas*, pp. 65–74. Canberra: Aboriginal Studies Press.

Bauer, F.H. 1964. *Historical Geography of White Settlement in Part of Northern Australia. Part 2. The Katherine-Darwin Region.* Canberra: CSIRO. Division of Land Research and Regional Survey, Report No. 64/1.

Berndt, R.M. 1959. 'The Concept of the Tribe in the Western Desert of Australia'. *Proceedings of the Royal Geographic Society of Australasia, South Australian Branch* 15: 57–212.

Bomford, M. and Caughley, J. (eds). 1996. *Sustainable Use of Wildlife by Aboriginal Peoples and Torres Strait Islanders.* Canberra: Australian Government Publishing Service.

Breeden, S. 1994. *Uluru: Looking after Uluru-Kata Tjuta the Anangu way.* East Roseville, NSW: Simon & Schuster Australia.

Brereton, K. 1990. 'Advertising Uluru. Commercial Images of Ayers Rock'. University of Technology Sydney. Unpublished PhD thesis.

Bryson, J. 1985. *Evil Angels.* Ringwood, Vic.: Viking.

Burbidge, A.A. 1985. 'Fire and Mammals in Hummock Grasslands of the Arid Zone'. In J.R. Ford (ed.). *Fire Ecology and Management of Western Australian Ecosystems*, pp. 91–4. W.A.I.T. Environmental Studies Group, Report No.14.

Burbidge, A.A, Johnson, K.A, Fuller, P.J. and Southgate, R.J. 1988. 'Aboriginal Knowledge of the Mammals of the Central Deserts of Australia'. *Australian Wildlife Research* 15: 9–40.

Cartwright, M. 1996. *Lasseter's Gold, Tanami Gold. A Story of Two Highways.* Adelaide: Flinders Press.

CCNT (Conservation Commission of the Northern Territory). 1981. *Annual Report July 1980 to June 1981.* Darwin: Government Printer of the Northern Territory.

Centre for Leisure Research. 1992. *Visitor Crowding Study of Uluru Kata Tjuta National Park.* Brisbane: Griffith University.

Clune, F.P. 1957. *The Fortune Hunters; An Atomic Odyssey in Australia's Wild West, and things seen and heard by the way in a jeep jaunt.* Sydney: Angus and Robertson.

Craig, D. 1992. 'Environmental Law and Aboriginal Rights: Legal framework for Aboriginal joint management of Australian National Parks'. In J. Birckhead, T. DeLacey, and L. Smith (eds). *Aboriginal Involvement in Parks and Protected Areas*, pp. 137–48. Canberra: Aboriginal Studies Press.

DeLacey, T. 1994. 'The Uluru-Kakadu Model: Anangu-Tjukurrpa – 50,000 years of Aboriginal law and land management changing the concept of National Parks in Australia'. *Society and Natural Resources*, 7: 479–98.

Department of Environment, Sport and Territories. 1994. *Renomination of Uluru–Kata Tjuta National Park by the Government of Australia for Inscription on the World Heritage List*. Canberra: Commonwealth of Australia.

—— 1995. *Australia's World Heritage*. Canberra: Department of Environment, Commonwealth of Australia.

Ditton, P. (ed.). 1990. 'Mutitjulu: A unique community'. Volume 1 and Volume 2. Report to the Minister for Arts, Sport, the Environment, Tourism and Territories. Alice Springs: Pamela Ditton.

Dunlop, S. R., and Anson, K. 1991. 'The Mutitjulu Community Today'. In T. Rowse (ed.). *Sharing the Park. Anangu Initiatives in Ayers Rock Tourism*, pp. 1–28. Alice Springs: Institute for Aboriginal Development.

Durbin, J. 1992. 'People and Protected Areas: A major theme of the IVth World Congress on National Parks and Protected Areas, Caracas, Venezuela, February 1992'. *Biodiversity and Conservation*, 1: 209–10.

English, P. 1986. *Storm over Uluru: The greatest hoax of all: a resume of events leading up to the questionable hand-over of Australia's most famous National Park to Aboriginal claimants*. Bullsbrook, W.A.: Veritas Publishing.

Environmental Science and Services. 1991. *Uluru Visitor Survey Report*. Canberra: Australian National Parks and Wildlife Service.

Finlayson, H.H. 1952. *The Red Centre. Man and Beast in the Heart of Australia*. New Edition. Sydney: Angus and Robertson.

Fiske, J., Hodge, B. and Turner, G. 1987. *Myths of Oz: Reading Australian Popular Culture*. Boston and London: Allen & Unwin.

Fleming, P.M. 1978. 'Types of Rainfall and Local Rainfall Variability'. In M.P. Austin and H.A. Nix (eds). *Studies of the Australian Arid Zone. III. Water in Rangelands*, pp. 18–28. Melbourne: CSIRO.

Fourmile, H. 1998. 'Using Prior Consent Procedures under the Convention on Biological Diversity to protect Indigenous Traditional Ecological Knowledge and Natural Resource Rights'. *Indigenous Law Bulletin*, 4: 14–15.

Gale, F. and Jacobs, A. 1987. *Tourists and the National Estate. Procedures to Protect Australia's Heritage*. Canberra: Australian Heritage Commission, Special Australian Heritage Publication Series No. 6. Australian Government Publishing Service.

Gibson, S. 1994. 'This Rock is Sacred. The Northern Territory Government and the Handback of Uluru Ayers Rock – Mount Olga National Park. November 1983 – May 1986'. Northern Territory University. Unpublished MA Thesis.

Giles, E. 1875. *Geographic Travels in Central Australia from 1872 to 1874*. Melbourne: McCarron, Bird & Co.

Gill, W. 1968. *Petermann Journey*. Adelaide: Rigby.

Goddard, C., and Kalotis, A. 1988. *Punu: Yankunytjatjara Plant Use. Traditional methods of preparing foods, medicines, utensils and weapons from native plants*. North Ryde, N.S.W.: Angus & Robertson.

Gosse, W.C. 1874. *W. C. Gosse's Explorations, 1873: report and diary of Mr. W. C. Gosse's central and western exploring expedition, 1873*. Adelaide: Government Printer.

Gould, R.A. 1969. 'Subsistence Behaviour among Western Desert Aboriginals of Australia'. *Oceania*, 39: 253–74.

—— 1977. *Puntutjarpa Rockshelter and the Australian Desert Culture.* Anthropological papers of the American Museum of Natural History. 54.

Gower, R. 1991. 'Uluru Off-limits. National Park guidelines ban photography'. *Industrial and Commercial Photography*, July/August: 45–6.

Griffin, G.F. and Friedel, M.H. 1985. 'Discontinuous Change in Central Australia: Some implications of major ecological events for land management'. *Journal of Arid Environments*, 9: 63–80.

Griffin, G.F, Price, N.F. and Portlock, H. 1984. 'Wildfires in the Central Australian Rangelands, 1970–1980'. *Journal of Environmental Management*, 17: 311–23.

Haines, K. 1992. 'The 1992 Variety Club Bash – a Second Invasion?' Northern Territory University. Unpublished MA Thesis.

Harney, W.E. 1957. *The Story of Ayers Rock told by Uluritdja Tribesmen Kadakadeka and Imalung to W.E. Harney.* Melbourne: Bread and Cheese Club.

—— 1963. *To Ayers Rock and Beyond.* London: Hale.

Hartley, J. B. 1988. 'Maps, Knowledge and Power. In D. Cosgrove and S. Daniels (eds). *The Iconography of Landscape*, pp. 120–34. Cambridge: Cambridge University Press.

Hartwig, M.C. 1965. 'The Progress of White Settlement in the Alice Springs District and its Effects on the Aboriginal Inhabitants'. University of Adelaide. Unpublished PhD Thesis.

Head, L. 1994. 'Aborigines and Pastoralism in north-western Australia: Historical and contemporary perspectives on multiple use of the rangelands'. *Rangeland Journal*, 16: 167–83.

Hill, B. 1994. *The Rock. Travelling to Uluru.* St. Leonards NSW: Allen & Unwin.

Hodges, J. C. 1977. *The Management of Ayers Rock–Mt Olga National Park: fourth report.* Canberra: Australian Government Publishing Service.

Hollingsworth, B, and Downing, J. 1987. *Aboriginal Christians and Denominationalism. Uluru – A National Park for all Australians or a National Tragedy.* Casuarina, N.T.: Nungalinya College.

Idriese, I.L. 1932. *Lasseter's Last Ride: An Epic of Central Australian Gold Discovery.* Sydney: Angus and Robertson.

Keon-Cohen, B. 1980. 'Aboriginal Land Rights in Australia: Beyond the legislative limits'. In R. Thomasic (ed.). *Legislation and Society in Australia*, pp. 382–411. Sydney: Law Society of New South Wales.

Latz, P.K., and Griffin, G.F. 1978. 'Changes in Aboriginal Land Management in Relation to Fire and to Food Plants in Central Australia'. In B.S. Hetzel and H.J. Frith (eds). *The Nutrition of Aborigines in Relation to the Ecosystem of Central Australia*, pp. 77–85. Melbourne: CSIRO.

Layton, R. 1986. *Uluru, an Aboriginal History of Ayers Rock.* Canberra: Aboriginal Studies Press.

—— 1993. *Australian Rock Art: A new Synthesis.* Cambridge: Cambridge University Press.

Lester, Y. 1993. *Yami: The Autobiography of Yami Lester.* Alice Springs: Institute for Aboriginal Development.

Lockwood, D. 1964. *The Lizard Eaters.* Melbourne: Cassell Australia.

Mabbutt, J.A. 1984. 'The Desert Physiographic Setting and its Ecological Significance'. In H.G. Cameron and E.E. Cogger (eds). *Arid Australia*, pp. 87–109. Sydney: Surrey Beatty and Sons.

Marcus, J. 1988. 'The Journey out to the Centre. The cultural appropriation of Ayers Rock'. In A. Rutherford (ed.) *Aboriginal Culture Today*, pp. 254–74. Sydney: Dangaroo Press.

Meggitt, M. J. 1962. *Desert People*. Chicago: Chicago University Press.

Messer, J. and Mosley, G. 1983. *What Future for Australia's Arid Lands?* Hawthorn, Vic.: Australian Conservation Foundation.

Mosley, G. (ed.). 1978. *Australia's Wilderness: Conservation Progress and Plans: Proceedings of the first National Wilderness Conference, Australian Academy of Science, Canberra, 21–3 October, 1977.* Hawthorn, Vic.: Australian Conservation Foundation.

Mountford, C. P. 1965. *Ayers Rock, its People, their Beliefs and their Art*. Sydney: Angus and Robertson.

Notzke, C. 1993. 'Aboriginal Peoples and Natural Resources: Co-management, the way of the future'. *Research and Exploration*, 9: 395–7.

Nutting, M. 1994. 'Competing Interests or Common Ground? Aboriginal participation in the management of protected areas'. *Habitat*, 22: 28–37.

Perry, R.A., Mabbutt, J.A., Litchfield, W.H., Quinlan, T., Lazarides, M., Jones, N.O., Slatyer, R.O., Stewart, G.A., Bateman, W. and Ryan, G.R. 1962. *Lands of the Alice Springs Area, Northern Territory, 1956–57. Land Research Series No. 6.* Melbourne: CSIRO.

Peterson, N. (ed.). 1976. *Tribes and Boundaries in Australia*. Canberra: Humanities Press.

Peterson, N., and Langton, M. (eds). 1983. *Aborigines, Land and Land Rights*. Canberra: Australian Institute of Aboriginal Studies.

Pianka, E.R. 1969. 'Habitat Specificity, Speciation, and Species Density in Australian Desert Lizards'. *Ecology*, 50: 498–502.

Porter, G. and McCormack, R. 1997. 'Uluru'. In *Outback Adventures*, Compact disk. EMI Music Australia.

Reynolds, P. 1989. *The Azaria Chamberlain Case: Reflections on Australian Identity*. Brisbane: University of Queensland.

Rowse, T. 1987. 'The Centre: A limited colonisation'. In A. Curthoys, A.W. Martin, and T. Rowse (eds). *Australians from 1939*, pp. 15–22. Sydney: Fairfax, Syme and Weldon Associates.

—— 1992. 'Hosts as Guests at Uluru'. *Meanjin*, 51: 247–59.

Smith, M.A. 1989. 'The Case for a Resident Human Population in the Central Australian Ranges during full Glacial Aridity'. *Archaeology in Oceania*, 24: 93–105.

Spencer, B. (ed.). 1896. *Report on the Work of the Horn Scientific Expedition to Central Australia. Parts 1–4.* Melbourne: Melville, Mullen and Slade.

Taylor, J. 2000. 'Mutitjulu Community Population Dynamics and Future Growth: 2000–2021. A Report to Parks Australia North'. Centre for Aboriginal Economic Policy Research, The Australian National University, Canberra.

Terry, M. 1930. *Hidden Wealth and Hiding People*. London: Putman.

Thompson, G. 1991. 'Regulations Close the Shutters'. *The Bulletin*, July: 28.

Thorley, P.B. 1998. 'Pleistocene Settlement in the Australian Arid Zone: Occupation of an inland riverine landscape in the central Australian ranges'. *Antiquity*. 72: 34–45.

Thornhill, M. 1984. *The Disappearance of Azaria Chamberlain*. Film. Produced by E. Sullivan. Australia.

Tindale, N. B. 1974. *Aboriginal Tribes of Australia: Their terrain, environmental controls, distribution, limits, and proper names with an appendix on Tasmanian tribes by R. Jones*. Canberra: Australian National University Press.

Tjamiwa, T. 1992. 'Tjunguringkula waakaripai: joint management of Uluṟu National Park'. In J. Birckhead, T. DeLacey, and L. Smith (eds). *Aboriginal Involvement in Parks and Protected Areas*, pp. 7–11. Canberra: Aboriginal Studies Press.

Toohey, J. 1980. *Uluru Ayers Rock National Park and Lake Amadeus/Luritja Land Claim. Report by the Aboriginal Land Commissioner to the Minister for Aboriginal Affairs and to the Minister for Home Affairs.* Canberra: Australian Government Publishing Service.

Toyne, P. 1994. *The Reluctant Nation: Environment, Law and Politics in Australia.* Crows Nest, N.S.W.: ABC Books.

UKTBM and PA (Uluṟu Kata Tjuṯa Board of Management and Parks Australia). 1999. 'Uluṟu Kata Tjuṯa National Park. Draft Plan of Management'. Canberra: Commonwealth of Australia.

Warumpi Band. 1988. 'Big Name, No Blanket'. Cassette music tape. Sydney: Festival Records Pty Ltd, Australia.

Wells, M, Brandon, K. and Hannah, L. 1992. *People and Parks. Linking Protected Area Management with Local Communities.* Washington: The World Bank, The World Wildlife Fund, and U.S. Agency for International Development.

Whittaker, E. 1994. 'Public Disclosures on Sacredness: The transfer of Ayers Rock to Aboriginal ownership'. *American Ethnologist*, 21: 310–34.

Williams, O.B. and Calaby, J.H. 1985. 'The Hot Deserts of Australia'. In M Evanari (ed.). *Hot Deserts and Arid Shrublands*, pp. 269–312. Amsterdam: Elsevier.

Williamson, J. and Williams, W.H. 1999. 'Raining on the Rock'. In *The Winners 7.* Compact disk. EMI Music Australia.

Willis, J. 1992. 'Two Laws, One Lease: Accounting for traditional Aboriginal law in the lease for Uluṟu National Park'. In J. Birckhead, T. DeLacey, and L. Smith (eds). *Aboriginal Involvement in Parks and Protected Areas*, pp. 159–66. Canberra: Aboriginal Studies Press.

Woenne-Green, S. 1994. *Competing Interests: Aboriginal Participation in National Parks and Conservation Reserves in Australia.* Melbourne: Australian Conservation Foundation.

Young, E., Ross, H., Johnson, J. and Kesteven, J. 1991. *Caring for Country: Aborigines and Land Management.* Canberra: Australian National Parks and Wildlife Service.

Index of Subjects

Index of Names

Bold pagination indicates an entry in the Note on Contributors and a chapter in this volume.